THE NORTHERN SECTION

Seattle to Eureka

Not to be used for navigation

For southern section see rear endpaper

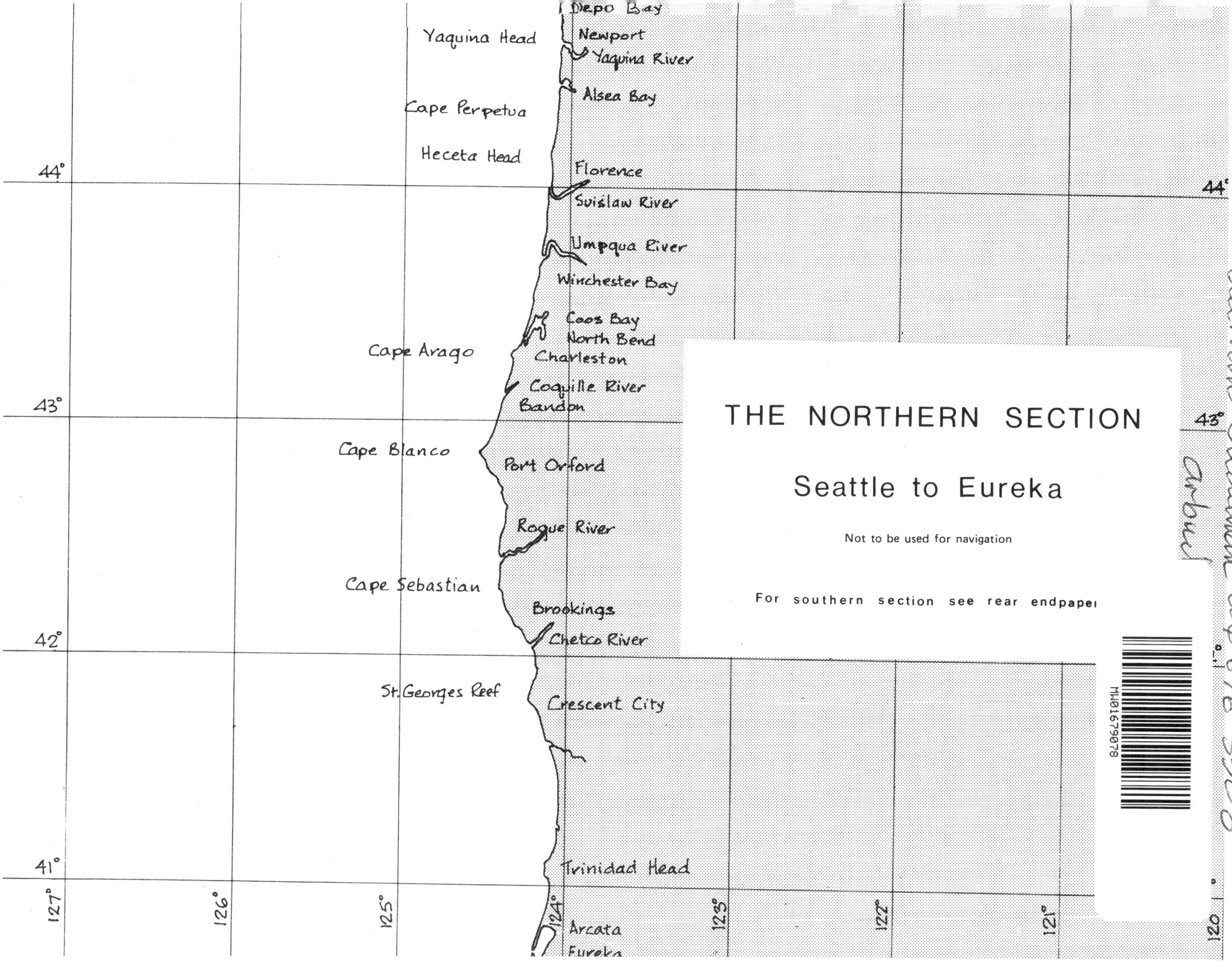

Adrian Buchanan 604 676 3308

arbuc

CHARLIE'S CHARTS CRUISING GUIDES

CHARLIE'S CHARTS of the Western Coast of MEXICO
CHARLIE'S CHARTS of POLYNESIA
CHARLIE'S CHARTS NORTH to ALASKA
CHARLIE'S CHARTS of the HAWAIIAN ISLANDS
CHARLIE'S CHARTS of the U.S. PACIFIC COAST
CHARLIE'S CHARTS of COSTA RICA

Other books:
BUILDING YOUR DREAM BOAT by Charles Wood
A PRAIRIE CHICKEN GOES TO SEA by Margo Wood

SOUTHEAST FARALLON, FROM N., distant abt. 5 miles.

Canadian Cataloguing in Publication Data

Wood, Charles E. (Charles Edward), 1928-1987
Charlie's Charts of the Pacific Coast of the United States: Seattle, Washington to San Diego, California
Includes index. ISBN 0-9686370-4-3
1. Nautical charts - Pacific Coast (U.S.). 2. Boats and boating – Pacific Coast (U.S.) Maps.
3. Pacific Coast (U.S.) - Description and travel. I. Title
G1461.P5W65 1988 623.89'22979 C87-091438-7

Illustrated by Charles Wood and Richard Miller
Photographs by Charles and Margo Wood

ISBN 0-9686370-4-3

PRINTED IN CANADA

CHARLIE'S CHARTS
of the
U.S. PACIFIC COAST

Seattle, Washington to San Diego, California
Including the Channel Islands

By

Charles and Margo Wood

Published By
CHARLIE'S CHARTS
(A DIVISION OF POLYMATH ENERGY CONSULTANTS LTD.)

P.O. BOX 45064
OCEAN PARK RPO
SURREY, BC V4A 9L1
CANADA

P. O. BOX 1702
BLAINE, WA
98231-1702
U.S.A.

TEL/FAX: (604) 531-6292
E-mail: info@charliescharts.com

Website: www.charliescharts.com

ACKNOWLEDGMENTS

Years ago while working on the fifth CHARLIE'S CHARTS cruising guide, my captain and husband passed away. I completed the book he had started and there are many people whose help I relied on in order to finish his project. I would like to acknowledge the encouraging words of John Guzzwell when I was searching for a author to write the text for this book. He said, “You can do it, you know Charles’ style and approach, just do it!” Such encouragement gave me the boost I needed and I have continued to update and enlarge these guides ever since. I wish to thank several people for their assistance in the production of this updated and expanded edition, in particular,

Captain Ann Kinner provided current information for the Southern California section,

Margaret Fago whose meticulous research has provided enhanced and up-to-date material for more complete coverage of San Francisco Bay,

Mike Skulsky took the time from a busy schedule to collect a mass of updating material covering the coast from northern Washington to Port San Luis, California and

Jo Bailie, co-author of the popular *Gunkholing* guides of Puget Sound and the San Juan Islands who kindly updated coverage of the Seattle area when I was unable to do so.

DISCLAIMER

The word, "CHARTS," in the title of this publication is not intended to imply that these hand-drawn sketches are sufficiently accurate to be used for navigation. They and the accompanying text are meant to act solely as a handy **cruising guide** to local harbors and marina facilities. As a result of marina development, shoaling in the vicinity of some coastal regions and other changes in the areas covered by this book it is inevitable that some of the material is out-of-date and inaccurate by the time it is used.

The safe navigation and operation of a vessel is entirely the responsibility of the skipper whose decisions must be based on the use of current NOAA charts and the U.S. Coast Pilot combined with weather and ocean conditions at the time. A prudent sailor makes decisions with careful consideration of all factors affecting the safety of the vessel and crew.

DO NOT USE ANY OF THE DRAWINGS IN THIS BOOK FOR NAVIGATION. Only the latest edition of NOAA charts and Light List for the Pacific Coast and Pacific Islands should be used for navigational purposes. The authors and publisher are in no way liable, directly or indirectly, for any loss or damages to persons and/or property resulting from the use or interpretation of any information within this book

UPDATING INFORMATION

CHARLIE’S CHARTS cruising guides are constantly being updated as new information becomes available. The material is printed and inserted in each guide and is posted on our website: www.charliescharts.com where it can be down-loaded from the Addenda section.

If you discover a significant change to facilities that would be worthwhile to pass on to fellow boaters please e-mail us at info@charliescharts.com

Coming in with the day's catch at Crescent City, California (Page 70)

The vastness of the Pacific Ocean is interrupted by a vessel on the horizon.

TABLE OF CONTENTS

SYMBOLS

Recommended approach	Range or leading direction
Anchorage	Reported anchorage
Rock or underwater reef	Dangerous rock
Shoal area	Shoal, dries at low water
Steep rocky edge	Sand or pebble beach
Land	Underwater patches
Lighthouse	Kelp
Navigation buoy (may be lit)	Navigation buoy (may not be lit)
Depth in fathoms	Heavy weather flag
Depth contour in fathoms	Buildings (indication only)
Surf, where strong, absence does not mean lack of surf	Coast Guard or commercial mooring buoy

NOTE: All depths are given in fathoms, except where feet are indicated. For metric conversion, approximately 1 fathom equals 2 meters and exactly 1 fathom equals 1.8 meters.

Introduction

The western coast of the U.S. from Seattle to San Diego covers a distance of over 1,500 miles. Within this distance a cruiser is treated to an ever-changing landscape including the snow-capped peaks of the Olympic Mountains and coastal ranges to the lower, often smog-shrouded, dry hills of the south and wind-blown, sometimes barren, Channel Islands. Coastal travel too, offers a kaleidoscope of conditions ranging from the quiet, protected anchorages of the San Juan Islands to challenging river entrances and well-found marinas offering respite from a tiring passage or just a relaxing stopover.

There is much of historic and natural interest ashore for those who have the time and inclination to visit along the way. But approach to any coast has its navigational challenges and decision-making to assure the safety of the vessel. Between Cape Flattery in the north and San Francisco in the south there are few protected harbors with easy access in stormy conditions. In many cases the only shelter from open water is found by transit of river entrances prone to shoaling bars. Even with man-made improvements of breakwaters, protecting jetties and other coastal works, the entrances continually suffer change. This can make entry challenging or even hazardous under some conditions of weather and tide.

In the past the Columbia River entrance became known as "Graveyard of the Pacific," because of the hundreds of lives lost from vessels foundering when attempting to transit the bar. In addition, the numerous vessels wrecked off the coasts of Washington and Oregon helped to form the perception that the only safe passage of this coast was via an offshore route. This principle has modified over the years, but care must still be taken. While this coast must be passed, sailors cannot consider it a recreational cruising area.

Coastal passage from harbor to harbor is one good alternative to consider. Many improvements have been made and continue to be made at most harbors, and the U.S. Coast Guard offers advice and assistance. This, together with a judicious selection of the time to travel during calm, settled weather, makes the coastal passage quite a feasible alternative.

South of San Francisco the number of harbors suitable for small craft increases. But it is not till south of Point Conception, in the relatively benign conditions of Southern California, that the intensity of the yachting scene is seen in the many well appointed facilities for pleasure craft.

Southbound travel along the coast can be an offshore passage, taking advantage of the favorable wind and current, a coastal passage using selected harbors, or harbor-hopping combined with a cruise of the Channel Islands. Though all choices should be approached with care, preparation and prudence, any of these routes can be interesting and exciting for both sail and powerboats.

Northbound travel involves fewer options. The coastal route becomes more favored for both sail and powerboats because it offers occasional respite against the prevailing winds of summer. For sailing craft the offshore route is possible, but it requires a long run offshore to achieve the needed northing if one is to make good headway against the prevailing northwesterly winds combined with the southbound current.

The following sections examine these options in greater detail. The guide describes the harbors, anchorages, points of interest and facilities useful to small craft along the coast, but the harbors are not all equally approachable, and caution should be used in selecting and using those chosen for stopovers.

PURPOSE and LIMITS OF THE GUIDE

This guide is intended to assist in selecting and identifying small boat harbors and anchorages between Seattle and San Diego for vessels transiting the coast. Descriptions of the harbors and facilities, local land form details, depth and land contour lines, lights and other navigational data are provided for this purpose but it is important to remember that the sketch charts are hand drawn and are NOT accurate surveys. Only proper nautical charts should be used for navigation. They are the basis for most sketches, augmented by such details or personal observations as may be of assistance to a small boat skipper. The chart sketches of this guide are intended to supplement proper harbor and coastal charts

All suitable small craft harbors and anchorages along the coast are presented, with the exception of a few that are very rarely used, and of those presented, some are not prudent selections for inexperienced visitors. This is particularly true of those harbors with bars subject to hazardous conditions at times of bad weather or an ebbing tide. Even those harbors deemed suitable as refuges for most conditions can become unapproachable if weather and sea states deteriorate. In such conditions we do not recommend any approach to the coast between Cape Flattery and San Francisco. Staying out at sea will be safer. Similar comments apply to some anchorages exposed to southerly winds.

For extended cruising of the Channel Islands Brian Fagan's *Cruising Guide to Southern California's Offshore Islands* is highly recommended. Herein, only the most popular spots are described, and the names of other anchorages given to facilitate identification as you move along the coast.

Every effort has been made to provide correct and current information, but errors can occur and change is constant on this populous coast. Information and lists, such as those of the charts required and telephone numbers etc., can only be current as of the date of the publication.

A good sailor is always prudent. It remains necessary for you to navigate and work your vessel safely from harbor to harbor. Obviously this guide does not relieve a skipper of the responsibility of decision-making and navigating his/her own vessel safely. It is essential not to rely only on these charts, or on any single aid to navigation. Check and verify all available data. and if variances occur use the greatest caution and make your own decisions in light of the conditions prevailing at the time.

PROCEEDING SOUTH

When leaving the Pacific Northwest, Puget Sound and the Juan de Fuca Strait to travel to southern U.S. ports, the sailing instructions could very simply be given as, "Go to Neah Bay, pass Cape Flattery, then turn left....." But there is much more to consider.

The passage past Cape Flattery opening to the Pacific Ocean is always exciting, no matter how often you cruise offshore, and it is even more thrilling if this is your first voyage to a distant destinations. However, serious thought must be given as to the choice of route to be followed.

The development of various sophisticated electronic navigational devices has supposedly reduced navigation to a matter of flicking switches to turn on the bank of instruments, then pushing a few buttons. Unfortunately this has spawned a new breed of cruisers who tend to be over-confident in their attitude toward navigation and in many cases, do not allow for a non-electronic back-up system. As long as the equipment operates satisfactorily it can provide an amazing amount of detailed information with a touch of a key, but when it fails this treasure chest of data becomes little more than a hollow shell.

The reality is that electronic equipment can and does fail, and there is a crucial need for a no-fail back-up system that will work without electrical power. It is essential that hourly positions be plotted on a current nautical chart, and the navigator must be able to plot a course to the destination in the event that the electronic systems fail. Prudent mariners will not rely solely on any single aid to navigation; charts, radar, depth sounder, compass, GPS and plotters should all be consulted, and when the final destination is approached your eyes will be the final arbiter.

The list of GPS positions given in Appendix II is provided only as a convenience and should be checked against the Current Light List for the position of the offshore buoy. Since this guide may be used for a number of years, the position given for offshore buoys may be out of date as they may change from time to time. Similarly, waypoints are not given for following entrance channels, since depths are subject to frequent change as a result of shoaling and dredging activities. After reaching the offshore buoy for any destination use the latest information available from all sources of navigational information to plot courses or set waypoints in order to follow the channel markers. If you are approaching a river entrance you are advised to contact the Coast Guard on VHF Ch. 16 to obtain information on changes in the channel and current bar conditions. Remember, your eyes can provide the most up-to-date information, so keep a sharp lookout.

OFFSHORE ROUTE

After standing well to the west of Cape Flattery the vessel's route turns to the south. The west coast of the United States tends to fall away to the east at a gradual rate till Cape Mendocino, then more rapidly thereafter. The vessel traveling south slowly leaves the coast further away to leeward as it carries the advantage of the California current and a favoring, prevailing northwest wind in summer months.

The California current has an average speed of 0.2 knots. It is stronger closer to the coast and its effect may be felt from about 10 miles off the coast to 300 miles offshore. It is a southerly proceeding stream of the main North Pacific current, curving clockwise around the northern Pacific basin.

The prevailing northwesterly winds curve around the North Pacific High which is at its most northerly position during summer. However the pressure gradient between the offshore high and the continental low pressure system sometimes becomes very tight, causing stronger winds and higher seas to develop in offshore waters more than 60 miles from land. Conditions are generally a less intense towards the coast. During summer months gales and storms can still occur, but are less violent and are of shorter duration.

The warmer winds passing over the cool water of the California current, together with coastal upwellings of colder water, are responsible for reduced visibility and fog in summer and fall months. The coldest water is nearer the coast, so the densest and most frequent fog occurs closer to shore, with the far offshore route being less affected. However, there are times when the fog banks can be extensive and occur for hundreds of miles.

As the coast recedes, a vessel following the rhumb line is further offshore with each mile traveled south. At Cape Mendocino it is common to be more than 60 miles offshore. Though this is beneficial in allowing sufficient distance from this Cape with its fog banks, wind and disturbed conditions, there is a drawback in that a vessel is in the higher wind (and resulting seas) caused by the pressure gradients. Once past Cape Mendocino courses are usually altered to begin to close the coast towards a destination such as San Francisco or San Diego.

Since north-south traffic lanes are usually densest towards the coast a skipper might assume that the vessel is clear of the shipping lanes. However the north-south shipping lanes are not restricted to a narrow corridor. Ships are widely dispersed and vessels can be found well beyond 60 miles off the coast. Therefore, a good watch is needed when traveling this route.

Briefly recapping-—a full offshore route appears more direct and less at risk from a lee shore, but strong winds and corresponding higher seas can be expected. Ships can still be met far offshore, though they are less prevalent than closer in. Advantages of this route lie in the clear sea room, less chance of fog and in the possibility of a relatively fast passage when southbound.

INSHORE ROUTE

A route that stays about 20 miles off the coast or roughly following the 60-fathom line offers several advantages, but has its own drawbacks. The effect of the California current is stronger and the prevailing westerlies of summer are often less forceful here than further offshore, and heavier seas can be avoided. Crab traps and associated gear are not found 20 or more miles offshore.

While being closer to a lee shore offers some danger, keeping sufficiently distant a well handled vessel need not be at risk and is close enough to find shelter if an emergency situation arises.

The main drawback of this route is increased sea borne traffic, for beyond 25 mile this is the area of the main coastal traveling zone. Vessels of all types, from the largest container ships and tankers to small fishing vessels, tugs and other yachts can be met. In addition this area is over the continental shelf, so that the fishing fleets of several nations are also found here.

The other drawback is fog. The strength of the California current in this zone gives any south-going vessel a boost, but also brings the densest and coldest water. The cooled moist air of the westerlies over this cold stream produces dense fog and usually overcast conditions. This band of deep fog and weather can extend over 100 miles offshore, often moving seaward during the day and rolling shoreward at night. Though the areas of densest fog can vary along the coast, the entire coast can experience fog, often for up to 10 days per month. Point Arguello in California is the foggiest spot on the coast with dense fog up to 20 days per month. South of Cape Conception there is often haze that reduces visibility and dense fog can also be experienced.

From Cape Flattery to the Columbia River the westerly winds are steadier and skies are a little less overcast than south of the Columbia River. However, fog, remains a problem and becomes more frequent after the middle of the summer. The entrance to the Strait of Juan de Fuca in particular is made more difficult by the presence of fog at Cape Flattery.

Between the Columbia River entrance and San Francisco, fog or haze can occur up to 25% of the time in summer and overcast or obscured skies can be expected up to 40% of the time. Gales, though infrequent, can still occur, up to 10% of the time.

The quickest passages can be made along this inshore route, as the weather conditions and seas (though not the fog) are more benign. In addition, trolling for tuna and other fish can be most productive.

Recapping, the inshore route is usually the fastest when northbound and is less bothered by seas and high winds (except near Cape Mendocino and Cape Conception), but heavier sea borne traffic will be met and fog is an annoying factor. A constant and careful look-out is essential.

COASTAL, HARBOR-HOPPING ROUTE

The bars that occur at the entrance to the harbors of the west coast, particularly in Washington and Oregon, make transit hazardous unless there has been an extended period of light to moderate wind combined with little swell from offshore weather disturbances. **Crossing bars should only be done during a flood tide in order to pass during the least turbulent conditions. If there is any question as to the advisability of crossing a bar contact the Coast Guard on VHF Ch 16 to obtain a report on current bar conditions.** When following this route one must allow sufficient time to wait out bad weather until settled conditions resume and the next leg can be taken. If a cruiser has the time to visit some of the coastal towns they will be treated to an interesting variety of communities, each with their own character and charm. When weather and sea conditions are benign this route is probably the least exhausting one to follow when a vessel is northbound.

While the rocky islands that extend all along the coast make this a scenic delight for the shore-based photographer, they make it essential that most vessels travel well off the land. For most of the coast, a berth of 5 to 10 miles suffices, closing in to the offshore harbor buoy at the end of a particular run. Most offshore harbor buoys are laid one or two miles out from the breakwater jetties and provide a point from which an entry line can be taken. It is always important to pick up the offshore buoy and not attempt to short-cut or otherwise avoid the proper approach.

San Francisco is a common destination point for a major stop-over (even when coasting). The next leg of the passage culminates in rounding Point Conception to arrive in the warmer and less boisterous conditions of Southern California, where harbor-hopping is easier. Harbors are more numerous, and access to them is generally less restricted by weather than those north of San Francisco.

The choice of route—whether fully offshore, inshore, or harbor-hopping—is usually determined by time and preference. The offshore routes, being direct, are shorter but are exposed to weather. The usual time from Cape Flattery to San Francisco varies from 6 to 10 days. The coastal route, on the other hand, is interesting but much slower, as the harbor entrances north of San Francisco have bars which are greatly affected by the state of the tide and weather. Thus, the harbor-hopping route could take as much as a month to complete.

While this guide describes and shows sketches for almost all useable harbors, not all of them are suitable for those unfamiliar with the area since under certain combinations of sea and tides some of them can become extremely dangerous. The Coast Guard maintains rough bar advisory signs, which must be heeded. They are diamond-shaped, painted white with an orange border and with the words "Rough Bar" in black letters. The signs are equipped with two alternating quick-flashing yellow lights, which are activated when seas exceed 1.2m (4 ft.) in height and are considered hazardous for small boats. The Pilot cautions that if the lights are not flashing, it is no guarantee that sea conditions are favorable. When conditions become sufficiently hazardous the Coast Guard may close an entrance.

In addition to the Rough Bar advisory sign a Heavy Weather Flag (a square red flag with a square black center) may also be displayed at several entrances along the Washington and Oregon coasts. The flags are displayed from one hour before sunrise to one hour after sunset and are visible from both directions to indicate winds 48 knots and above are forecast.

ENTRY PROCEDURES

Vessels of foreign registry entering the United States must report to U.S. Customs at a Port of Entry and make formal entry for which there is a small charge. Failure to report properly can result in a fine of $5,000 US for each offense. Except for the person reporting the vessel's arrival no one is allowed to board or leave the boat until customs processing is completed. Passports are required for all persons aboard a visiting vessel.

Canadian vessels transiting the coast will be issued a Cruising License, which is valid for one year. U.S. and Canadian vessels (with a Cruising License) may report up to one hour prior to departure from Canada, while enroute or immediately after stopping in the U.S. Pacific Northwest vessels clearing US Customs in Washington are issued a Personal Identification Number (PIN) number (free) which provides ease of clearing customs. The PIN number is not recognized in Alaska or the Hawaiian Islands. Clearance (costing about $10) must be obtained when leaving the U.S. before proceeding to a foreign port

Canadian vessels 9.14m (30 feet) or longer moored in U.S. waters must pay an annual fee of US$25 for a user fee decal; and clearance is not required when departing. Vessels with a user fee decal are issued a Cruising License at no charge. Current vessel registration is required at many marina facilities.

COAST GUARD INSPECTIONS

The US Coast Guard may board US and foreign flag vessels in US or International waters at any time. The usual reason for boarding is labeled a "Safety Inspection" but since September 11, 2001 national security concerns have increased the frequency of these "visits." Appendix V lists recommended safety equipment for cruising vessels.

WEATHER

The most important factor to consider when deciding the time of year to travel is the weather. The Pacific Ocean can be benign and seem to live up to its name, but it is a very large body of water that has significant interlocking relationships with global weather.

The major meteorological feature that concerns travel along the Pacific Coast is the North Pacific High—a high pressure area that is resident off the west coast of North America. At the perimeter of the high, and continuing radially beyond, the wind velocity increases as it rotates in a giant clockwise direction resulting from the Coriolis forces of the Earth's spin. The North Pacific High does not remain stationary, nor is it the same size all the time. It moves as the overall planetary weather systems affect it, enlarging, building, moving, weakening and reducing, and rebuilding again. Statistically, the High shows a trend of annual migration from near 30°N, 130°W in January and February (winter) to further northwest, about 38°N, 150°W, in July and August (summer) before retracing southwesterly again. During any season, the day-to-day position of the high can vary from these average positions, and its perimeters are reported in weather broadcasts. Sources of current weather information are listed in Appendix VI.

The stable and normally larger highs of the summer period (i.e. June to August) provide the best time for travel along the coast. Though heavy weather can be encountered at any time of year, the generally settled weather across the Pacific at this time also means that fewer gales are encountered near the coast. May and September are almost as good as the best summer period. As the advent of winter occurs, the frequency of storms and the parade of lows producing bad weather increases and voyages made during these times must expect to meet inclement conditions. High pressure systems dominate the weather in California offshore waters, although an occasional storm can disrupt settled weather, particularly in winter and early spring. From San Francisco northward the pressure gradient between highs and lows is often very tight, creating strong northerly winds during spring months.

The prevalence of fog from the Strait of Juan de Fuca to San Diego is another common feature of the weather throughout the summer months. The reduced visibility makes the entry to strange harbors not just difficult, but often dangerous unless the vessel is equipped with radar. A night entrance to a strange harbor is never a good idea, but to do so during dense fog is extremely foolhardy. Remaining offshore until visibility improves and daylight arrives, or continuing along the coast would be safer and wiser courses of action to follow. One's navigation must be accurate, and current NOAA charts must be used to plot positions and courses in the event that electronic navigation equipment fails.

Santa Ana (Santana) Winds are local, violent winds that may occur from Santa Barbara to Newport. These hot, desert winds may increase from 0 to 50 knots in twenty minutes with little or no warning. Prior to their arrival there is often good visibility, the air seems noticeably dry and a dark-brown dust cloud may appear. And then the wind . . . . .

CURRENTS

The presence of the massive ocean movement known as the California Current sweeps south along the Pacific Coast of the United States at about .2 knots. The direction of surface winds can accelerate or retard its speed to a considerable extent. Extending offshore more than 300 miles it gives a nice boost to southbound vessels, while northbound boats must work harder for each mile gained against the prevailing northwesterly winds.

A weak counter-current—the Davidson Inshore Current—flows in a northerly direction from San Diego to Point Conception from July to February and from Point Conception to Cape Flattery from November to February.

Tidal currents vary greatly depending on the inequality of the tidal differential (two high waters and two low waters) each day, the area of the waters being drained (or filled) by the tidal flow and the size of the channel. Tidal Current Tables must be referred to before transiting a harbor entrance in order for passage to be made at the optimum time. Do not confuse high or low water with slack water for land-locked harbors, narrow channels, or tidal rivers, as there are several factors determining slack water. For the strength of the current and estimated time of slack water refer to Tidal Current Tables.

DANGERS

Aside from the obvious and standard dangers of rocks, shoals and heavy weather there are some additional dangers peculiar to these waters.

Rogue Waves

A very large breaking wave can sometimes suddenly appear out of nowhere. During heavy weather a particular pattern of waves can produce an immense, steep wave known as a rogue or freak wave. It can also be created by the wake of large commercial or naval craft when combined with a heavy swell; rogue waves of more than 4.5m (15 ft.) can be experienced up to 2 miles from the wake of a large ship traveling at 20 to 30 knots. Rogue waves should not be confused with sneaker waves which sometimes form in shoaling waters along the coast as a result of a combination of factors: the speed of the ebb current, the size of the swell and depth of the water.

Kelp

Kelp grows on rocky bottoms and reefs, its tendrils steaming out in line with surface currents. Spotting it in or on the water indicates caution for it is a sign of danger and usually indicates depths of less than 10 fathoms. Along this coast kelp is particularly heavy in parts of the Santa Barbara Channel, adjacent to parts of the Channel Islands and in the San Diego area. Dead or detached kelp seen floating freely on the surface should cause no alarm.

Deadheads and Log Debris

Logging is a major industry in Washington and Oregon. Much of this timber is transported by log booms, which periodically lose some of the logs. In addition, spring freshets and heavy rains often add to the debris from logging operations. Thus it is no surprise to find logs, deadheads (almost completely waterlogged logs that float vertically in the water) and other detritus floating in coastal waters.

Dredges

Dredging is necessary to combat the natural shoaling of many harbor entrances and channels. But when a dredge is operating in a narrow channel, the space available for passage seems small indeed. Judgment and skill are required when passing a dredge, for its anchor lines must be given adequate clearance. To pass an operating dredge in a channel you must await safe conditions. Some dredges have a barge nearby for collecting dredged material, while others have a long system of large pipes for pumping the silt outside the channel area. Never pass between the dredge and the buoys or lights marking the pipeline.

Pipe-laying Barges and Jet-barges

Clearance of a mile in all directions should be given to pipe-laying barges and jet-barges as they have anchors that extend outwards for as much as 1,525m (5,000 ft.). Vessels associated with the pipe-laying operation may be contacted on VHF Ch. 16 if advice is needed for safe passage through the vicinity.

Fish Havens

Fish havens can be found anywhere along this coast but are more common in Californian waters. Usually marked by buoys and relatively close inshore, they must be given adequate clearance.

Crab Traps

From December to September 15 crab traps can be found along the coast at depths from 5 to 40 fathoms. Though most crab traps are taken out of the water about April 15, a few remain and must be avoided by giving safe clearance to their white or orange floats.

Oil Wells

Many offshore operating oil wells and exploratory wells are to be found in Santa Barbara Channel and other coastal Californian waters. Though the main structure is normally well lit (resembling a huge Meccano set creation strung with lights) any subsidiary structures within 91m (100 yards) of the main platform are marked only by reflective material. A vessel should give good clearance to such operations.

Congested Traffic

The hundreds of pleasure craft in and around the large cities along this coast can be intimidating to vessels from smaller population centers. All one can do is to obey the rules of the road and be ready to take evasive action in the event that other boaters are not operating their boats safely. Moving with the flow of traffic and maintaining a safe distance from other vessels helps to prevent collisions. In theory, this sounds reasonable but there is no doubt that freeway boat traffic adds an element of stress by day's end.

Collisions with Commercial and Naval Shipping

Offshore Vessel Traffic Management recommendations to commercial traffic are as follows: Commercial vessels of 300 gross tons or more transiting the coast are expected to stay a minimum of 25 nautical miles offshore; tankers are expected to travel at least 50 nautical miles off the coast.

It is the small craft skipper's responsibility to keep a sharp lookout at all times to avoid closing with commercial and naval craft. Fast ships beyond the visible horizon of small craft can close for a possible collision in the time taken for a yachtsman to make a cup of coffee. Many a merchantman's watch keeper will trust the radar equipment, and consider the first determined closest point of approach of another vessel as valid and consistent, even when the contact is a sailing yacht (if it displays an effective radar reflector). But if a puff or lull in the wind occurs, the yacht's speed can change considerably and the calculation could become inaccurate with deadly consequences for the yacht. In addition, modern container vessels have an arc of visibility limited to 10° on either side of the bow because of highly stacked containers. So a small vessel is virtually lost in the range of visibility of the larger vessel when within 1.5 miles. Furthermore, only passenger and naval vessels have an appointed lookout making it imperative that yachts look out for their own safety.

Crossing Bars

As ocean swells approach shoaling water, the increased resistance of the ocean bottom causes the water to pile up, forming increasingly steeper seas. When these seas become so steep that they become unstable and cannot support the water at the top of the swell, they collapse in the form of breaking seas. Thus, when crossing a bar, a vessel may encounter anything from a gentle swell to towering, breaking seas only a few seconds apart.

The condition of a bar can change dramatically within a few minutes. This fact was demonstrated recently when I was walking beside the breakwater to view the breakers at the entrance to Humboldt Bay (Eureka). For some time the swells rolled into the entrance. with five or six in a half-hour period creating a wave that broke on the south side of the entrance channel, leaving at least three-quarters of the channel relatively smooth. Suddenly, an eight-foot breaker formed across the entire width of the channel, crashing over the northern breakwater with a cascade of cold water. The Coast Guard officer had predicted that conditions on the bar were about to change with the outgoing tide and a strong southeasterly. Trust recommendations of the Coast Guard for their experience in analyzing weather and sea conditions provide warnings that are to be believed.

Boat-handling problems under these conditions are similar to those experienced in heavy following seas, the main difference being that the waves are steeper and closer together. The vessel often responds by slewing about and tending to broach. Broaching must be avoided because of the potential damage, and also because of the possibility that the vessel may be carried out of the channel on to jetties or shoal areas. Controlling a vessel in such conditions demands skill at the helm and puts tremendous loads on the rudder and steering mechanism. Sailboats are easier to control than powerboats in these conditions because of their straighter keels, larger rudder area and greater weight below sea level.

The decision to enter a harbor involving the crossing of a bar cannot be taken lightly. Preparation of the vessel is necessary and careful planning by the skipper is essential. Most pleasure boaters are unprepared for the experience of crossing a bar under turbulent conditions no matter what is read prior to the passage. Let us examine a worst case scenario to visualize what can go wrong, so that preventive measures may be taken ahead of time.

The most common vessel failures are loss of steering from either shearing of the rudder stock or from steering gear failure. This leads to loss of control of the vessel's direction in heavy swells and/or breaking seas. Broaching of the vessel makes it vulnerable to the full force of the seas on the boat's hull. The cabin sides may be stove in, ports and windows broken or the companionway damaged and the vessel flooded. If batteries are stowed in the bilge, this usually means the loss of engine power and restart capability. Another possibility is the reduction in available power if the engine becomes starved for fuel. The erratic motion while crossing the bar can cause the fine deposits in fuel tanks to plug strainers and screens, thereby restricting the passage of fuel, especially if the boat has only a single fuel filter. In addition, the fine rigging of modern light displacement sailboats often suffers failure under heavy surf conditions, and this can be another casualty.

Assuming that one has survived the worst of the bar there is still a possibility of engine failure occurring upstream from the bar. Fuel may run out because more has been used (or spilled through overflow piping) in making the passage. Many vessels are not equipped with sufficiently heavy ground tackle to set an anchor against strong tidal and river currents; as a result, they may be swept onto a shoal or breakwater by the current.

To avoid these calamities, the skipper must plan his entrance so that it is attempted only when the seas are benign and during a flood tide—the last of the flood being the best. Therefore it is critical that Tide and Current Tables be consulted to determine when the optimum time has arrived for starting the entrance.

The Coast Guard has lookout towers and stations near all bar entrances. Contact them (Ch. 16) prior to committing the vessel to crossing a bar for the first time or if sea conditions are at all in doubt. Remember, viewed at a distance from seaward, especially when seen from the backside, seas appear to be more benign than they actually are. It is far better to get on-the-spot advice regarding the safety of attempting an entrance, than to run into problems and have to call for help. The Coast Guard will often escort a vessel into a harbor; they would rather aid a boat to make port than look for bodies following an unwise entry that results in tragedy.

Prior to crossing a bumpy bar, double-check that all deck gear is securely fastened so that damage will not be done by its coming adrift, nor will it be lost overboard. The risk of sending someone forward to secure a loose object must be avoided. All crew on deck should wear safety harnesses and preferably a survival suit. No short-cuts should be taken where safety is involved, as all resources may be taxed to the limit.

Hypothermia (Exposure)

The cold California Current sweeping the coast and brisk, chilly winds are common phenomena for the entire Pacific Coast until south of Point Conception. Sailboat crew are particularly exposed to the full effect of wind, spray, air temperature and wind chill factor.

Cool winds can cause considerable chilling, while strong cold winds can cause heat loss greater than the replacement rate and can significantly lower body temperature. As wind chill is moderated by a barrier to the wind's effect, the value of a good dodger behind which to shelter, and windproof clothing combined with several layers of underclothing is obvious. Except when the sun's rays beat directly down on calm days, it is worthwhile to wear one's foul weather suit as an outer windproof covering, with several layers consisting of sweaters, flannel shirts and longjohns underneath. Gloves are necessary at times—wool being the best in cold and fisherman's rubber working gloves best in wet conditions. Wool toques help to reduce major heat loss from the head; neck scarves and sea boots with woolen socks are also recommended.

Such cold weather clothing is needed even in summer because one may be standing or sitting for long periods without sufficient activity to generate heat. When traveling on cloudy or foggy days convective cooling can cause body heat loss, particularly when augmented by a strong wind.

Every effort must be made to stay on board. Safety harness should be used whenever it is warranted. If a member of the crew falls into the water, every minute counts in the recovery. It is especially important to realize that a person in cold water quickly goes into shock and as hypothermia advances may become disoriented and unable to assist himself.

A periodic refueling of body heat for those on watch greatly helps to keep the energy level high and prevent chilling. By keeping one or two thermos bottles full of boiling water, hot drinks can be made by any of the crew without having to start the stove. But keep the Mexican coffee and hot toddies until one is off watch, as alcohol gives an artificial feeling of warmth and can cause confusion and impaired judgment.

DISPOSAL OF REFUSE

For centuries the oceans of the world have been considered as limitless, not only in size but also as a cheap and convenient pump-out station. Only recently has the ecological movement, sparked by the research of Jacques Cousteau and others, begun to publicize a need for concern. Cruising boaters now may go ashore on any beach in the world and find bottles, plastic debris and trash of every kind. Since the advent of the plastic age in the 1950s the ocean's shores have been fouled and defaced by the indestructible evidence of the "disposable" attitude.

Every boater has a duty not to discard non-organic garbage at sea. Since there was sufficient space on board to stow the containers at the beginning of the trip, there obviously is the same capacity available for holding them for their return to shore-based recycling depots or garbage containers. It is hoped the photographs on the following page will encourage the reader in this regard. The following poem hopes to make people think:

IN YOUR WAKE

What do you leave in your wake?
The choice is many, but few stop to think.
A sparkling trail on a shining sea
Gone in a trice to eternity?
Or a bear can track that staggers along
To mar and foul a virgin beach?
Or do you leave the plastic debris of the packaged world
To smother a bird, kill a fish, foul a prop or litter a coast?
The wake you leave tells it all.
What is the legacy that you bequeath?

Carey Paige

WHALE WATCHING

Boaters who travel along the coast in the areas covered by this guide will probably experience the joy and excitement of seeing whales from the deck of their own boat. Since there is a growing population of both whales and boaters, it is essential that some guidelines for viewing be understood.

Whales have never gone out of their way to follow, chase, or kill human beings, or to separate a mother from her child. Unfortunately, man does not have the same unblemished record. Man's actions have at times resulted in injured whales attacking boats and frantic "mother whales" (cows) reacting in an erratic manner when separated from their calves. Therefore, a few basic rules for viewing whales will make it a pleasurable, worry-free experience for both parties:

- Stay at least 100 yards away from a whale
- Approach slowly, from the whale's side, never "head-on"
- Travel parallel to and at the same speed as the whale
- Avoid abrupt changes in direction or speed of the vessel
- Do not cross the whale's path or get between two whales
- Never "box inn" a whale, i.e. have a boat on either side
- If a curious whale approaches your vessel, put the engine in neutral

Debris Harms Our Marine Life

Rex Herron

Endangered species also suffer. Sea turtles mistake plastic bags for jelly fish (a major source of food) and choke to death.

Brian Lawhead

Some 30,000 Northern Fur seals die yearly from entanglement in netting, a 50% decline in population in 30 years has been noted.

Common items like six-pack rings, fishing line, and strapping bands entangle and kill sea birds, fish and mammals. Plastics can last many hundreds of years, harming even large mammals like the gray whale.

Michigan Dept. of Natural Resources

Fishermen and boater safety is jeopardized when debris fouls propellers or causes engines to overheat. Heavy losses of time and money are reported from debris damage to vessels and equipment.

Pierce Harris

George Antonelis

This page and the one following marked ⚡ have been reprinted (with permission) from a brochure published by the National Marine Fisheries Service, Maine Refuse Disposal Project, Newport, Oregon.

Plastics can last for hundreds of years. A careless moment lasts generations. The best method of disposal is at a recycling depot.

Dale Snow

Frans Lanting

Birds, fish and mammals mistake plastic for food. Some birds even feed it to their young! With plastic-filled stomachs, animals die of starvation or poisoning.

Bill Keay

Six-pack rings can result in starvation and death to seabirds such as this Great Blue Heron and other marine life.

DON'T TEACH YOUR TRASH TO SWIM!
WGM

Western Canada Wilderness Committee

Oil spills such as the disasters caused by the Exxon Valdez in 1989 (which ran aground in Prince William Sound) and the leaking oil barge (which was deliberately hauled out to sea from near Grays Harbor, Wash. in 1988) caused the deaths of thousands of seabirds and damage to marine habitats along the coasts of Washington, British Columbia, and Alaska.

Public pressure is needed to have a moratorium on offshore drilling, require the transport of oil in double-hulled vessels and establish a Law of the Sea making the owners of oil shipments liable for all clean-up costs and damages anywhere in the world. This would encourage oil companies to act responsibly toward the fragile marine environment.

Jim Boeder

You can help!

* Make boat policy that no trash is discarded, washed, or blown overboard.
* Minimize the amount of non-degradable products on board. Provision your vessel using bulk-refillable containers.
* Stow trash for disposal in port. Encourage your port to provide convenient refuse disposal facilities.
* Where possible retrieve trash found on the ocean.
* Share your concern with friends, fellow mariners and politicians.
* Participate in beach clean-ups, and leave the beach clean after visits.

The gray whale migrates from the summer feeding grounds in Alaskan waters to the birthing and breeding havens in Mexican waters, following a route usually within 10 miles of the coast. Pregnant females commence the southbound trip in the fall, followed by the rest of the herd making their passage as late as February. The northbound trip of the herd of about 17,000 whales begins in April, with arrival in Arctic waters in June or July.

Gray whales are stocky, have no distinct dorsal fin, may reach 12m (40 ft.) and weigh up to 40 tons. Females are larger than males. They are a slate or mottled gray, with scattered whitish patches of barnacles or whale lice. Unique among baleen whales, they feed by scooping great mouthfuls of silt and filtering out worms, crabs, shrimp and tiny fish.

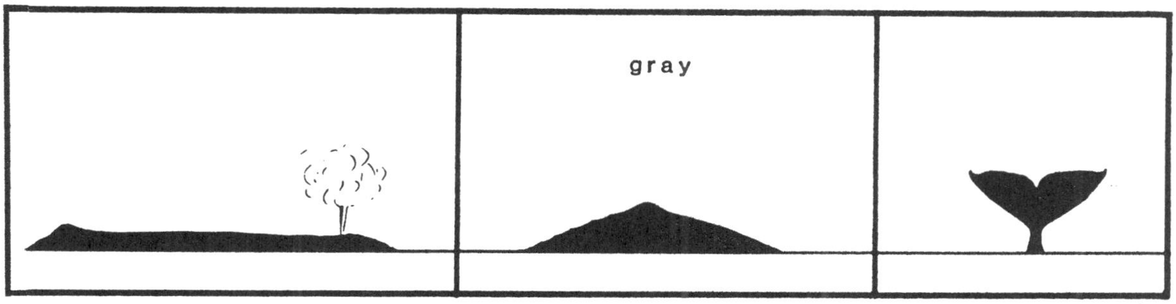

During the summer months great blue whales and humpback whales can be seen in the Santa Barbara Channel feeding on krill, which they filter from the water by means of baleen plates growing down from either side of the roof of the mouth. It is believed that these whales avoided these waters during the period from the 1940s to the 1990s because of noise and other disturbances caused by seismic exploration for oil. Fortunately this kind of exploration ceased in the 1980s and these magnificent creatures have returned to their customary feeding grounds.

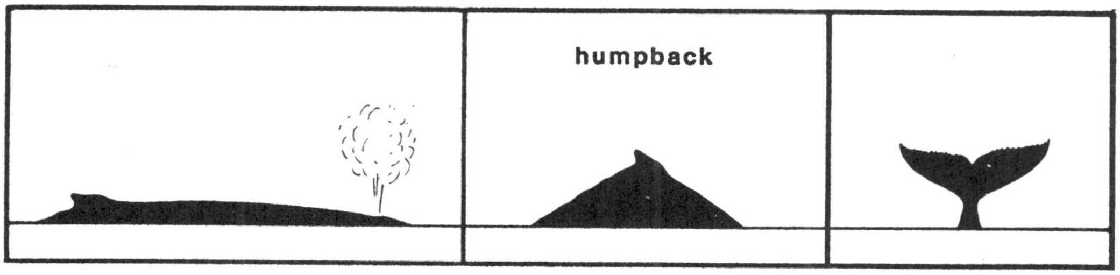

Both humpback and great blue whales have grooves on their throats and chest and have a small dorsal fin and are known as rorquals. Humpback whales may reach 15m (50 ft.) and weigh up to 45 tons. They are often seen in Hawaiian waters during the winter and are known for their acrobatic breaches and haunting underwater songs. Great blue whales can grow up to 29m (95 ft.) and weigh up to 150 tons.

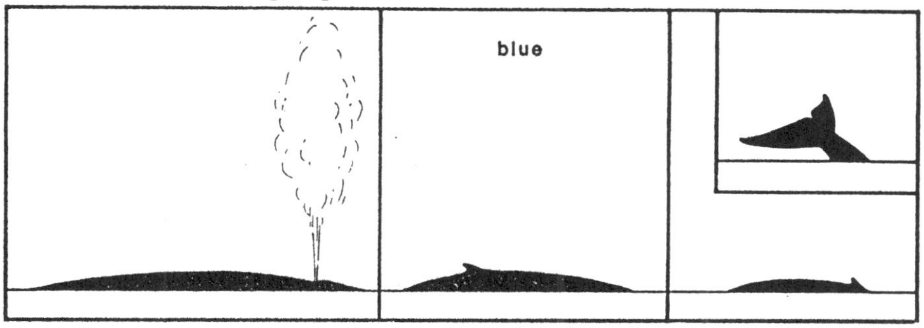

Drawings by Larry Foster, courtesy of General Whale.

SEATTLE

The bustling city of Seattle is well situated for the enjoyment of protected boating on both salt and fresh water, as well as mountaineering, skiing and hiking on the nearby Cascade Range and Olympic Mountains. To the east, Mount Rainier provides a spectacular alpine backdrop to this historic city. Surrounding Salmon Bay and Lake Union, it shares the shores of Lake Washington with the cities of Kirkland, Bellevue and Renton. Seattle's western shoreline borders Elliott Bay and Shilshole Bay on Puget Sound.

West Point lighthouse

Since adjoining fresh water areas are about 6m (20 ft.) above sea level it is necessary to transit the Hiram M. Chittenden Locks before proceeding via the 8-mile long Washington Ship Canal to Lake Washington. This trip takes you boating literally through the heart of the city, a feature making Seattle unique among west coast ports.

The city boasts a wealth of first-class facilities and services for the boating public. Two large moorage facilities operated by the Port of Seattle provide moorage for over 1,500 boats at Shilshole Marina on Shilshole Bay (as well as 150 guest slips) and Fishermen's Terminal on Salmon Bay is a large mooring basin for the fishing fleet based here. When fishing boats are out (late spring to early fall) guest moorage for pleasure boats is available with Med-style mooring (bow or stern tie). In addition, there are two privately owned marinas: Elliott Bay with 1,200 slips and Bell Harbor Marina catering to primarily to visitors with over 80 slips. Seattle is a busy commercial port with 18 piers servicing deep-sea vessels. Private marinas, ship chandlers, repair and hauling facilities can be found lining the shores of Lake Washington. Many residences enjoy the luxury of a private dock extending from their waterfront property.

After passing the deep-sea port facilities in Elliott Bay south of Shilshole Bay Marina there are two well appointed facilities, Elliott Bay Marina and the newest marina, Bell Harbor. Vessels moored in these marinas have the luxury of not having to line up to wait their turn to pass through the locks as do boats moored on Lake Washington's shores. Information regarding these facilities is provided on pages 16 and 17.

Other marinas with guest slips are:

Nautical Landing at 2500 Westlake Avenue North, Seattle, WA 98109. This facility features superyacht moorage catering to vessels 25m (80 ft.) to 60m (200 ft.). The large piers permit convenient vehicular shipside access when provisioning, taking on fuel and other deliveries. For information telephone (206) 284-2308, Fax (206) 285-6998; www.nauticallanding.com

Parkshore Marina, 9050 Seward Park S., Seattle WA 98118. Limited transient moorage is available making it advisable to call in advance at telephone (206) 725-3330; www.parkshoremarina.net

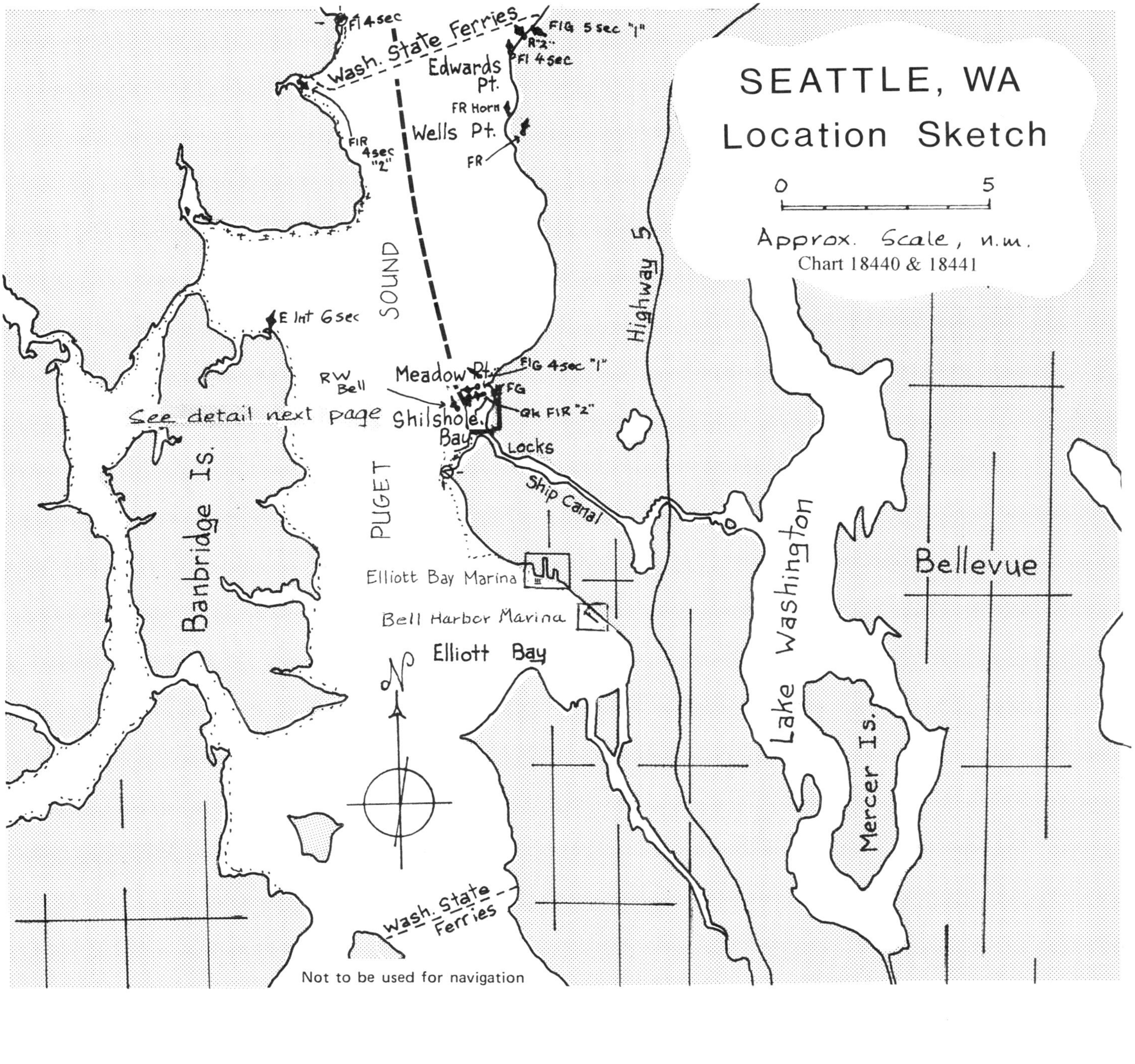
SEATTLE, WA
Location Sketch
0
5
Approx. Scale, n.m.
Chart 18440 & 18441
Fl 4sec
Wash. State Ferries
FlG 5 sec "1"
R"2"
Fl 4sec
Edwards Pt.
FR Horn
Wells Pt.
FR
FlR 4sec "2"
SOUND
E Int 6sec
Highway 5
RW Bell
Meadow Pt.
FlG 4sec "1"
FG
Qk FlR "2"
See detail next page
Shilshole Bay
Locks
Ship Canal
PUGET
Banbridge Is.
Lake Washington
Bellevue
Elliott Bay Marina
Bell Harbor Marina
Elliott Bay
N
Mercer Is.
Wash. State Ferries
Not to be used for navigation

Salmon Bay Marina, 2100 W. Commodore Way, Seattle, WA 98199. Guest moorage is available for vessels up to 18m (60 ft.). For information call (206) 282-5555 or Fax (206) 282-8482m. www.salmonbay-marina.com

Elliott Bay Marina is just around West Point in a southeasterly direction toward the deep-sea commercial docks in Elliott Bay. The marina can accommodate 1,200 vessels in sizes ranging from 10m (32 ft.) to 19m (62 ft.) along with end ties that can handle much larger vessels. Facilities are available for monthly and overnight moorage. The office is open daily with 24-hour service and VHF Ch 78A is monitored. For information call (206) 285-4817 or Fax (206) 282-0626. The entrance to the marina is at 47°37.84'N and 122°23.47W.

The usual amenities of fuel, pump-out, restroom and laundry services are available and showers are free. Propane is sold at a U-Haul dealership about five blocks from the marina on Fifth Avenue. Wireless internet is available in the marina but the nearest internet café is downtown at I-Table located at Pike Place. Bus service links the marina to the downtown area and various points of interest in the city.

A unique feature of the marina is that although it is located conveniently close to the downtown area it is away from the noise and activity associated with a large city. In addition, the panoramic view of the snow-capped peaks of the Olympic Peninsula and beautiful Mount Rainier add to its unique ambience.

Bell Harbor Marina, about three miles south of Elliott Bay Marina, is the newest facility providing moorage for pleasure craft in the Seattle area. The entrance to the marina is at 47°36.6'N, 122°20.9 W. The marina caters to visiting boaters and has a 7-day maximum stay during the peak summer season and 30 days in the off season. Vessels from 9m (30 ft) to 36m (120 ft) can find accommodation on a first-come first-served basis as reservations are limited. The office is open from 7 a.m. to 10:30 p.m. during the summer and from 7 a.m. to 5 p.m. in the winter months. To arrange moorage call the office on VHF Ch 66A or telephone (206) 615-3952. Reservations can also be made on line. Internet services are available in the Business Center at the Marriott Hotel just across Alaskan Way from the marina.

The marina is located conveniently close to downtown with just a 5-minute walk to Pike Place Market (open daily) and numerous restaurants and interesting shops. Just across Alaskan Way is street car service with connections to buses serving Pioneer Square (where the famous underground tour of Seattle commences), the Space Needle, Seahawks Stadium and Exhibition Center and Safeco Field. Northwest of the marina is the Odyssey Maritime Museum with an observation deck on the fifth floor. On Pier 62-63 summer concerts are regularly held. Although the marina is close to the hustle and bustle of the downtown area boat security is assured with keyed access gates and security cameras.

In addition to the marinas described there are many other facilities in the Seattle area. Many of these allow a limited number of livaboards; the services offered and rates charged vary. For information on a few of these such as Swiftsure Marina, Westlake Marina and University Marina contact **Boatworld Marinas** at telephone (206) 284-4420; www.boatworldmarinas.com

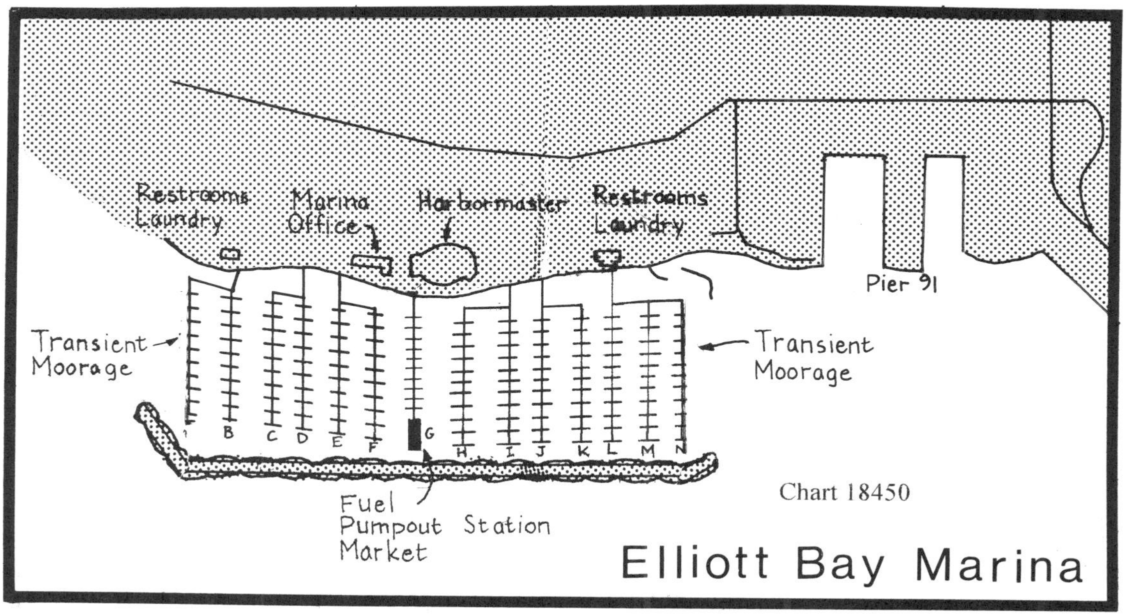
Restrooms Laundry
Marina Office
Harbormaster
Restrooms Laundry
Pier 91
Transient Moorage
Transient Moorage
B C D E F G H I J K L M N
Fuel
Pumpout Station
Market
Chart 18450
Elliott Bay Marina

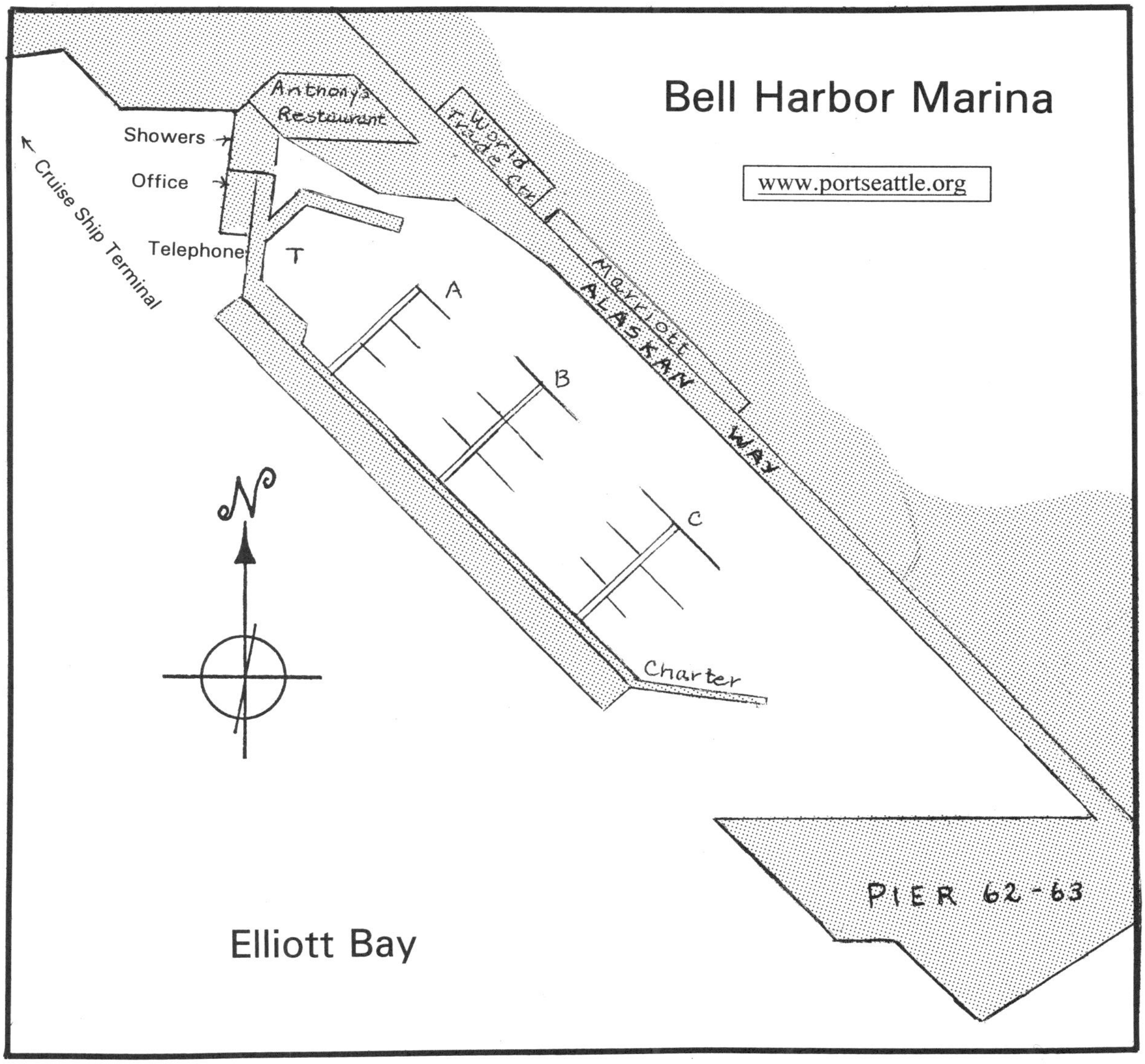
Bell Harbor Marina
www.portseattle.org
Anthony's Restaurant
World Trade Ctr
Showers
Office
Telephone
Cruise Ship Terminal
T
A
B
C
Marriott
ALASKAN WAY
Charter
N
PIER 62-63
Elliott Bay

Shilshole Bay Marina is the largest pleasure craft facility in Seattle and the first one seen after passing the extensive outer limits of this large city. It is eevident when traveling south in Puget Sound just before the entrance to the Salmon Bay waterway and prior to reaching deep-sea port facilities in Elliott Bay. For slip assignment call the marina on VHF Ch. 17 or telephone the harbormaster at (206) 728-3006; after hours call (206) 601-4089. Entrance may be made at either end of the breakwater. Transient berths are on J dock near the fuel dock and on the north and south sides of the central pier.

In addition to gasoline and diesel the fuel dock sells bio-diesel (20 – 80 blend) and has a convenience store selling snacks, beer and various sundries. Pump-out stations are available in several locations. For major grocery shopping one must take a taxi to the nearest supermarket which is located about one mile away. A bus stop on the road next to the marina provides access to downtown attractions. Restroom, laundry and shower facilities are located throughout the marina. A boatyard at the southern end can haul vessels up to 13m (42 ft.).

The **Hiram M. Chittenden Locks** consist of two locks for boat traffic—a large one 250m by 25m (825 ft. by 80 ft.) and a smaller one 45m by 9m (150 ft. by 28 ft.). The lift varies from 2m to 8m (6 to 26 ft.) depending on the lake level and tidal fluctuations. Approximately 100,000 commercial and pleasure craft pass through the locks annually. On the south side of the small lock is a 21-weir fish ladder which allows migrating salmon and trout to swim upstream; an underwater viewing gallery is provided to watch the action as about 500,000 salmon and trout use the fish ladder

You will need at least two 15m (50-ft.) mooring lines with an eye of at least 30 cm (12-inch) when passing through the locks. The crew must be alert and attentive to taking in or paying out the line as the vessel is raised or lowered. Several fenders should be tied to each side of the vessel since the sides of the locks are rough cement and other boats may be touching your hull. A free booklet available at the Shilshole harbor office and most marine stores entitled, "*Guidelines for Boaters,*" should be studied prior to locking through for the first time. During busy holiday weekends you may have to wait as much as five hours until you can enter a lock. After entering it takes an hour or more by the time you tie up, lock through and exit the large lock, less for those using the small lock. The lock tenders are very helpful and will tell you where to go and how to tie your vessel. It is essential that you secure the vessel properly as currents are quite strong when the gates of the lock open.

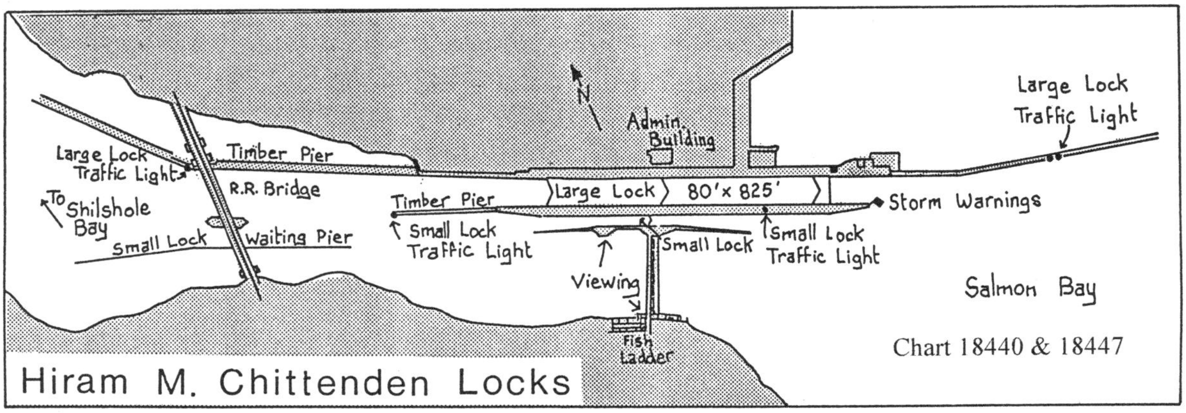

Hiram M. Chittenden Locks

Shilshole Bay Marina

Approx. Scale n.m.

Not to be used for navigation

Chart 18447

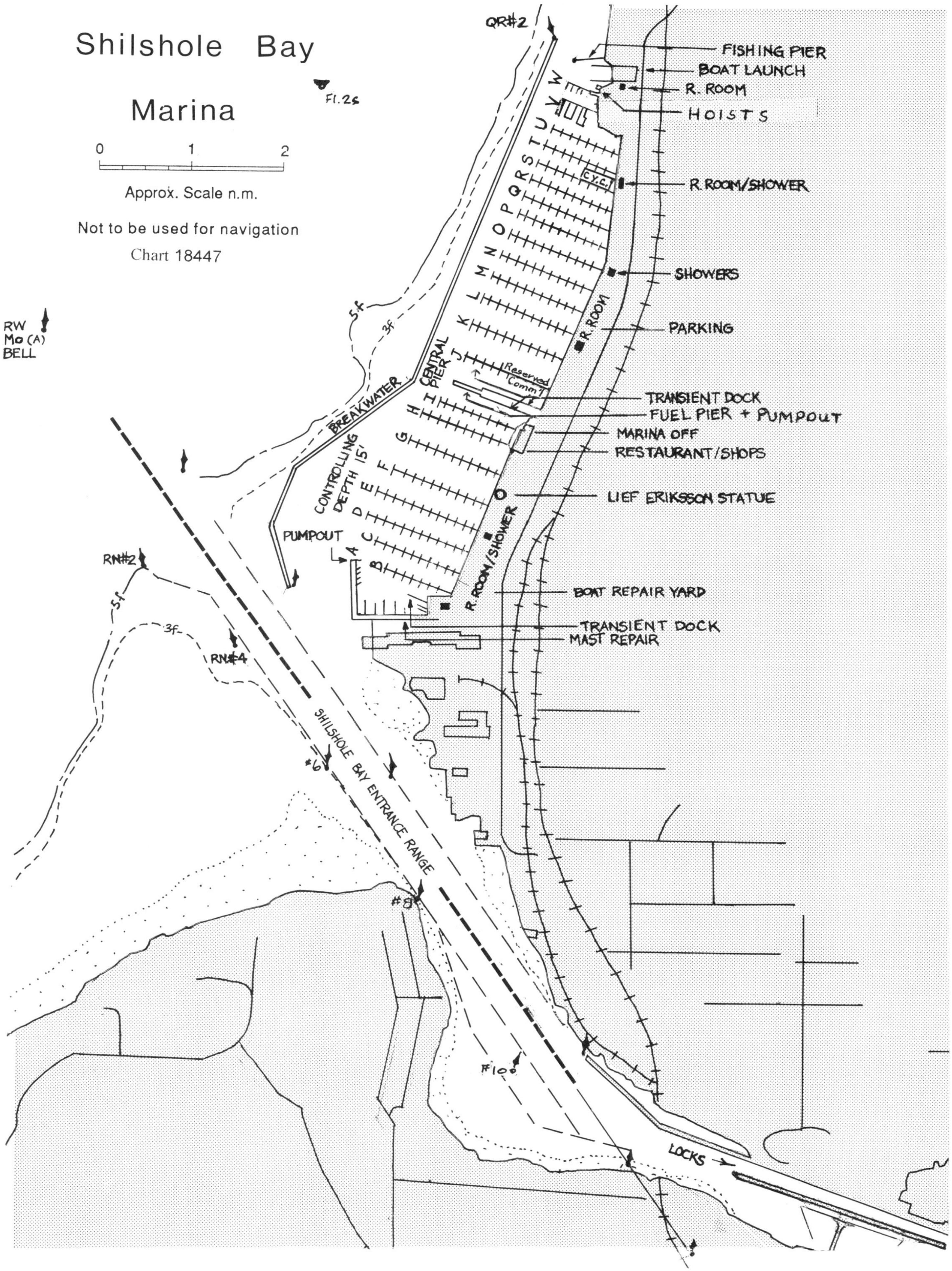

LAKE WASHINGTON SHIP CANAL & LAKE WASHINGTON

The Lake Washington Ship Canal from Puget Sound passes through Shilshole Bay, Salmon Bay, Lake Union, Portage Bay and Union Bay before entering Lake Washington. These heavily traveled waters are lined with repair yards, yacht clubs, moorages and used for a variety of sports. The Ship Canal is crossed by five bascule bridges and two fixed bridges. Clearances of the draw spans vary from 4m to 14.5m (14 to 43 ft.). Tenders of drawbridges monitor VHF-FM Ch. 13 and it is their working channel.

The Burlington Northern Railroad Bridge is west of the locks and its clearance varies from 12m to 17m (40ft. to 56 ft.) depending on the state of the tide. When the bridge is down it indicates that a train is about to pass over the bridge and it will not be opened until the train has passé. Blowing your whistle/horn for it to open at this time is futile since train traffic has priority over marine traffic. The amount of clearance will vary somewhat since the level of water in the lake is lower during the winter months and higher in the summer.

Telephone numbers and clearance heights for the bridges are as follows:

Ballard Bridge (15th Avenue) . . (206) 282-9525: 13.4m (44 ft.) clearance

Fremont Avenue Bridge . . . . (206) 386-4234: 9.1m (30 ft.) clearance

University Bridge . . . . (206) 684-4765: 13.1m (43ft.) clearance

Montlake Bridge . . . . (206) 720-3048: 14m (46 ft.) clearance

Hours of operation for bridge tenders are as follows: Monday to Friday from 9 a.m. to 4 p.m. and from 6 p.m. to 11 p.m.; on weekends, from 7 a.m. to 11 p.m.; between 11 p.m. and 7 a.m. bridge openings are by appointment only. Since there is only one bridge tender on duty from 11 p.m. to 7 a.m. and he is responsible for all four bridges some delay may be experienced. You can call the bridge shop at (206) 386-4251 to let them know approximately when you wish to transit; one hour of notice of your desire to transit during this time is required. If you don't reach a person, leave a message on the answering machine. It is still necessary to use the horn signals (one long and one short blast) during this period even though arrangements have been made. Highway 520 (Floating Bridge) opens when an appointment has been made 24 hours in advance.

Whistle signals are the preferred method of contacting bridge tenders since their use reduces radio chatter. Alternatively you can call on a cell phone or VHF Ch 13 (the ship to ship channel). Normally, the bridge tender responds within 10 minutes.

The Seattle area has an abundant supply of full service boatyards, chart dealers, marine supply stores and specialty shops. Some of the long-established larger yards are indicated on the sketch on the opposite page; showing them does not indicate that they are superior to the many smaller yards that have not been listed as a result of space limitations. A number of yacht clubs are based here and have reciprocal docking privileges with yacht clubs scattered along the coast. The Seattle Yacht Club is the only one which has a number of slips devoted to guest moorage at the present time.

It's a tight squeeze when week-end traffic passes through the Hiram M. Chittenden Locks in Seattle, Washington. (Page 16)

Fighting the tidal current in Deception Pass, Washington, can be exciting. (Page 24)

Left: The snow-capped Olympic Mountains make a beautiful back- drop for Port Angeles, Washington. (Page 34)

Pinnacle rocks south of Cape Flattery, Washington discourage close inspection. (Page 38)

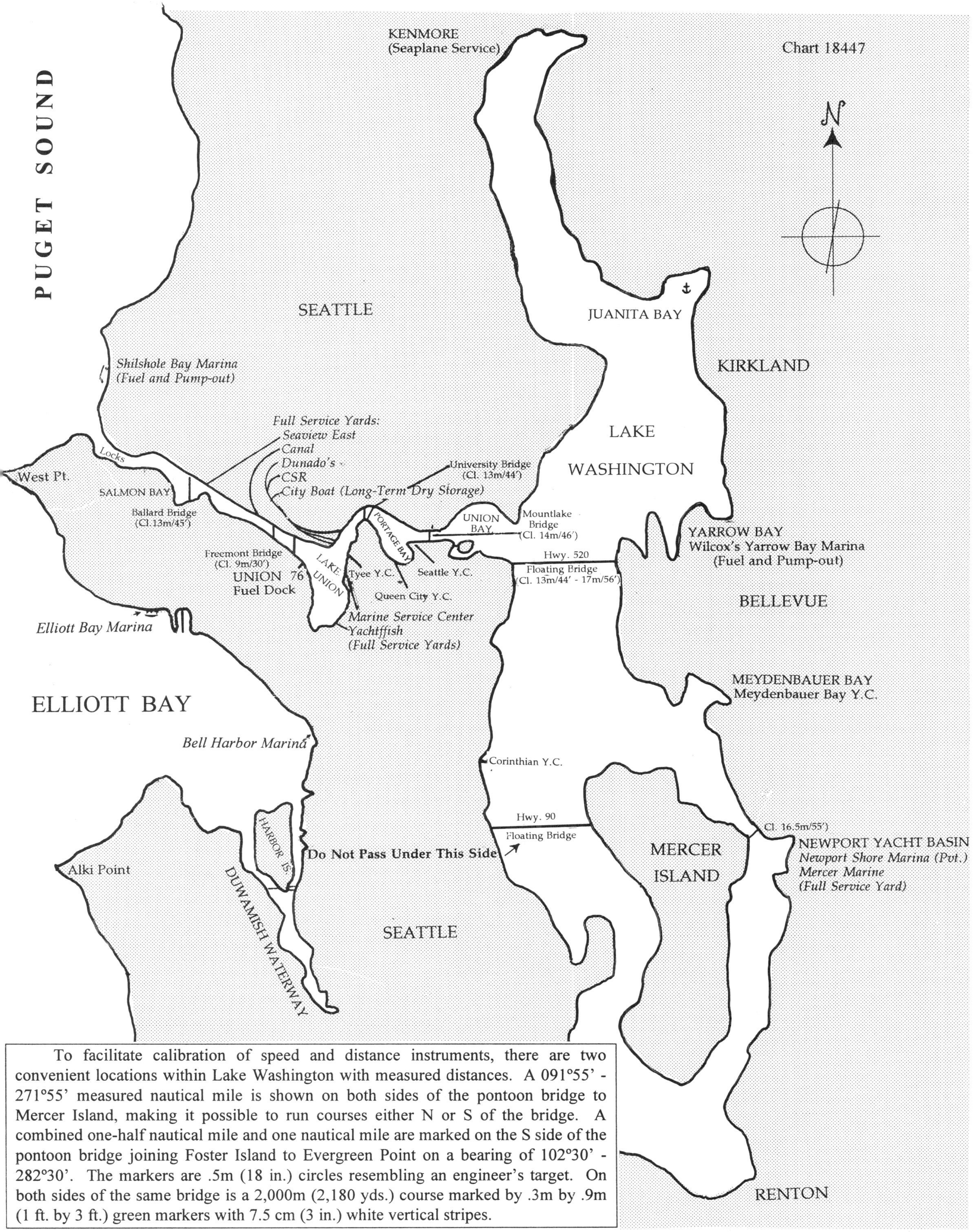

To facilitate calibration of speed and distance instruments, there are two convenient locations within Lake Washington with measured distances. A 091°55' - 271°55' measured nautical mile is shown on both sides of the pontoon bridge to Mercer Island, making it possible to run courses either N or S of the bridge. A combined one-half nautical mile and one nautical mile are marked on the S side of the pontoon bridge joining Foster Island to Evergreen Point on a bearing of 102°30' - 282°30'. The markers are .5m (18 in.) circles resembling an engineer's target. On both sides of the same bridge is a 2,000m (2,180 yds.) course marked by .3m by .9m (1 ft. by 3 ft.) green markers with 7.5 cm (3 in.) white vertical stripes.

EVERETT

This city of over 60,000 is strategically located on the doorstep of two cruising areas: Puget Sound and the San Juan Islands. The popularity of boating has made necessary the development of small craft facilities second only to the massive complex at Marina del Rey in southern California.

Everett is located about 25 miles north of Seattle on the edge of a fan-shaped shallow area in the eastern side of Possession Sound. The snow-capped mountains of the Olympic Peninsula make a stunning backdrop to the waters of Puget Sound while to the east are the tree-covered slopes of the Cascade Range.

The shoal area near Everett extends south from Tulalip Bay to flashing buoy "1" and is marked further to the south by bell buoy "3". After rounding this buoy to port the vessel can proceed toward the entrance that is protected by two breakwaters. There are lights at the ends of each of the breakwaters that protect the dredged channel leading to the small craft harbor. The channel is marked by lights, buoys and lighted and unlighted ranges.

The Snohomish River has a navigable channel that runs through the western side of the small boat harbor and continues for some distance to port facilities upriver. The huge smokestack of a pulp mill located upriver is prominent for miles. Though the river current becomes more evident during low tide, it is not normally a major problem when navigating in the small craft harbor area. Log booms are often moored to the eastern side of the long rock breakwater protecting the basin and channel from westerly winds. From fall to spring California sea lions congregate on this breakwater (Jetty Island) making their presence known with a continuous babble of barking.

The **Port of Everett Marina** has accommodation for over 1,500 boats including some covered moorage. Transient berths are available on the breakwater though water and power is currently not available. The harbormaster may be called on Ch. 16 for berth allocation or by telephone at (425) 259-6001. Self-registration and moorage payment stations are located on the northern end of the southern breakwater, at the north end of the fuel dock and outside the Marina Office. Repair facilities, hoists, ship chandlers and yacht brokers are located on the north side of the docks, while on the south side is Everett Marina Village, a collection of specialty shops and restaurants. As propane is no longer available at the fuel dock one must inquire about availability at an RV outlet in the city.

Points of interest in the area include a Firefighters' Museum on 13th Street which offers through-the-window viewing, while a self-guided walking tour of the Port Gardner district can be obtained from the Public Library. Transportation can be arranged to reach the Boeing 747/767 plant where 90-minute tours are held Monday through Friday. Here one can see the largest (by volume) building in the world that is 11 stories tall and covers 62 acres. For more information call (206) 342-4801. Children under 12 are not allowed to participate in these tours.

EVERETT, WA

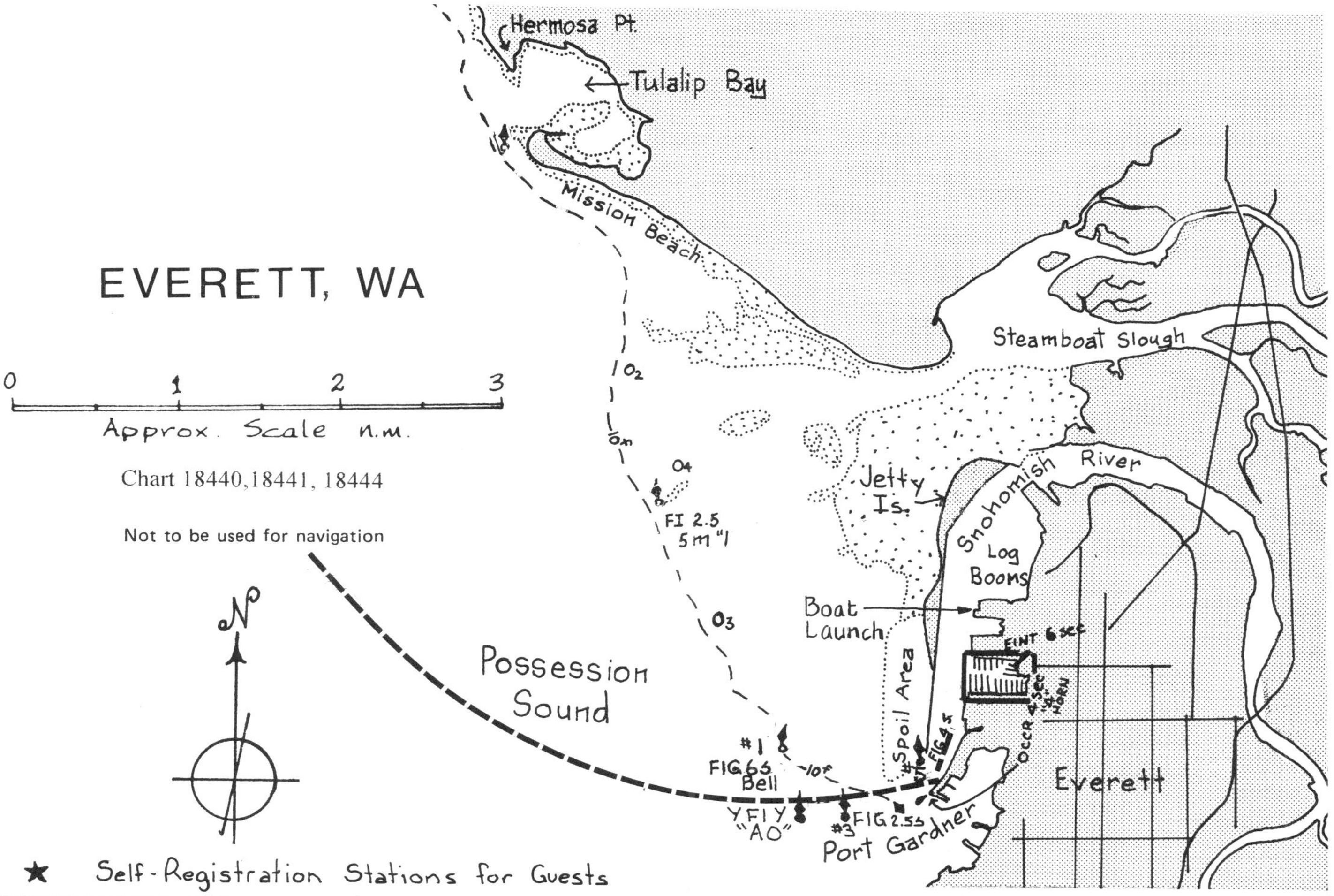

★ Self-Registration Stations for Guests

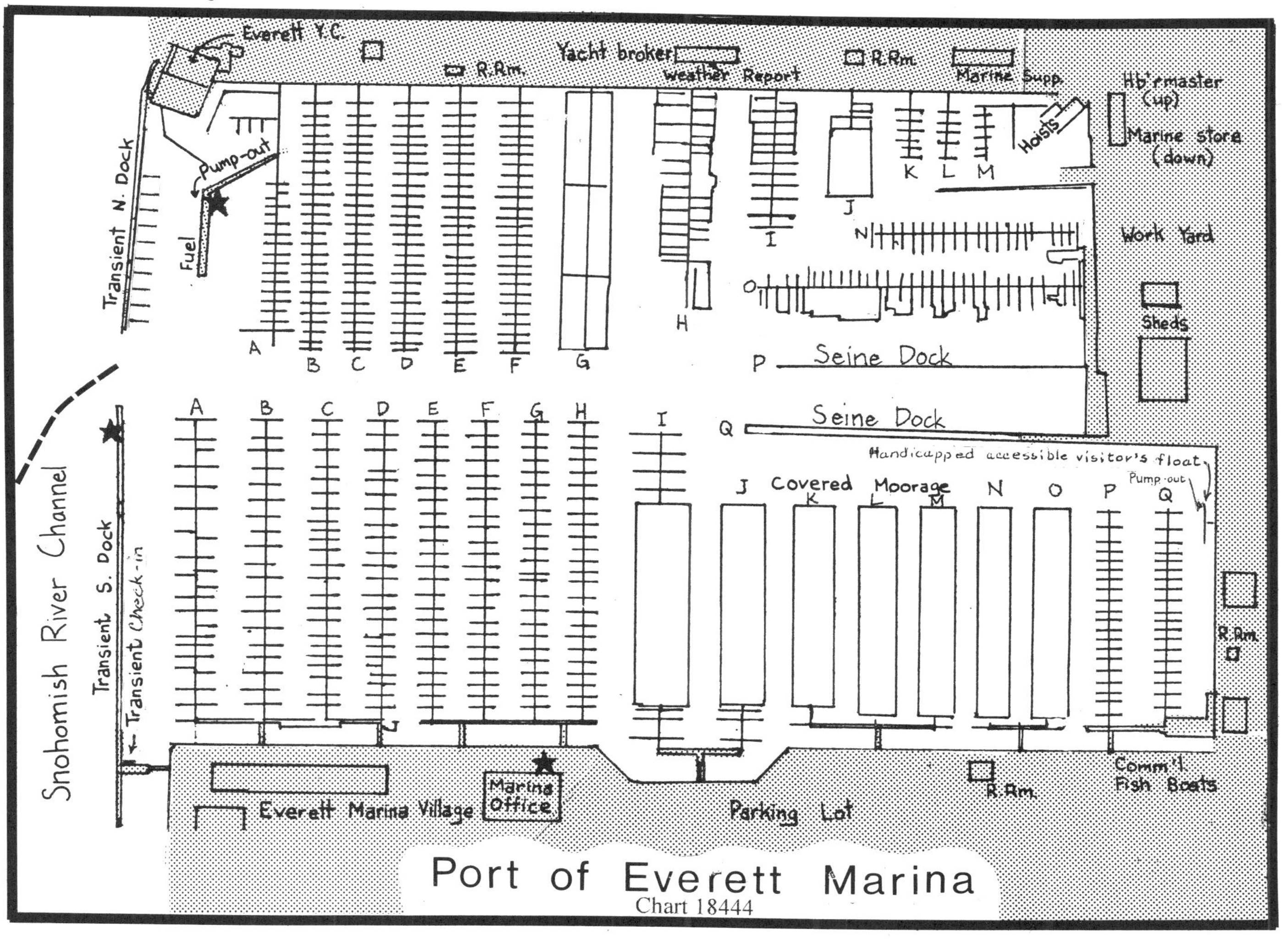

Chart 18444

BELLINGHAM

Whatcom County Museum

Bellingham is a Customs **Port of Entry** on the northeastern side of Bellingham Bay and is a major wood and wood products exporting port. Prominent features include a water tank on top of a furniture building, a stack at the cement plant and a church spire near the waterfront. A lighted HERALD sign on the newspaper building is conspicuous at night. The city contains an interesting mix of well maintained buildings featuring architecture from the 1800's combined with various modern designs.

Squalicum Harbor is located next to the Squalicum Creek Waterway and is protected by breakwaters on the southern side. The harbor is operated by the Port of Bellingham and now consists of two large basins that can accommodate over 1,400 vessels. The outer and older basin has a breakwater with two entrances and depths inside of 1.5 to 2 fathoms. The harbormaster's office is on the north side, just to the east of the main float walkway and the Bellingham and Squalicum Yacht Clubs also have quarters nearby. A popular café known for its tasty hamburgers is the Web Locker, just a few steps north of the harbor office. Visiting cruisers are welcome on a first-come, first-serve basis with a three-day limit in effect. The transient berths are just ahead of the sea wall. The harbor office monitors VHF Ch 16 from 8 a.m. to 5 p.m. daily. A Coast Guard Station is located in the harbor.

A block behind these shore buildings is Harbor Mall, with several small marine restaurants and coffee shops. Nearby are large marine hardware chandlers such as LFS (Lummi Fish Supply) and Redden Marine Supply catering to both pleasure and fishing craft. West Marine Products is located on Meridian Street in the northern part of the city. A 25- to 30-ton Travelift and dry storage yard is at Hilton Harbor a short distance to the south. For information call (360) 733-1110.

The newer harbor basin is slightly to the east and behind the older basin and is entered at the southeast corner. Several large areas of floats are enclosed within. The transient berths lie off the Harbor Center on the northeast side. Nearby are boat launching ramps and a Coast Guard Station. A short walk takes one to a powerful statue on a base lined with beautiful detailed friezes honoring fishers who make their living from the sea.

Roeder Avenue runs roughly east-west behind the harbor areas, as does the railroad. The downtown Bellingham area and shopping areas are to the east along this avenue, far enough away to require a taxi or bus ride. Bus service is hourly except Sundays and holidays.

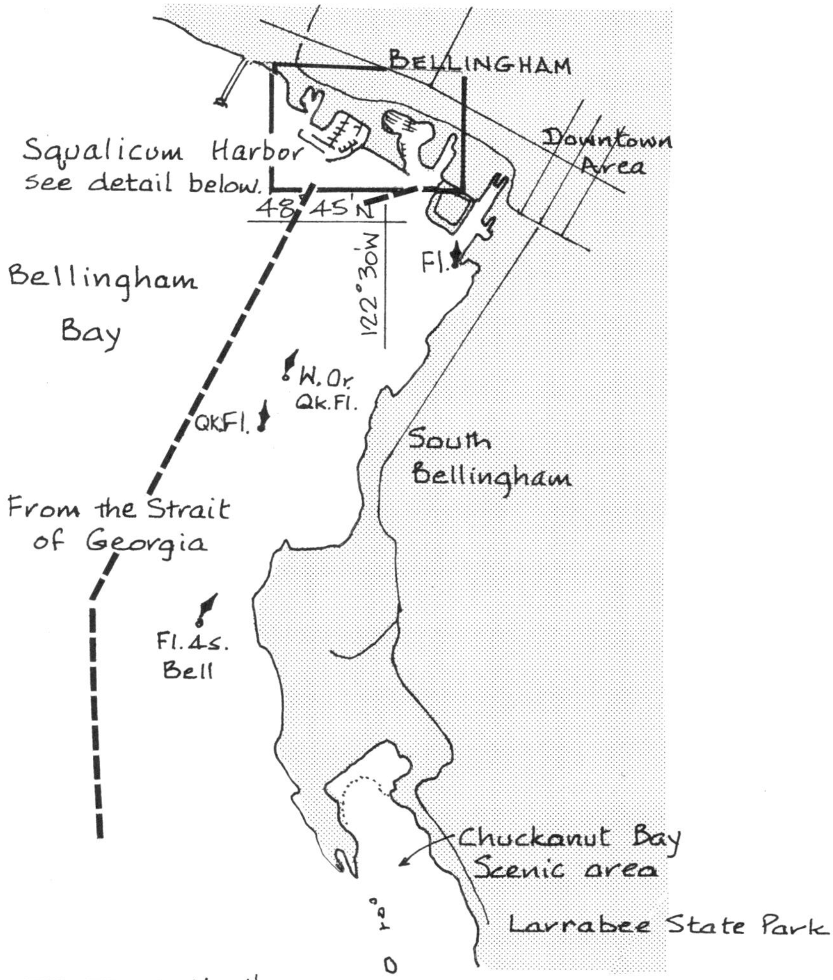

BELLINGHAM, WA

0 1 2n.m.

Approx. Scale, n.m.

Not to be used for navigation

Chart 18423

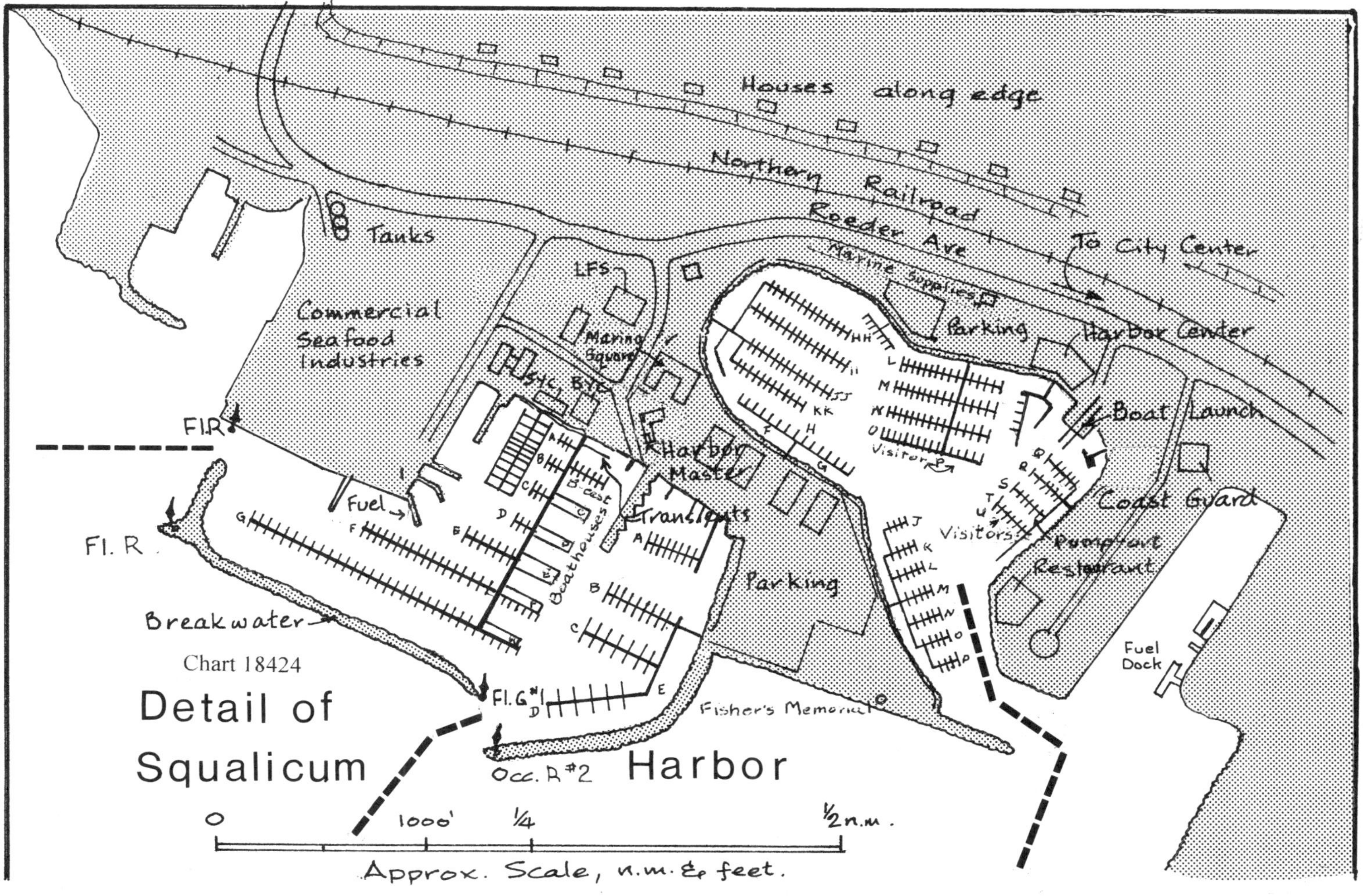

BLAINE

Adjacent to the US - Canada border, this small town is a **Port of Entry**. It lies on the eastern side of Drayton Harbor that is a large, mostly shallow bay protected by Semiahmoo Spit. A protective dike and breakwater encloses the port of Blaine, which is entered through an opening between Semiahmoo Spit and the breakwater. Entry to the U.S. can be made from either the Public Marina or Semiahmoo Yacht Basin. Call U.S. Customs at (360) 332-6318 from the telephone booths ashore to obtain clearance either by phone or having an inspector come to check the vessel and crew. After 5:00 p.m. or on Sundays and holidays, call 1-800-562-5943.

Blaine harbor is a controlled by the Port of Bellingham. **Blaine Harbor** is a first-class facility, occupying the northeastern part of Drayton Harbor. The western portion of these floats is for commercial fishing vessels and nearby is a repair yard with haul-out facilities and fish unloading docks. The eastern floats, occupied by Blaine Harbor Marina, offer moorage for over 650 pleasure craft where transients can usually find space. The Harbor office is in the large one-story building with a clock tower at the head of the gangway leading to the docks. The harbormaster monitors VHF 16; the working channel is VHF 68. The main part of the town is about .25 mile to the east. In addition to Blaine Marine Services and a West Marine store in town, a marine supply store shares space with a furniture store in the vicinity of the westernmost fuel dock. The nearby restaurant caters to fishermen and serves substantial meals at a reasonable cost.

Another marina is **Semiahmoo Yacht Basin**, which can be entered through an opening in the floating breakwater behind Semiahmoo Spit. Immediately to starboard on the inside of the breakwater is a fuel station where temporary moorage can be taken until moorage has been assigned. The Harbor Office is in a large building at the head of the main ramp. Also enclosed in this building are a convenience and gift store, small marine supply store and service department. Next door is a dry storage area, travel lift and work yard. Facilities include a laundromat, showers and restrooms. The yacht basin is adjacent to Semiahmoo Resort, which features a hotel, restaurant, housing development and an Arnold Palmer designed golf course. Though the town of Blaine is less than .25 mile away by water, it is about 7 miles distant by road. A short walk north of the town leads to a beautifully manicured park featuring the historic Peace Arch monument.

Anchorage can be taken in Drayton Harbor south of Semiahmoo Yacht Basin, taking care to anchor clear of extensive shoal area. Southerly or southeasterly winds can kick up a nasty chop in the harbor.

Blaine is a small border town, so if any major refitting is needed, better prices and a larger variety of services may be obtained in Bellingham or Seattle (a short trip by bus). Bus service connects Blaine to the city of Vancouver and surrounding cities of Surrey, Richmond Delta and White Rock.

WHITE ROCK, British Columbia

Across the bay is White Rock, a **Canadian Port of Entry**. A long pier protrudes from the shore with a short breakwater at right angles to the end. Proceed around the breakwater to the dock on the eastern side, taking care to clear the shoals extending well out from the shore. A call can be made from the telephone booth at the shore end of the pier to contact Canadian Customs to arrange for entry to Canada by telephoning (604) 531-7581.

BLAINE, WASH.; WHITE ROCK, B.C.
Ports of Entry

Chart 3313, 3463

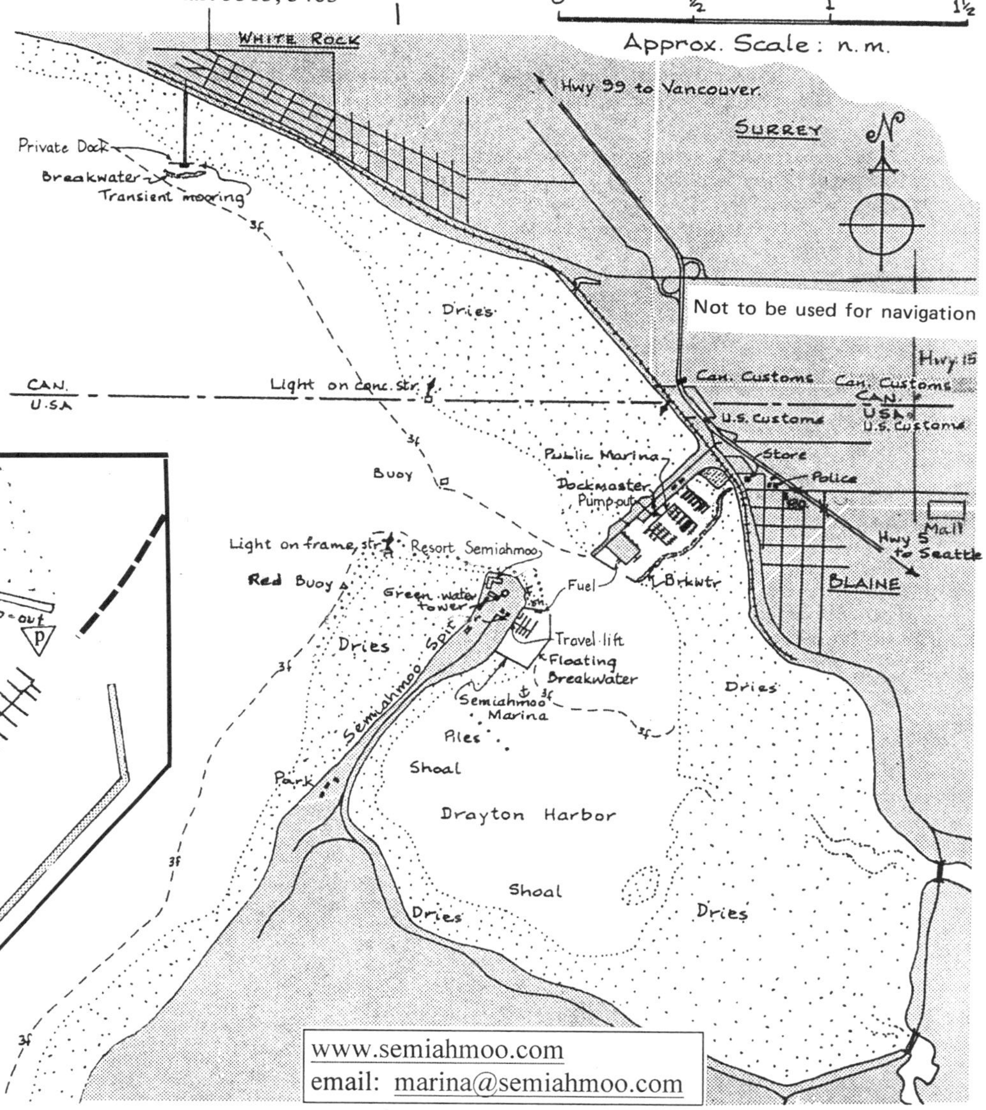

SEMIAHMOO BAY

Semi·ah·moo Marina

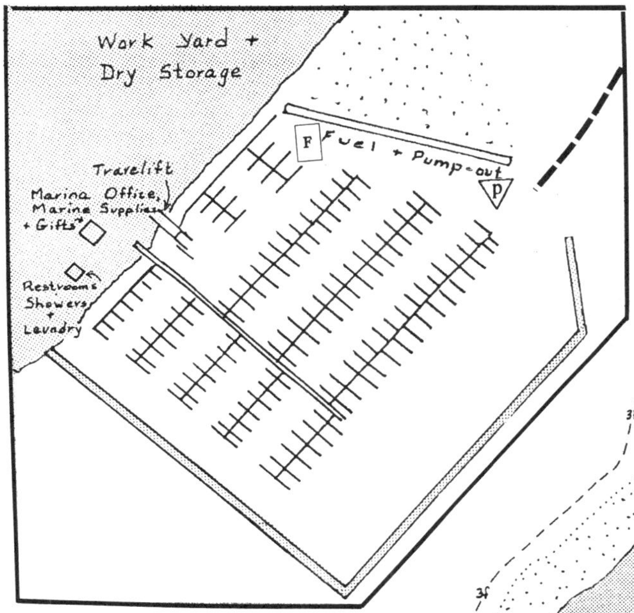

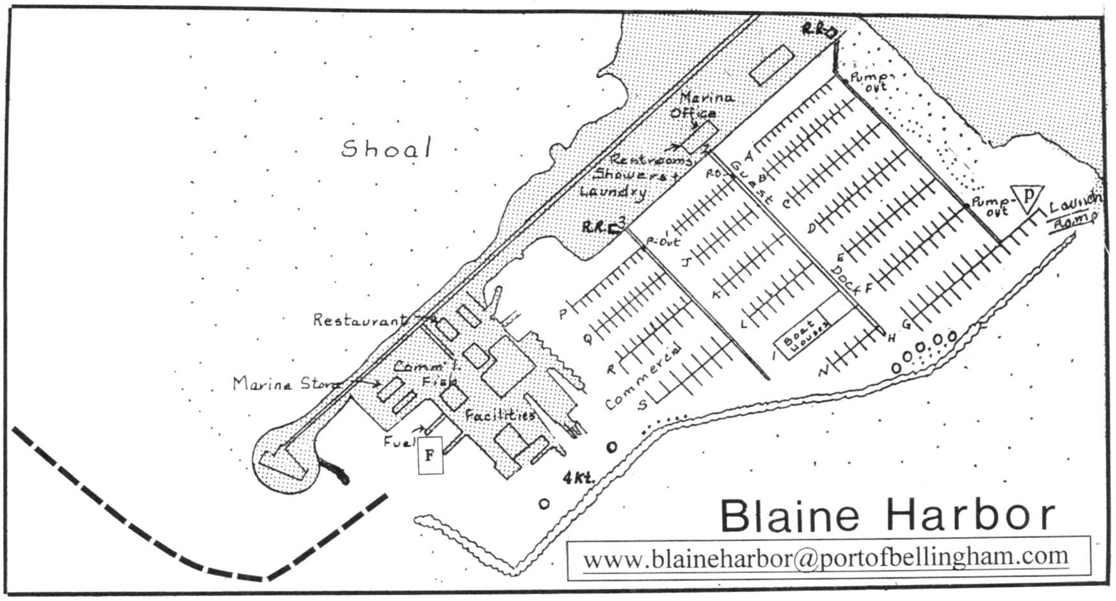

POINT ROBERTS

Point Roberts projects below the 49th parallel of latitude and thus the southern tip of the peninsula is US territory. The BC Ferry docks, super port loading facilities and mounds of coal at the Roberts Bank coal terminal are landmarks for the yacht basin a couple of miles to the southeast. An attractive marina is on the southern side of the peninsula, strategically placed for access from the heavily populated lower mainland (Vancouver and environs) to the Gulf and San Juan Islands.

Point Roberts Marina has a well thought out and effective design. It is arranged to allow the tidal flow to flush out the yacht basin, while affording moored vessels maximum protection. There is berthing capacity for over 1,000 pleasure craft from 6 to 22m (20 to 70 ft.) in length. This would be an ideal moorage if it weren't for the regular dusting of coal dust deposited on decks as it drifts southward from the huge Roberts Bank coal terminal. For slip assignment call (360) 945-2255 or Fax (360) 945-0927.

The basin is entered from the west, passing between the shore and the breakwater. Do not enter from the east as it is shoal. The entrance curves toward the west of north and the travel lift and fuel floats are on the starboard side when entering. These floats are also used for awaiting Customs. Call 332-6318 on weekdays (8:00 a.m. to 5:00 p.m.), or after hours and on Sundays and holidays call 800-562-5943 to arrange clearance. Clearance may be given by phone, so take paper and pen along to make a note of the clearance number in case any reference is needed later. Remember that only the skipper can go ashore to report, and no one else should leave the vessel until clearance is granted.

Marina facilities include a travel lift, mast tower and a small marine supply store. Restaurants and grocery stores are about .5 mile away where liquor and provisions are available. The US/Canadian border is about 2 miles away where Tsawwassen (a part of the city of Delta) has supermarkets, restaurants, hotels and a variety of services available.

From Point Roberts it is only 11 miles across the Strait of Georgia to Active Pass. However, since Point Roberts is US territory, you must enter Canada if you plan to sail across to the Gulf Islands. From Point Roberts you can travel directly northward to Vancouver for entry into Canada or about 12 miles east to enter at White Rock. If you want to visit the Gulf Islands you can take advantage of the prevailing northwesterly winds to reach down to Boundary Pass around the south end of Saturna Island, turning up and entering Canada at Bedwell Harbor on Pender Island. Bedwell Harbour (currently known as Poet's Cove) is a **Canadian Port of Entry** where the Customs office is maintained during the summer from 9:00 a.m. to 5:00 p.m., Monday to Saturday. You can then leisurely cruise in the protected waters of British Columbia's Gulf Islands.

If you are northbound and enter Canadian waters at White Rock, and wish to stop near Tsawwassen but not enter Point Roberts, there is a small area south of the B.C. Ferry Terminal mole that can be used as an anchorage. Boats of the Tsawwassen Yacht Club moor here in summer, though in winter it is open southeasterly storms. Enter close by the southernmost ferry docking structure where a relatively deep channel runs along the south side of the mole. After about half a mile the channel it widens into a basin suitable for anchorage. Painted oil drum buoys mark the passage.

POINT ROBERTS, WA
Port of Entry

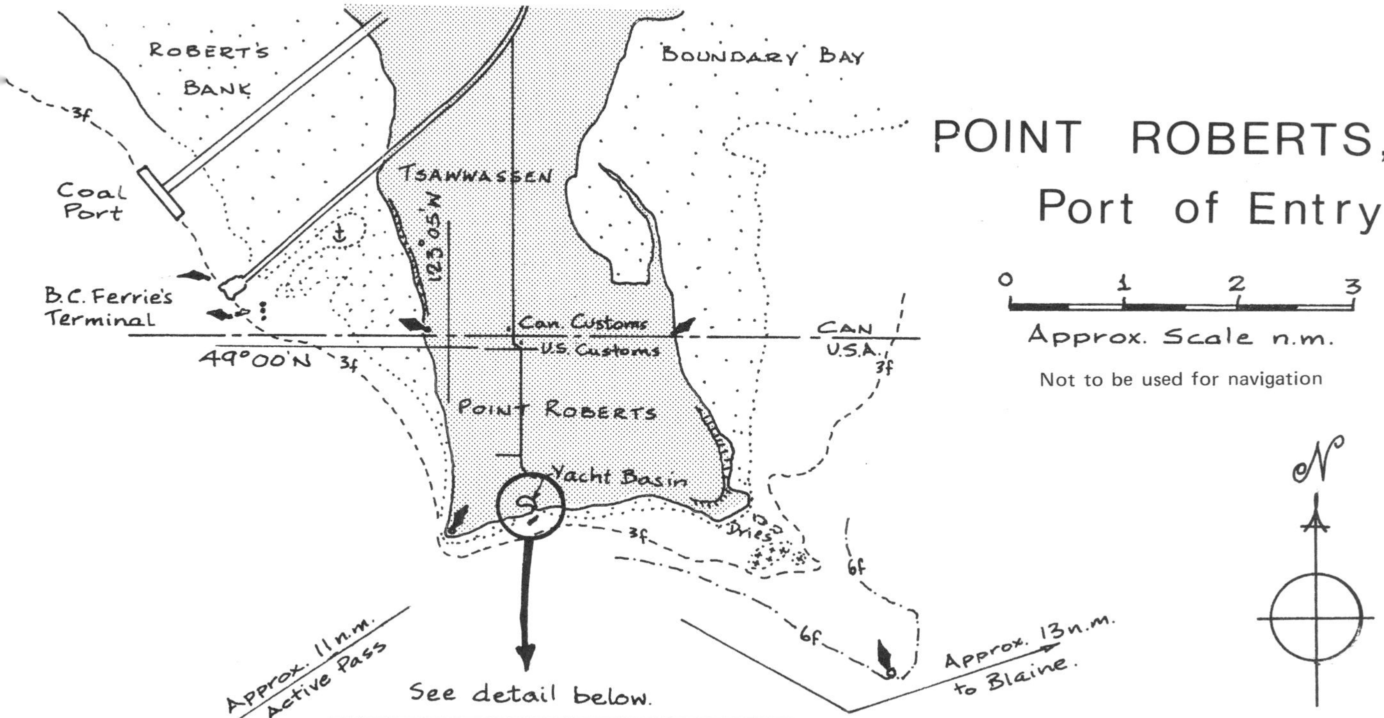

www.pointrobertsmarina.com
email: prmarina@robertsmarina.com

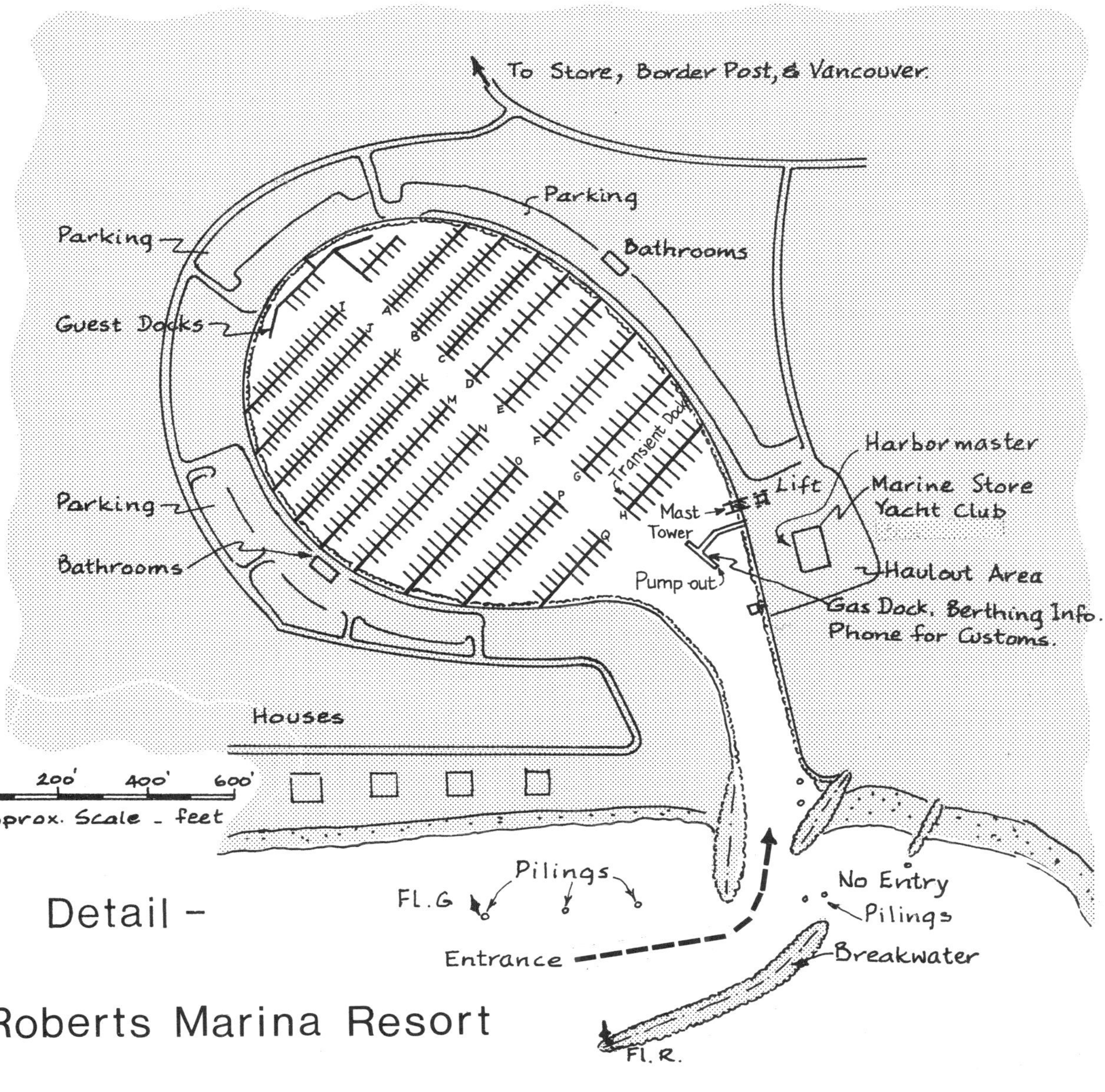

Detail -

Point Roberts Marina Resort

FRIDAY HARBOR

Haro Strait, trending north from the Strait of Juan de Fuca, separates the San Juan Islands in the US from the Gulf Islands in Canada, with the International Boundary passing roughly midway up the Strait. San Juan Island is one of the largest islands in the archipelago and has **Ports of Entry** in Friday Harbor and Roche Harbor.

Friday Harbor is the principal town of the San Juan Islands. The harbor is in a cove about a mile wide, with Brown Island set in the middle of the cove. The north entrance is the wider one and is marked by lights; it leads directly to the town. At the northern entrance the University of Washington Oceanographic Laboratories are prominent, where a pier extends from the shore. The southern entrance is narrower but is clear and can also be used.

The town of Friday Harbor is in the western part of the cove and has a ferry terminal where ferries connect the island with Anacortes on Fidalgo Island and other islands in the San Juan Islands. Adjacent to the ferry terminal is a public marina operated by the Port of Friday Harbor where docks are allocated for transient moorage. The building at the head of the main dock houses the harbor office with restrooms and showers nearby. The Customs office is on the ground floor of the 2-story building west of the harbor office; the San Juan Island Y.C. occupies the second floor. The Customs office is open 8 a.m. to 5 p.m. daily. If a Customs inspector is not on duty call 1-800-562-5943 for telephone reporting.

A walk around the town is worth the exercise for there are a number of interesting shops and eating spots. Kings Market ground floor offers sundries and a good selection of groceries (which they will deliver to the marina). The second floor of the market houses a well-stocked marine supply store. Laundry facilities are in town. This is a very popular destination for yachts, and reservations are highly recommended from May 15 to Sept. 15; call the harbor office at (360) 378-2688.

In the southern end of the bay is Shipyard Boat Harbor, and next to it is Jensens Shipyard Marine where a vessels may be hauled and repair facilities are available. Numerous private docks, floats and buoys are scattered around the bay. Float plane and ferry traffic areas must be kept clear and anchorage should not be taken within 90m (100 yds.) of the breakwater to avoid fouling anchor cables. Good anchorage can be taken elsewhere in the bay in 6 to 7 fathoms, soft mud.

ROCHE HARBOR

Roche Harbor is an attractive destination on the northwest corner of San Juan Island, where Henry Island and Pearl Island give good protection to the harbor. The main entrance is a mid-channel course between the two islands. Various other entrances can be used, but they are narrower and obstructed in places by kelp beds and shoals, making them less convenient. A 5-knot maximum speed is in effect within the harbor, monitored by a patrol vessel.

Nestled against the base of wooded hills are a hotel, restaurant, general store, cafe and post office. The marina has slips for almost 400 vessels. During the summer months the Customs Office is clearly seen at the outer end of the marina, at other times it is next to the cafe. The office monitors VHF 78A from 9 a.m. to sunset and is open from 8 a.m. to 7 p.m. from May 15 to Sept. 15. Reservations are recommended from June to September at this busy marina, particularly on holidays and weekends. Roche Harbor Marina and Resort and can be made by calling (800) 451-8910 /Ext. 500 or (360) 378-2155.

Good anchorage may be taken in the extensive area in the lee of Pearl Island.

FRIDAY HARBOR

ROCHE HARBOR

SAN JUAN ISLAND

1000' 0 ½
Approx. Scale, n.m. & feet.

Not to be used for navigation

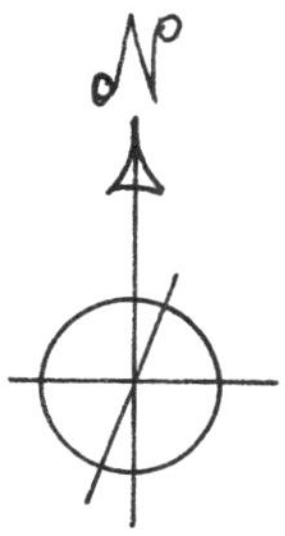

Ports of Entry

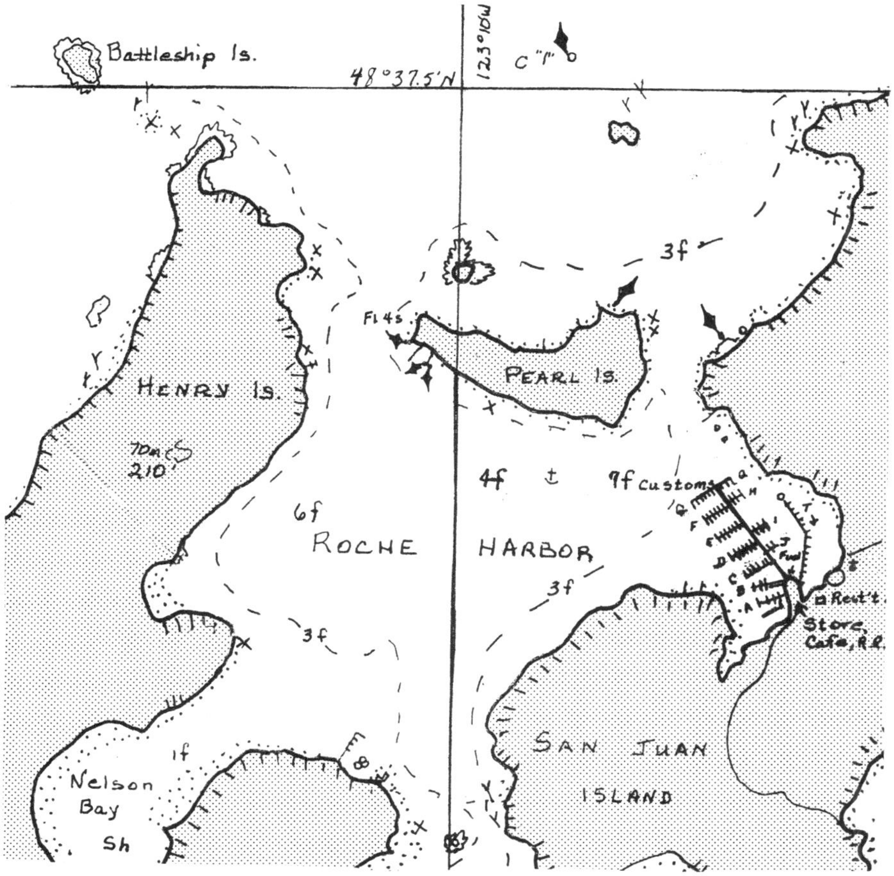

ROCHE HARBOR

www.rocheharbor.com
email: rhmarina@rockisland.com

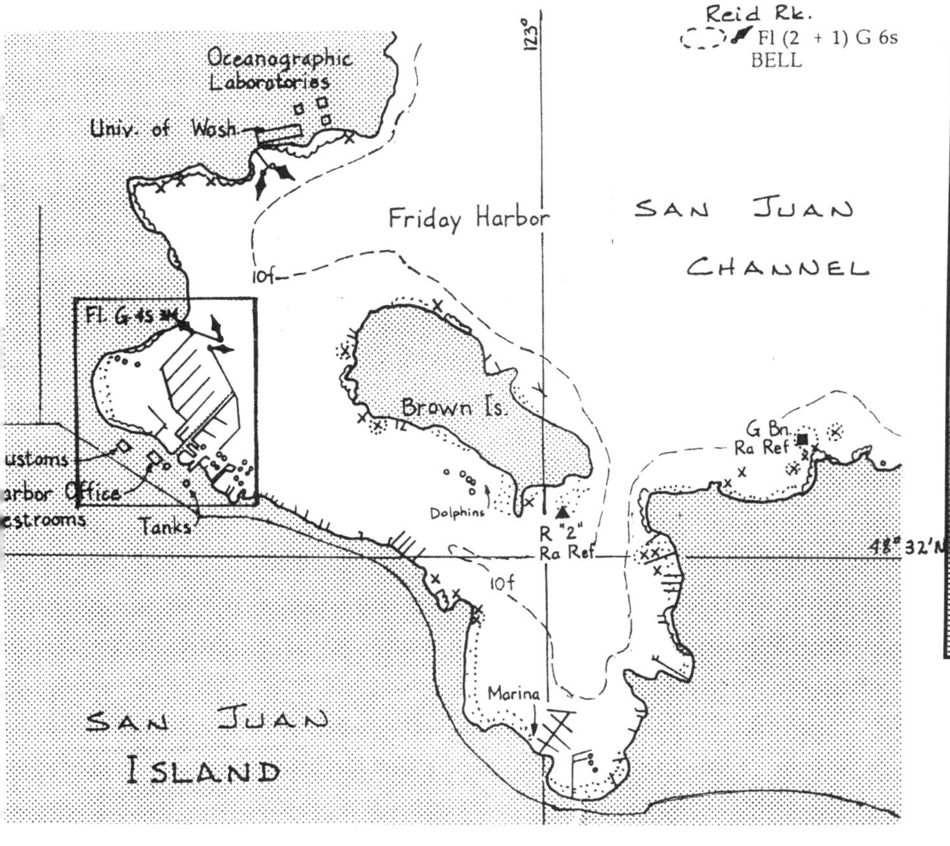

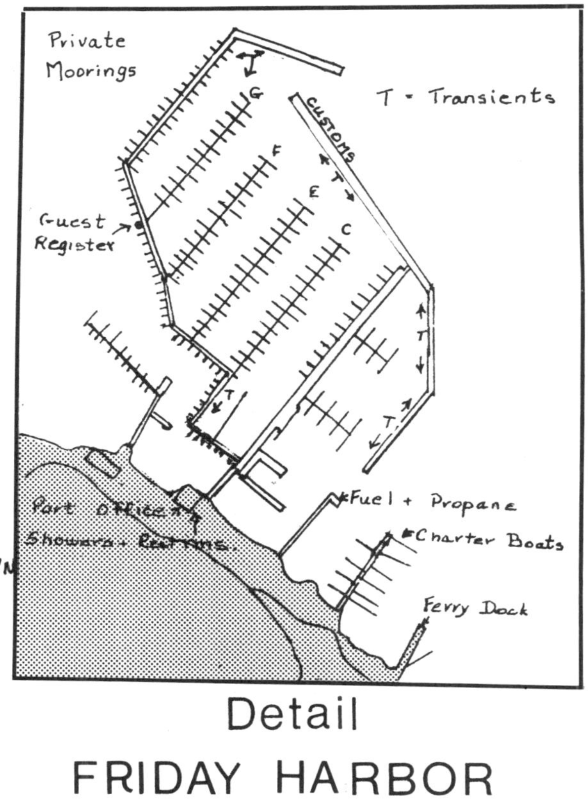

Detail

FRIDAY HARBOR

www.portfridayharbor.org

ANACORTES

Anacortes lies on the northern tip of Fidalgo Island and is connected to Rosario Strait by Guemes Channel. Rosario Strait in turn leads northward into the Strait of Georgia and thence to the extensive cruising areas of British Columbia, then around the top of the main group of San Juan Islands to the Strait of Juan de Fuca. This strategic location has a ferry terminal that provides access to the San Juan Islands as well as across to the town of Sidney, BC on Vancouver Island.

The original town was concentrated around its waterfront where older parts of the city are now characterized by brick and concrete buildings. The present city sprawls for some distance southward and westward along the highway.

There are three marinas in and near Anacortes, the two largest being on the eastern shore of Fidalgo Bay. Across the bay the tanker docking facilities, refinery works and massive storage tanks rise on March Point. The bay is shallow but dredged, and marked channels lead to the two marinas. This is an excellent place to haul a boat as there is a large work area and equipment and supplies are readily available. Commercial Avenue is lined with a variety of stores and restaurants.

Anacortes Marina is less than one-half mile south of Cap Sante, obvious by the wooden piling breakwater that surrounds it. Privately owned and operated, the marina offers long-term moorage but no facilities for transient boaters. Live-aboards are welcome and there are covered moorage areas, a fuel dock, facilities for haul-outs and maintenance, excellent security and plenty of parking spaces. For information telephone (360) 293-4543, Fax (360) 293-7013 or call on VHF Ch 66.

Skyline Marina at Flounder Bay is west of Anacortes. Though the marina is privately owned and fully occupied by local vessels, transients may be accommodated on a hot-berth basis when space is available; to inquire phone (360) 293-5134 or Fax (360) 293-2427. The approach channel is well marked; favor the port side when entering. A pump-out station and propane is available at the fuel dock where bins for recycling aluminum cans and cardboard are nearby. A full repair shipyard operates on the premises. VHF Ch 16 is monitored.

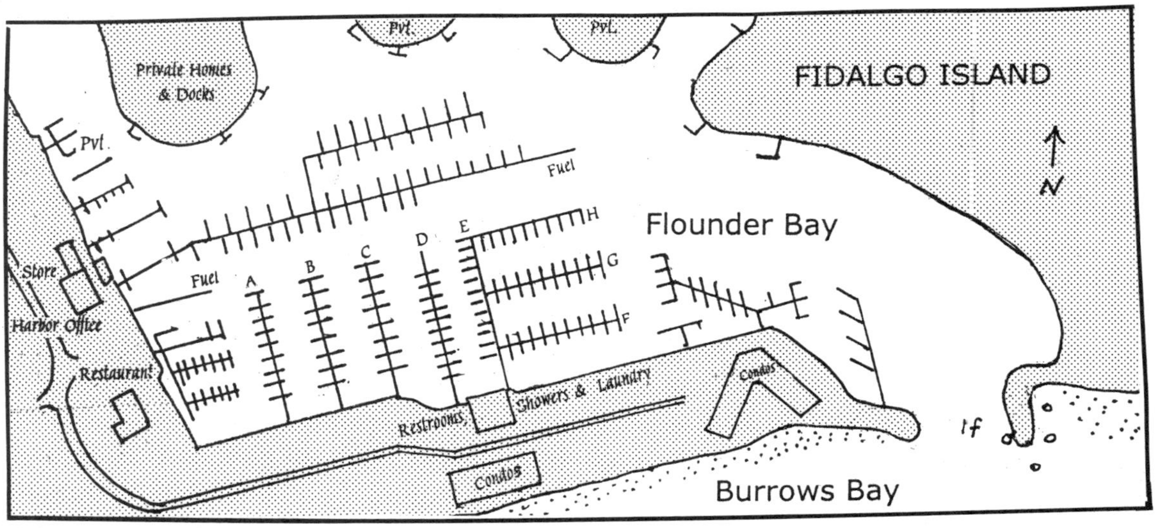

Skyline Marina

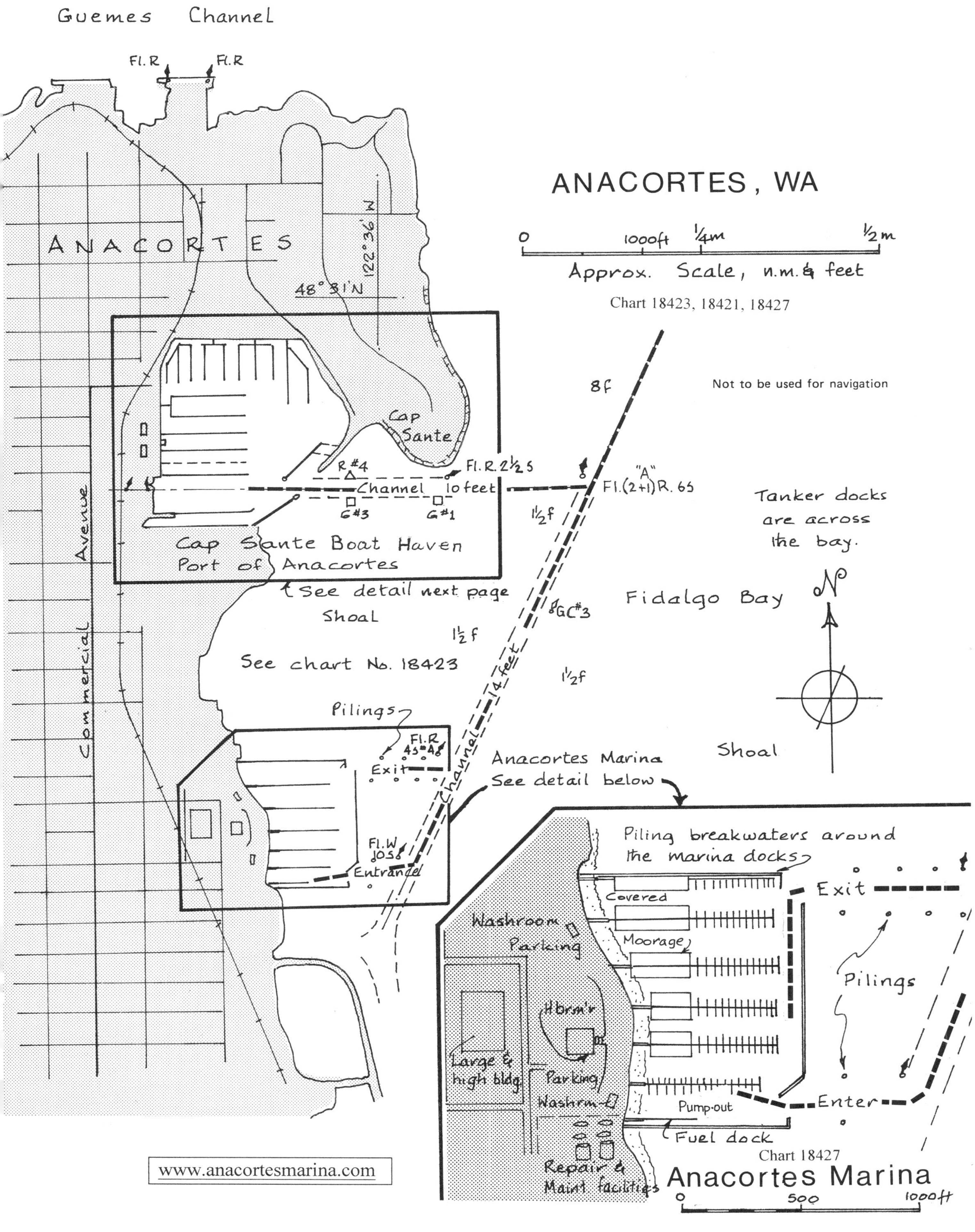

ANACORTES, WA
0 1000ft 1/4m 1/2m
Approx. Scale, n.m. & feet
Chart 18423, 18421, 18427
Not to be used for navigation
Guemes Channel
Fl. R
Fl. R
ANACORTES
122°36' W
48°31' N
Cap Sante
R#4
Fl. R. 2½ s
Channel 10 feet
Fl. (2+1) R. 6s
"A"
8f
G#3
G#1
1½f
Cap Sante Boat Haven
Port of Anacortes
See detail next page
Shoal
Tanker docks are across the bay.
Fidalgo Bay
GC#3
1½f
See chart No. 18423
1½f
Commercial Avenue
Channel 14 feet
Pilings
Fl. R 4s #4
Exit
Anacortes Marina
See detail below
Shoal
Fl. W 10s
Entrance
Piling breakwaters around the marina docks
Exit
Covered
Washroom
Parking
Moorage
Pilings
Hbrm'r
Large & high bldg.
Parking
Washrm
Pump-out
Enter
Fuel dock
Repair & Maint facilities
Chart 18427
Anacortes Marina
0 500 1000ft
www.anacortesmarina.com

Operated by the Port of Anacortes, **Cap Sante Boat Haven** lies below the rocky bulk of Cap Sante which affords it excellent protection while breakwaters protect the harbor entrance. This is a large marina with over 1,100 slips of which about 150 are for visitors. It is home to many of the fishing vessels that work the northern Pacific waters. Prior to entry call the harbormaster on VHF Ch. 16 or call (360) 293-0694. Space is usually available on C Dock where signs indicate temporary moorage. Register in the harbor office which is next to the C Dock shore walkway. Beyond the marina road is a marina supply store, restaurant and dry storage facilities.

SKAGIT BAY TO PADILLA BAY OR THE STRAIT OF JUAN DE FUCA

From Seattle or Everett two interesting routes are available which provide access to Padilla Bay or the Strait of Juan de Fuca: Deception Pass and Swinomish Channel. Each requires preparation and alertness, though only transit of Deception Pass is dependent on the tide.

Deception Pass is 5 miles south of Anacortes and separates Fidalgo Island on the north from Whidbey Island on the south. Two miles long, its narrowest section is 165m (200 yards) wide in the vicinity of Pass Island. Currents through the pass are very strong, reaching 8 knots, hence it is essential to travel at slack water. The pass is spectacularly narrow, steep-sided and crossed by a bridge with a vertical clearance of 44m (144 ft.) at the center; otherwise, 32m (104 ft.).

Docks to await the tide can be found at the north end of Whidbey Island in Cornet Bay. A dredged channel marked by private daybeacons leads to the marina. Marine supplies, fuel, launching ramp and a hoist are available. To the east of the marina is a state-run small craft facility where berthing and a launching ramp are available. Overhead power cables with clearances of 9.6m (56 ft.) cross the west end of the Bay.

Bowman (Reservation) Bay is a small bight between Reservation Head and Rosario Head on the west Bay side of the Pass. When awaiting the tide here, anchorage may be taken in 2 fathoms, mud. Docks maintained by the Deception Pass State Park may be used on a temporary basis.

When westerly winds blow in from the Strait of Juan de Fuca, the ebb tide has heavy rips that can make the passage difficult for small boats. A further obstacle one may encounter is log booms transiting the pass at slack water, effectively narrowing the channel for other boats.

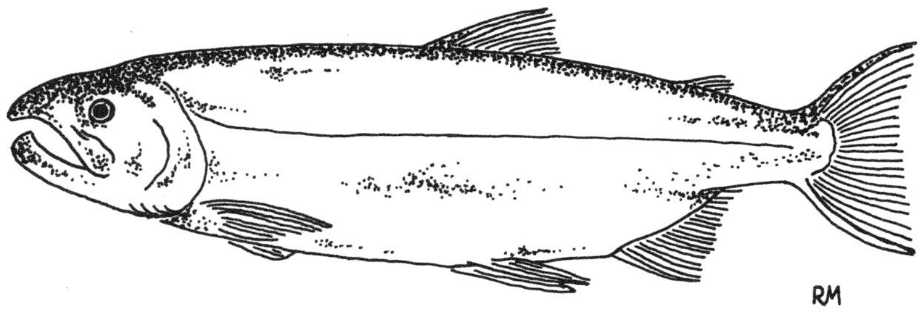

Chinook Salmon
(Oncorhynchus tshawytscha)

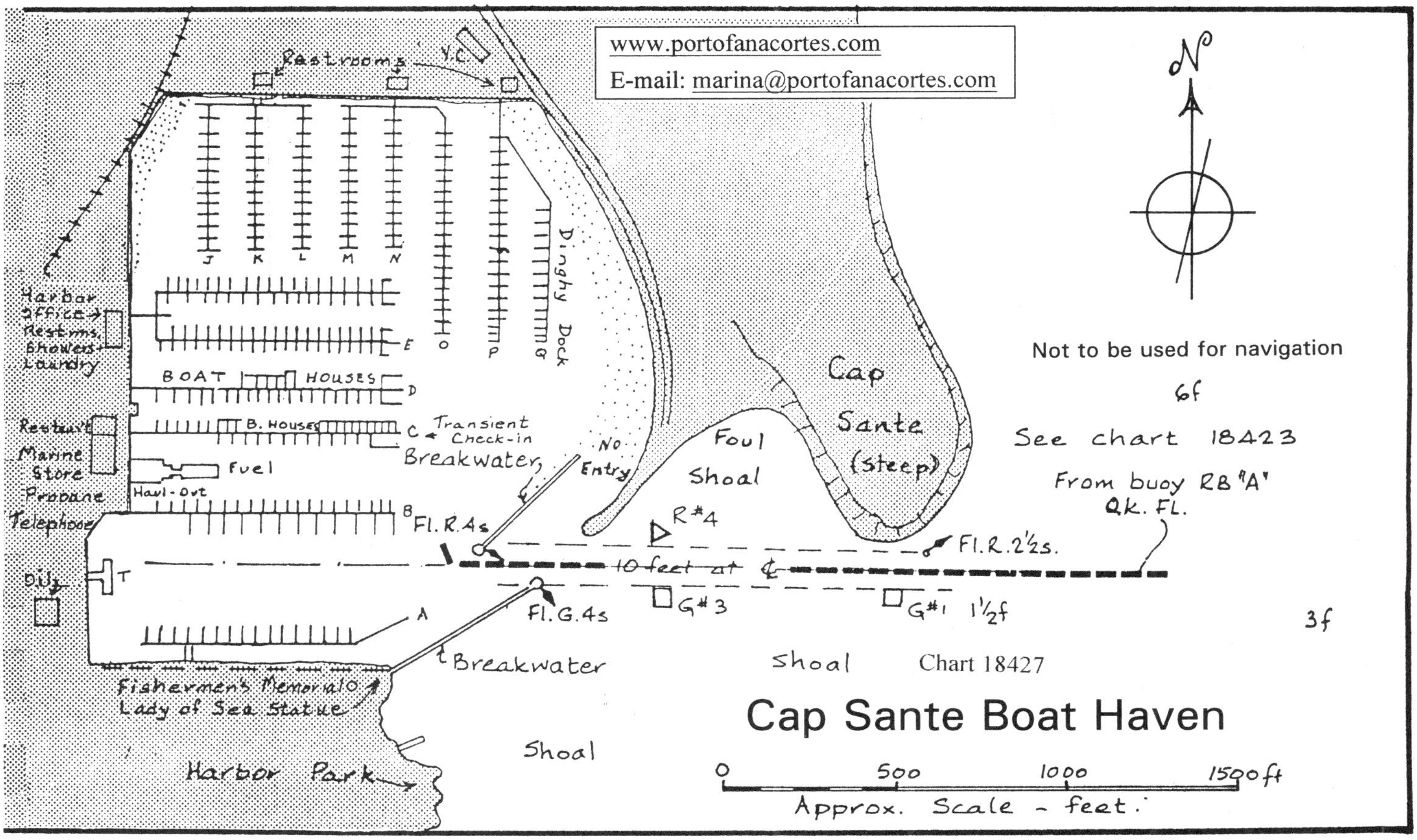
www.portofanacortes.com
E-mail: marina@portofanacortes.com
Restrooms
Y.C.
Harbor office
Restrms. Showers Laundry
Restaurant
Marine Store
Propane
Telephone
Oil
BOAT HOUSES
B. Houses
Transient Check-in
Fuel
Haul-Out
Dinghy Dock
Breakwater
No Entry
Fl. R. 4s
Fl. G. 4s
10 feet at ℄
R#4
G#3
G#1
1½f
Fl. R. 2½s.
Foul Shoal
Cap Sante (steep)
Not to be used for navigation
6f
See chart 18423
From buoy RB "A" Qk. FL.
3f
Breakwater
Shoal
Chart 18427
Fishermen's Memorial Lady of Sea Statue
Harbor Park
Shoal
Cap Sante Boat Haven
0 500 1000 1500 ft
Approx. Scale - feet.

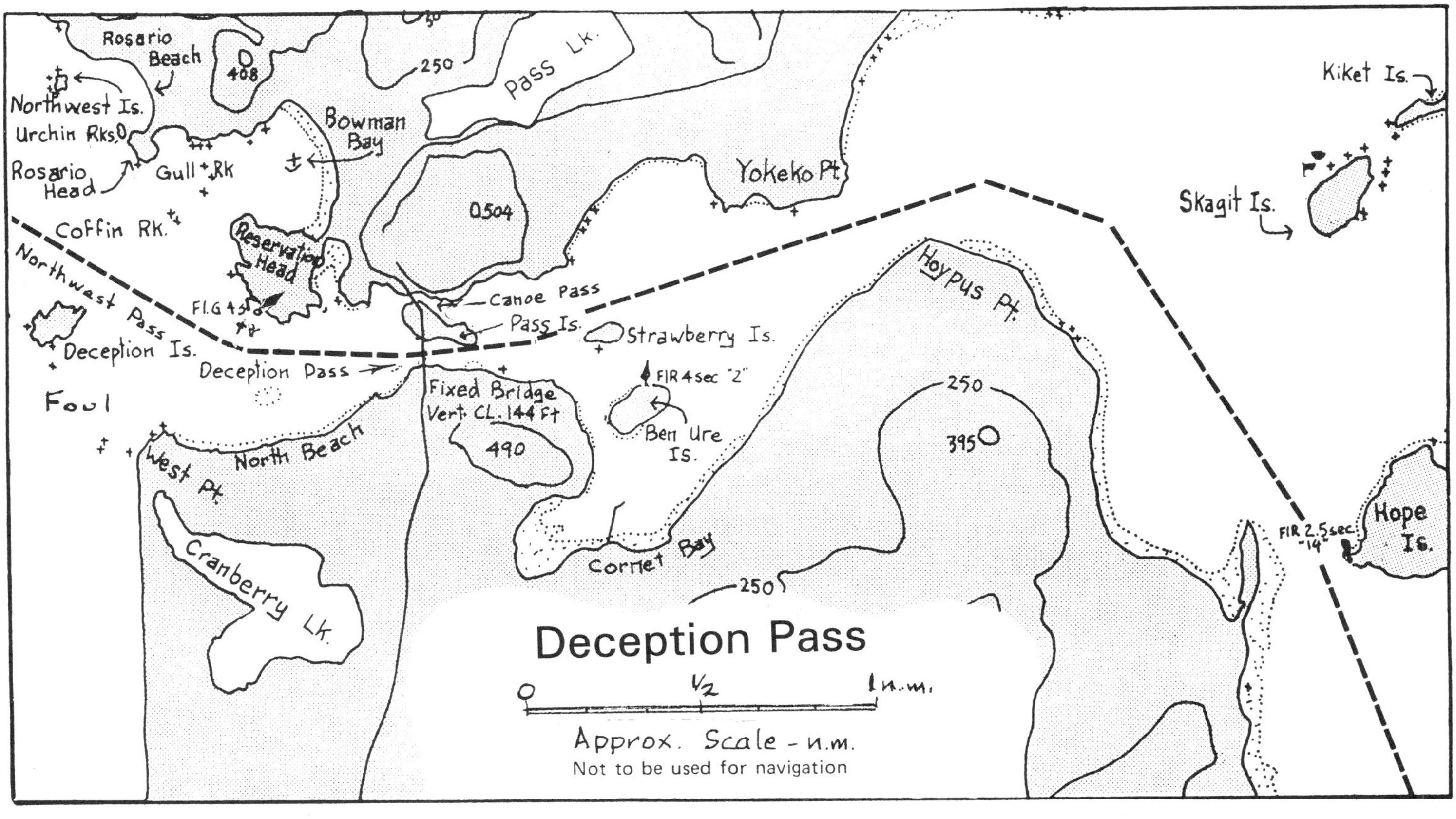
Rosario Beach
408
250
Pass Lk.
Northwest Is.
Urchin Rks.
Rosario Head
Gull Rk
Bowman Bay
0.504
Yokeko Pt.
Kiket Is.
Skagit Is.
Coffin Rk.
Reservation Head
Northwest Pass
Deception Is.
Fl. G 4s
Canoe Pass
Pass Is.
Strawberry Is.
Hoypus Pt.
Deception Pass
Foul
Fixed Bridge Vert. CL. 144 Ft
490
Fl R 4 sec "2"
Ben Ure Is.
250
395
North Beach
West Pt.
Cranberry Lk.
Cornet Bay
250
Fl R 2.5 sec. "14"
Hope Is.
Deception Pass
0 ½ 1 n.m.
Approx. Scale - n.m.
Not to be used for navigation

Chart 18427

Swinomish Channel is sometimes referred to as the "chicken route" between Seattle/Everett and Bellingham/Anacortes, since it provides a shorter, calmer run than the more exciting and beautiful transit of Deception Pass.

Swinomish Channel, connecting Skagit Bay to Padilla Bay, is regularly dredged and clearly marked. Entrance to the channel from Skagit Bay leads ENE between two jetties, then north of Goat Island. **Caution:** both jetties are submerged except for a short section, but they are well marked by daybeacons. A lighted range 72° - 252° marks the entrance from Skagit Bay. Take care to line up the range markers in Dugualla Bay before turning in to the channel and keep to the north side of the red buoy marking the southern boundary. Keep to the channel, which has a controlling depth of only 8 ft., for it is very shallow on either side. Where the channel turns to the north at Hole In The Wall, a fork of the Skagit River is seen on the east side. After passing an Indian settlement, Shelter Bay Development and logging operations you enter the town of La Conner.

The fixed highway bridge south of La Conner has a vertical clearance of 23m (75') for the center width of 95m (310') and 14m (45') for the remainder. Several power and telephone cables with clearance of at lease 22m (72 ft.) also cross the channel. A railroad swing bridge at the Padilla entrance is left open unless a train is approaching. As long as a vessel stays in the marked channel and keeps range markers lined up there are no problems in using this route.

La Conner is an attractive tourist-oriented town which is also busy with logging and smelt-fishing from October to March. Located in the midst of a rich agricultural area largely devoted to raising tulip bulbs, the country is a blaze of color during the spring. Several marinas are found in the channel. Large, city-operated marinas where transient moorage is available are in the basins on the east side of the channel roughly .6 miles and .8 miles north of the fixed highway bridge. Pilings on the north side of the south basin limit the entrance somewhat. It is just a short walk to the town where interesting gift and craft shops, restaurants and tourist facilities are in abundance.

Tulip fields near La Conner

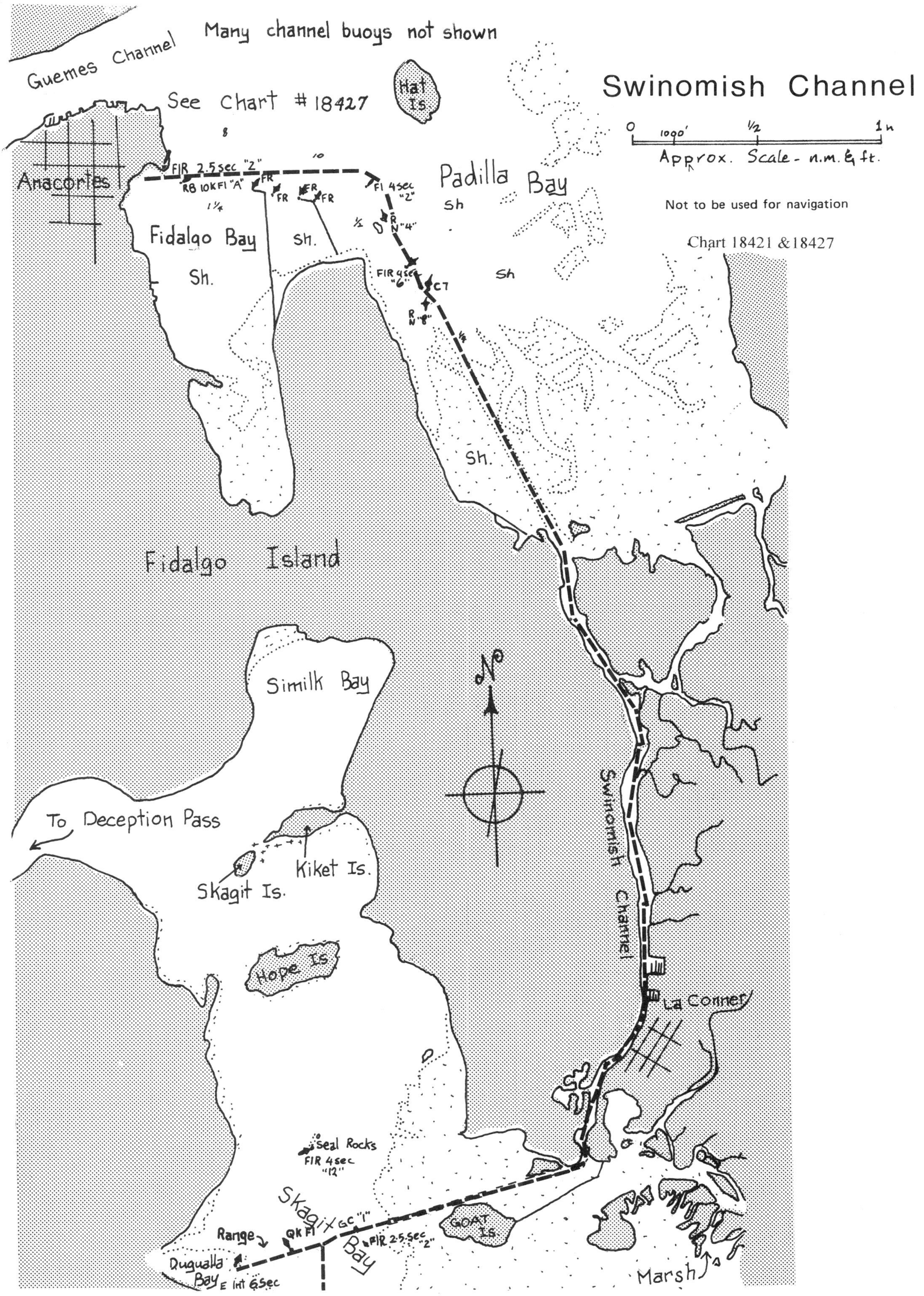

Swinomish Channel
Many channel buoys not shown
Guemes Channel
See Chart #18427
Hat Is.
0
1000'
1/2
1 n
Approx. Scale - n.m. & ft.
Not to be used for navigation
Chart 18421 &18427
Anacortes
Padilla Bay
Fidalgo Bay
Sh.
Sh
FlR 2.5sec "2"
RB 10k Fl "A"
FR
Fl 4sec "2"
R N "4"
FlR 4sec "6"
C7
R N "8"
Fidalgo Island
Similk Bay
To Deception Pass
Kiket Is.
Skagit Is.
Hope Is
Swinomish Channel
La Conner
Seal Rocks
FlR 4sec "12"
Skagit Bay
GC "1"
GOAT Is.
Range
QK Fl
FlR 2.5sec "2"
Dugualla Bay
E Int 6sec
Marsh

PORT TOWNSEND

The town of Port Townsend, a **Port of Entry**, lies at the mouth of an inlet with the same name. It lies at the threshold of the Strait of Juan de Fuca and can be recognized by the prominent light and foghorn at Point Wilson. A distinctive feature in the area is the long line of sand cliffs along Whidbey Island just across Admiralty Inlet from the Port.

Port Townsend is a National Historic Landmark City because of the countless Victorian houses and downtown buildings that give the town a unique and colorful character. The gray stone Customs/Post Office building, built in 1833, is prominent on the bluff above the sand cliffs behind the waterfront. The ferry terminal lies below, where a ferry connects Port Townsend to Keystone Harbor across the inlet on Whidbey Island.

The money and efforts of many dedicated citizens has resulted in the construction of the Education Dock which is the first stage of a Northwest Maritime Center that will eventually be a cultural and educational facility celebrating the maritime history of Puget Sound. For information call (360) 379-2629 or fax (360) 379-0885. Temporary moorage for very large pleasure craft may be arranged by calling the harbormaster on VHF Ch. 16

Port Townsend Boat Haven lies a short distance southwest of the ferry terminal. A stone breakwater encloses the harbor, with the entrance on the northeast corner. Within the entrance a mole divides the harbor into a smaller commercial basin and a larger pleasure craft section. The moorage office monitors VHF Ch. 9 and 66A and is located on the mole above the fuel float. Transient vessels can tie here until assigned a berth. If the harbormaster is away a list of vacant berths is posted near the door and vessels can fill out the forms supplied and take up one of these berths. For information call (360) 385-2355 or Fax (360) 385-3988. Facilities include showers, laundry, fuel, pump-out and haul-out services. About five blocks away is Café Internet providing internet service.

Point Hudson Harbor is a well protected indentation just southwest of Point Hudson and it is within easy walking distance of the main street of Port Townsend. The harbormaster's office is on the north side of the harbor and moorage is on a first-come, first-serve basis. Two restaurants, showers and laundry facilities are on the premises. For information call (360) 385-2828.

Anchorage in 7 to 8 fathoms, mud can be taken a .5 mile south of the railroad ferry terminal.

Every year the town hosts several popular events, including the Rhododendron Parade at the end of May. The Wooden Boat Festival in early September has become the most popular festival of its kind, attracting hundreds of classic wooden boats from all over the Pacific Northwest.

Detail of

Point Hudson Harbor

Chart 18464

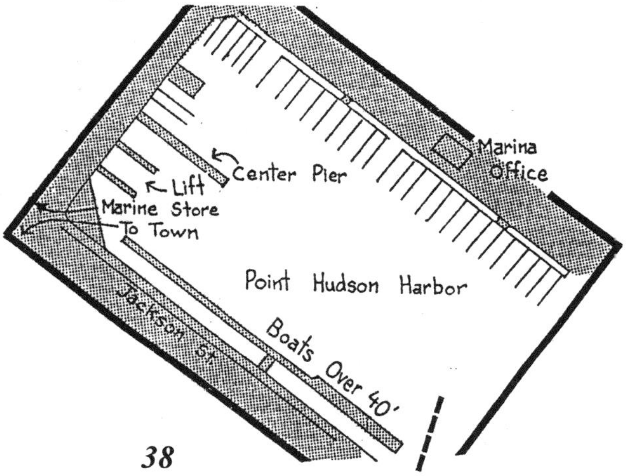

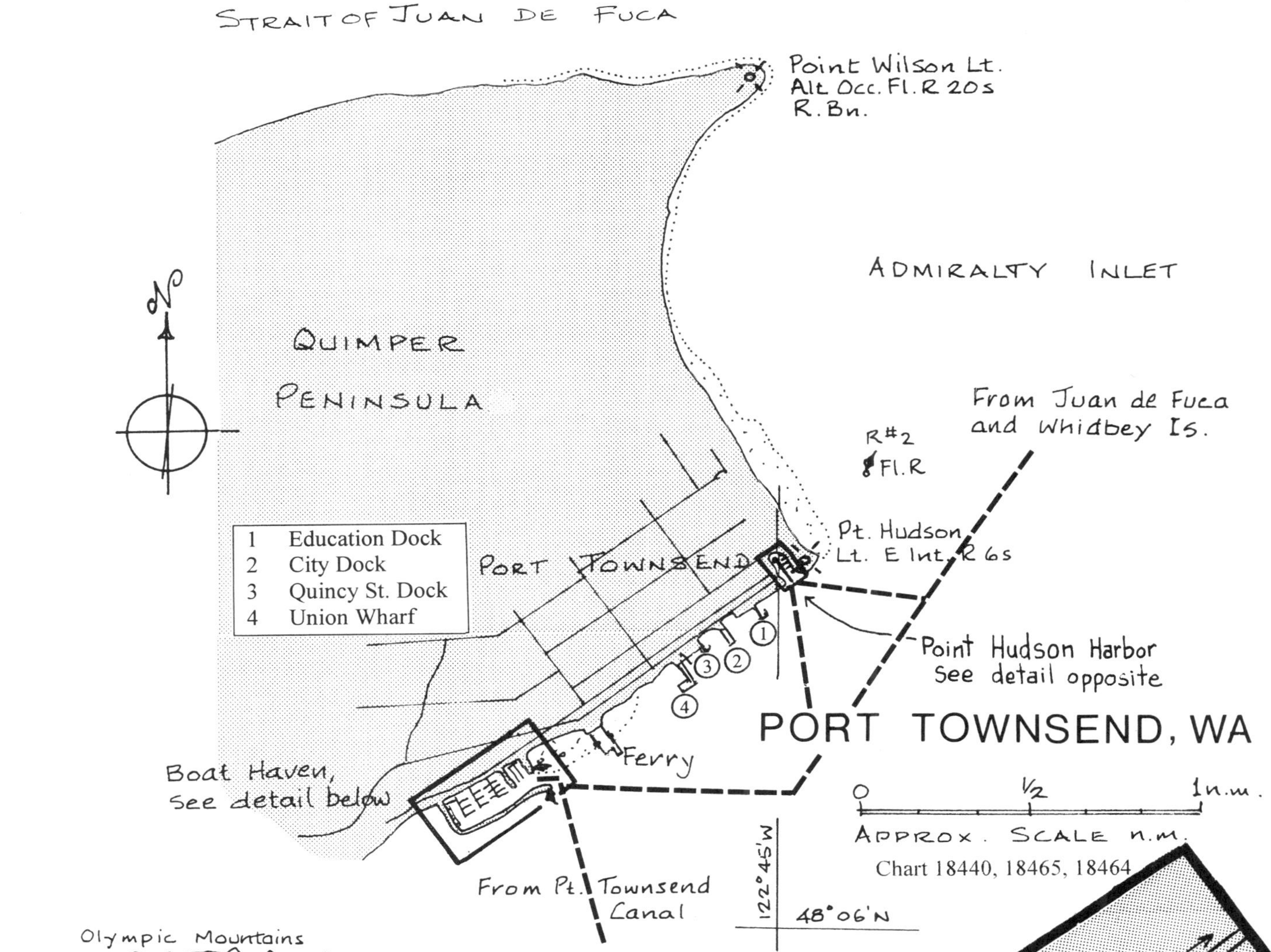

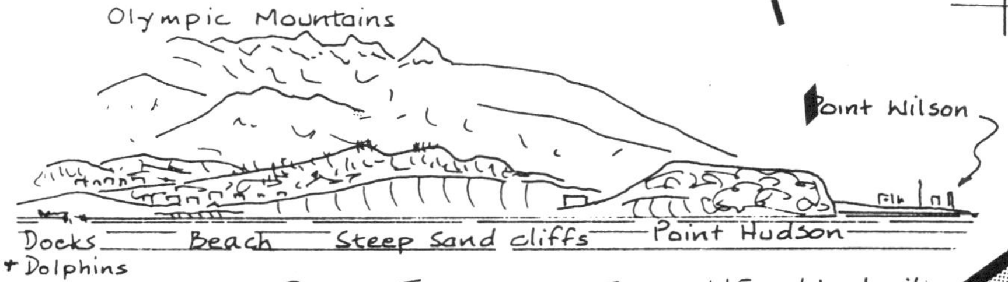

APPROACHING PORT TOWNSEND FROM NE abt. 1 mile

A very distinctive line of sand cliffs are seen on Whidbey Is. across Admiralty Inlet.

Detail of Port Townsend Boat Haven

Chart 18464

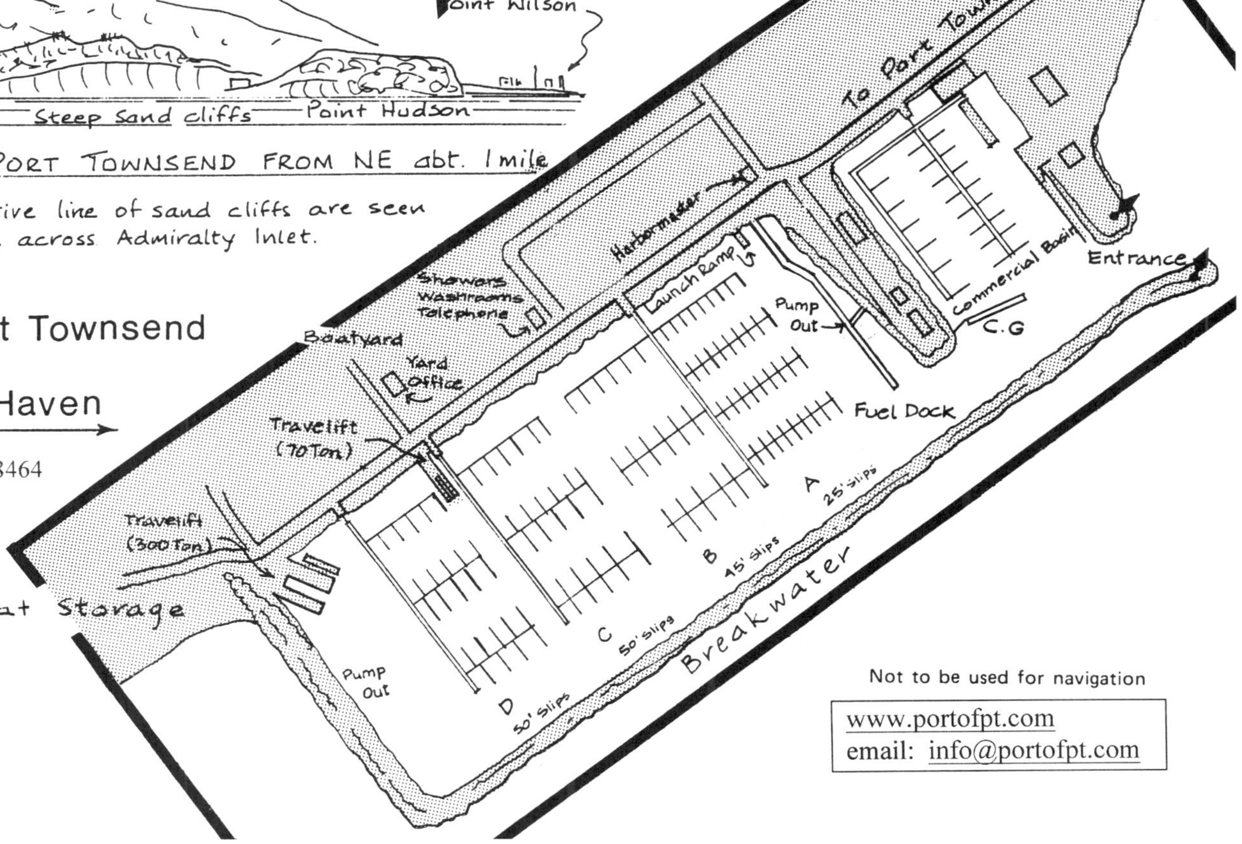

Not to be used for navigation

www.portofpt.com
email: info@portofpt.com

PORT LUDLOW

In the Admiralty Inlet part of Puget Sound the yachting center of Port Ludlow is tucked into a protected area off the Hood Canal south of Port Townsend. When approaching Port Ludlow give a safe clearance to Colvos Rocks, marked on the west by a flashing light and on the east by a red nun. At the eastern point of the entrance foul ground off Tala Point is marked by a red flashing buoy. The green slopes of second-growth timber blanket the Olympic National Forest where majestic trees were logged in the early 1900's. All evidence of the large sawmill that was once so active have long since disappeared, and the only legacy for sailors is the debris that rests on the bottom and sometimes causes concern for cruisers finding their anchor fouled on a bit of history.

The extensive development of the Port Ludlow Conference Center becomes evident as you approach the establishment. Facilities include a restaurant, gas barbecue, bar, golf course, condominiums and swimming pool.

Port Ludlow Marina has 300 slips that can accommodate boats up to 15m (50 ft.) and side ties where vessels up to 170 ft. can be moored. Shore power is 110/220 volt, 30 and 50 amp. and showers, laundry and restrooms are available. Both long-term and transient moorage are available, with free showers for overnight guests. The Marina office and store are open May 1 through September 30 from 8:00 a.m. to 7:00 p.m. on weekdays and 8:00 a.m. to 9:00 p.m. on weekends. From October 1 to April 30 the office is open from 8:00 a.m. to 5:00 p.m. The office monitors VHF 16 and 68. Moorage for holiday weekends must be booked several months in advance. For reservations call (360) 437-0513 or 1-800-308-7991.

Fuel is available at the dock. If the attendant is not at the dock use the blue phone box on the fuel dock building to call the marina store (dial 220) and someone will be sent to pump fuel. Separate bins for glass are near garbage containers.

At the southern end of the bay is a beautiful land-locked cove where anchorage may be taken. To enter, follow a mid-channel course between The Twins (two small wooded islands) at the entrance to the cove. The depth in the entrance channel between the islands is about 1.5m (5 ft.) at low tide. Well protected anchorage may be taken in the center of the cove in 4.5m (15 ft.) good holding mud. It is advisable to buoy your anchor for sometimes anchors are fouled on the logging debris on the bottom.

Anchorage may also be taken in an extensive area off the marina but holding in the soft bottom is not the best, calling for additional scope and extra care to set the anchor securely, especially if a brisk breeze from the southwest develops.

PORT LUDLOW

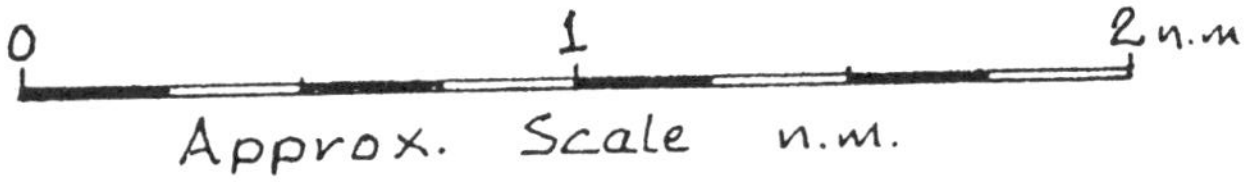

Not to be used for navigation

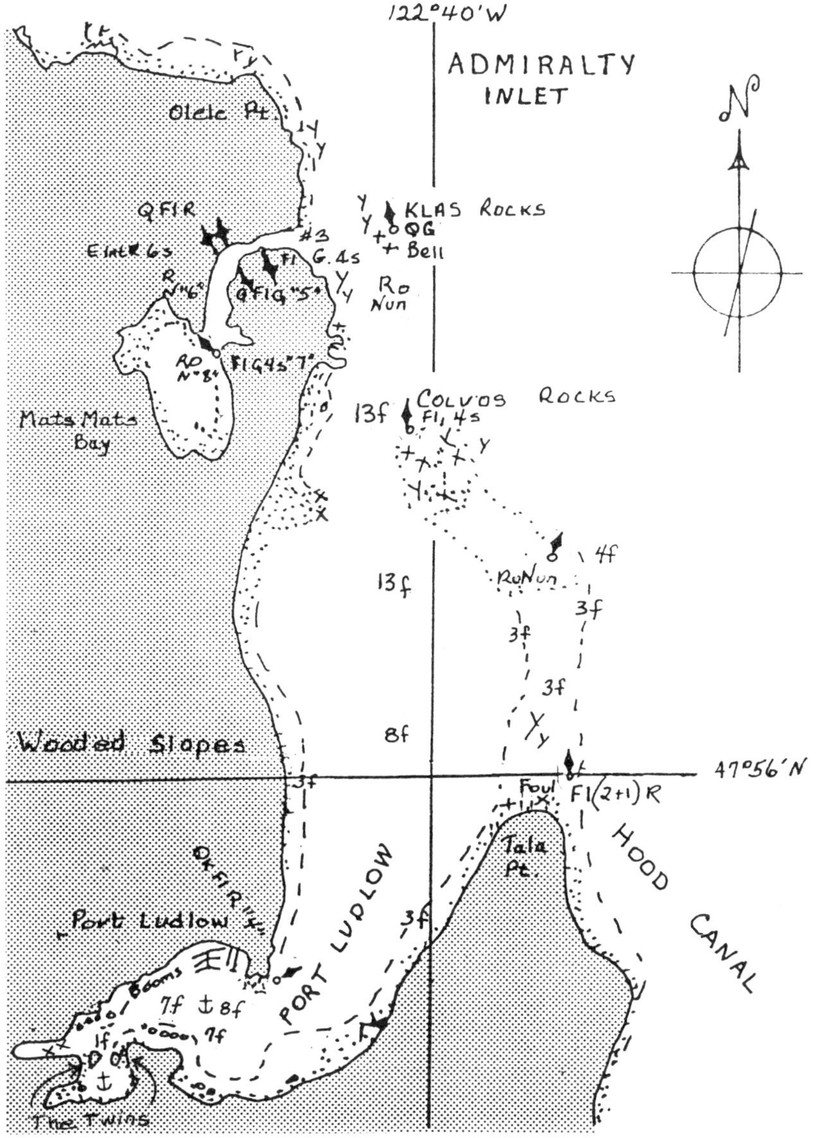

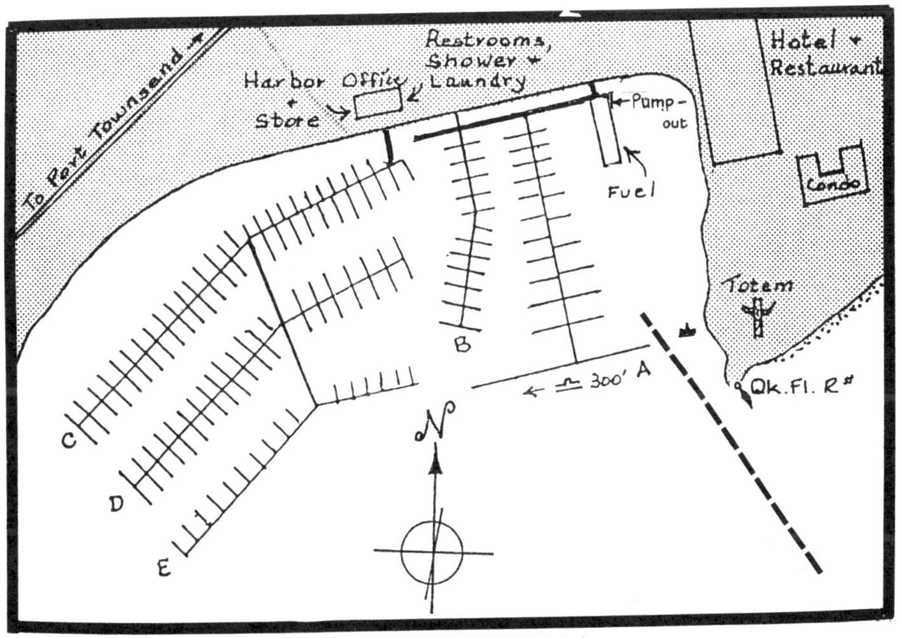

Harbor Details

www.visitportludlow.com
email: marina@portludlow.com

SEQUIM BAY

The Olympic Peninsula forms the south side of the Strait of Juan de Fuca. Several deep bays and indentations are found along the coast including Port Townsend at its eastern end. To the west, Discovery Bay winds SSE'ly for about 8 miles. Protection Island lies off the entrance and though there is a small harbor on the south shore of the island it has limited use. It is a National Wildlife Refuge reserved for birds and no one is allowed on shore or within 185m (200 yds.). Captain Vancouver anchored *Discovery* and *Chatham* here in 1792 while refitting his boats before probing further northward. The bay is too deep for most small cruisers except towards its head.

Sequim Bay is the next bay, about 6 miles to the west, penetrating inland for about 4 miles. At the entrance a long, low sand spit extending westward from Kiapot Point almost encloses the bay. Because of shoals extending almost half a mile from the western shore, vessels approaching the bay should aim towards Kiapot Point on the east side, then pick up the buoys marking the channel along the spit, turning in at the narrow passage on the west. Buoys mark the continuation of the channel to the west of Middle Ground Shoal. Once past the buoys the bay is open to anchorage in depths of 6 to 20 fathoms. Ebb currents can be strong, and use of Tide Tables is advised.

John Wayne Marina, operated by the Port of Port Angeles, is about a mile into the bay on the west side, off Pitship Point. An oval rock breakwater encloses the floats with a narrow, 15m/50 ft.-wide entrance at the northern end. The modern marina has first-class facilities for 280 permanent slips and 22 transient spots. A fuel dock, pump-out and launch ramp lie just within the entrance. The harbormaster's office is in the large marina building; office hours are 8 a.m. to 12 p.m. and 1 p.m. to 4 p.m. If transient slips are occupied check the bulletin board outside the marina office where the numbers of open slips are posted for after hours check-in. For more information call (360) 4417-3440 or fax (360) 417-3442.

The Sequim Bay Yacht Club meets in the marina building. Shore facilities include a restaurant, laundry, showers, a picnic area and access to the beach. Dock power (20 and 50 amps) is available at $5.70 per day. Some marine hardware, fishing supplies and limited groceries can be obtained in the store upstairs. Transient guests are welcome; on arrival tie up at the first dock seen upon entering the basin or you may be directed to the inner end of H dock. Large craft may side-tie to E dock. Check in with the harbormaster.

Years ago John Wayne used to visit the bay with his family who donated the land to the community in his name. Excavation of land where the present marina is located was completed before water from Sequim Bay was let in to flood the basin. The well-designed facility keeps clean with a natural flushing system with the deepest water 3.6m (12 ft.) at the mouth of the marina sloping to 1.5m (5 ft.) at the inner end. The success of this plan is evidenced by the clear waters in the marina where tidal flow has prevented the accumulation of sediment with a regular movement of fresh seawater.

A casino, tourist facilities and stores with a greater selection can be found in the town of Sequim, about two miles to the west. A propane outlet is about 3 miles away.

The unique combination of weather and soil makes this area successful in the growing of lavender which is available at U-pick operations from July to October 1. This feature is celebrated in early August with a Lavender Festival. Also of interest is a well-attended Jazz Festival in June.

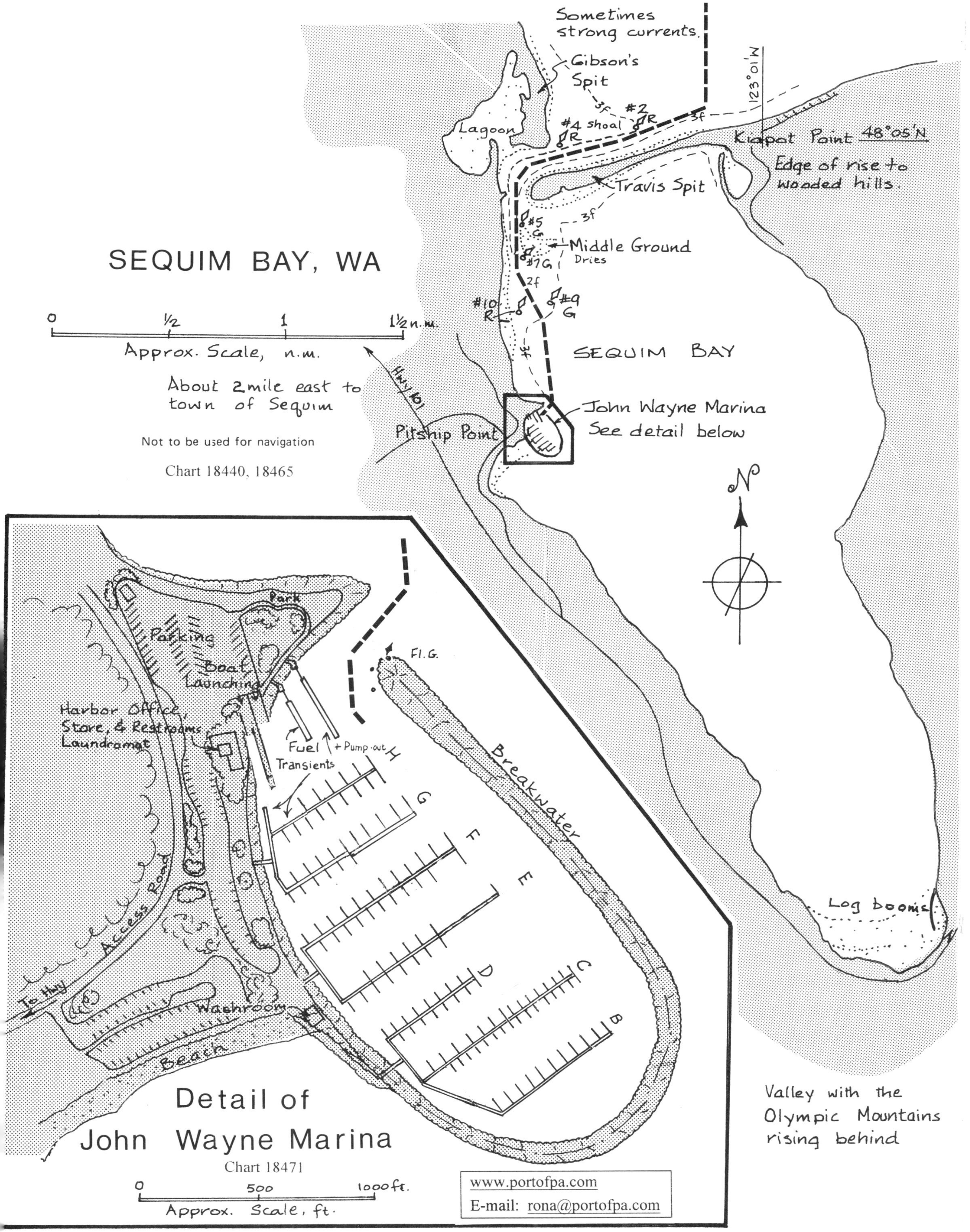

SEQUIM BAY, WA
0 1/2 1 1 1/2 n.m.
Approx. Scale, n.m.
About 2mile east to town of Sequim
Not to be used for navigation
Chart 18440, 18465
Sometimes strong currents.
Gibson's Spit
Lagoon
#4 shoal R
#2 R
3f
123°01'W
Kiapot Point 48°05'N
Edge of rise to wooded hills.
Travis Spit
#5 G
Middle Ground Dries
#7G
2f
#10 R
#9 G
SEQUIM BAY
Hwy 101
John Wayne Marina See detail below
Pitship Point
N
Log booms
Valley with the Olympic Mountains rising behind
Park
Parking
Boat Launching
Harbor Office, Store, & Restrooms Laundromat
Fl. G.
Fuel
Pump-out
Transients
H
G
F
E
D
C
B
Breakwater
Access Road
To Hwy
Washroom
Beach
Detail of John Wayne Marina
Chart 18471
0 500 1000ft.
Approx. Scale, ft.
www.portofpa.com
E-mail: rona@portofpa.com

PORT ANGELES

This important harbor is about 10 miles west of Sequim Bay and 56 miles from Cape Flattery. Ediz Hook is a low, 3-mile long, narrow and curving sand spit that shelters the harbor of Port Angeles. Ediz Hook light is shown from a 21m (70-foot) circular white tower .3 miles west of the eastern extremity of the spit and a 52m (170-foot) Coast Guard radar tower is just to the west of the light. When approaching Port Angeles give a safe clearance to the shoals extending about 69m (75 yds.) east of the end of the spit, marked by a lighted buoy. The harbor is exposed to easterly winds that occasionally blow during the winter, though strong winds from other directions can cause a sharp chop within its large expanse. Further caution calls for a look-out for deadheads and numerous logs adrift in this predominant logging center. The U.S. Coast Guard maintains an air and helicopter station near the end of the spit.

The U.S Navy advises that a precautionary area, located within a 1-mile radius centered around a point 48°15'36"N., 123°15'48" is used to conduct tests. When surface vessels and submarines are maneuvering in circles in this area reference lights consisting of a lazy "T" bar, 1 sec flashing yellow, 2/sec flashing red, and a high intensity spot will be lit. The group of lights is visible from the N side of Ediz Hook with the "T" bar to the W and spot light to the E. The Navy Test Facility Port Angeles will monitor VHF Ch 16 and 69. Mariners are advised to proceed with caution when transiting this area.

Port Angeles is the largest city on the north Olympic Peninsula and is a **Port of Entry.** It is a convenient place to complete last-minute provisioning and complete entry procedures before turning the corner at Neah Bay and heading south. It is recommended that vessels leaving Canadian waters complete entry procedures before transiting the coast so that if necessary, entry can be made at any location that is not a Port of Entry. With a Cruising Permit in hand and all paperwork completed you can then avoid problems with officialdom. The Customs Office is located about .5 miles east of the Boat Haven in a two-story, brick building at the corner of First Street and Oak. Office hours are from 8 a.m. to 5 p.m., Monday to Saturday; the telephone number is (260) 457-4311. At all other times one may call 1- 800-562-5943 to obtain information.

Port Angeles Boat Haven is a large rectangular small boat harbor maintained by the Port. A breakwater protects the northwest section and short breakwater extends in a northwesterly direction from the tip of the East Jetty providing protection to the entrance from some of the chop in the outer harbor. While there are floats and berths in both sections, the main transient area is on the central float of the southeast section. The harbormaster's office is in the building at the end of the east jetty, with a fuel and utility float directly beneath it in the harbor. For information call (360) 457-4505 or Fax (360) 457-4921. A boat launching area, shipyard with marine ways, mobile hoist and some chandlers are located in the harbor area. Laundry facilities are eight blocks from the marina and propane is two blocks. Wood and steel repairs are available at the New Day Boat Repair Shop.

The city center lies to the southeast of the basin and contains many stores with a wide variety of merchandise. Several buildings in the town have attractive murals depicting historic moments in the area's past. The headquarters of Olympic National Park is here and the town is a short distance from the spectacular and much photographed Hurricane Ridge. The ferry terminal, for service connecting Port Angeles to Victoria, B.C. is toward the center of the town.

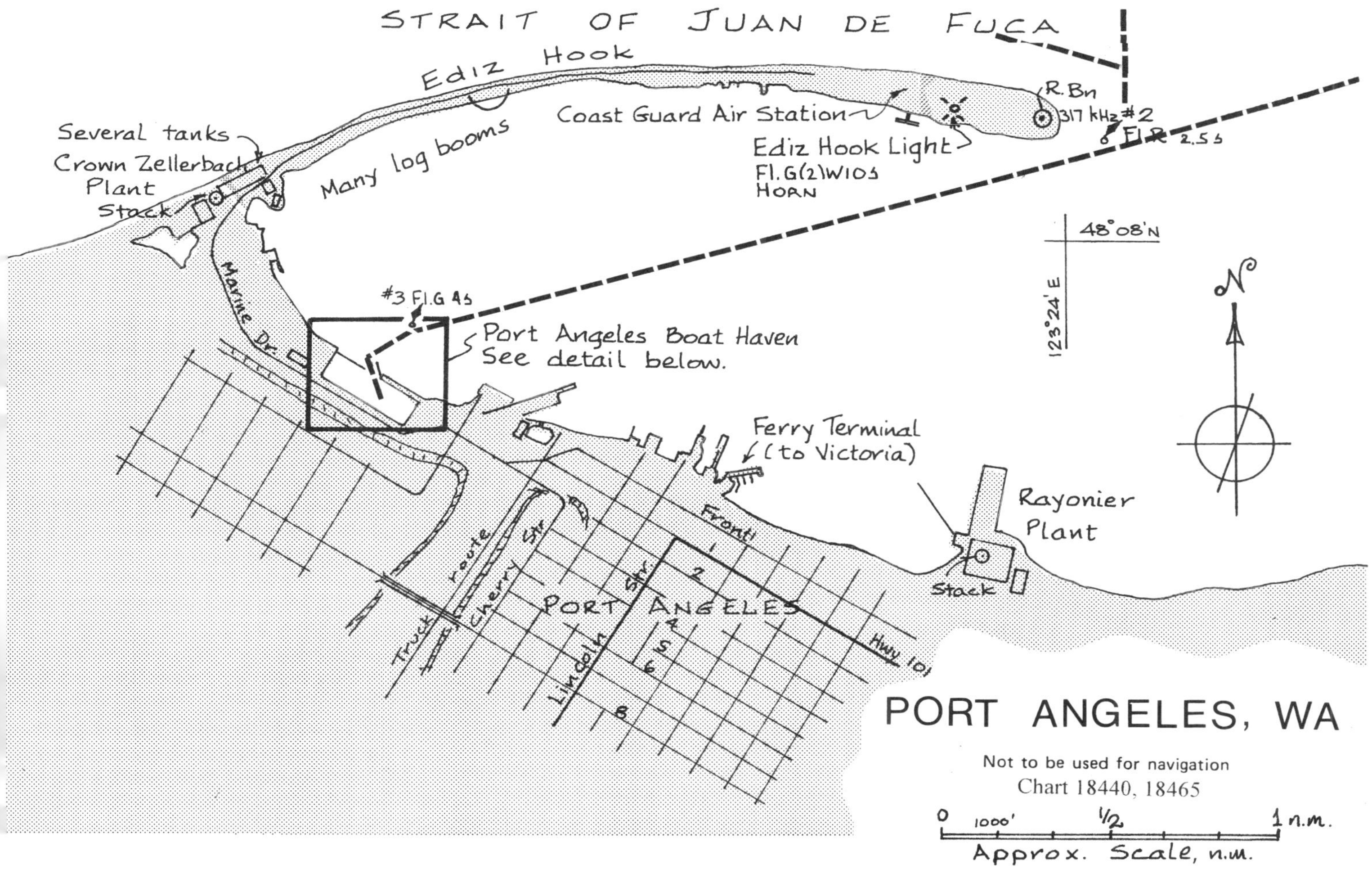

STRAIT OF JUAN DE FUCA
Ediz Hook
Several tanks
Crown Zellerbach Plant
Stack
Many log booms
Coast Guard Air Station
R.Bn 317 kHz
#2 Fl R 2.5s
Ediz Hook Light
Fl.G(2)W10s
HORN
48°08'N
123°24'E
N
Marine Dr.
#3 Fl.G 4s
Port Angeles Boat Haven
See detail below.
Ferry Terminal (to Victoria)
Rayonier Plant
Stack
Front
Truck route
Cherry Str
Lincoln Str.
PORT ANGELES
Hwy 101
PORT ANGELES, WA
Not to be used for navigation
Chart 18440, 18465
0 1000' 1/2 1 n.m.
Approx. Scale, n.m.

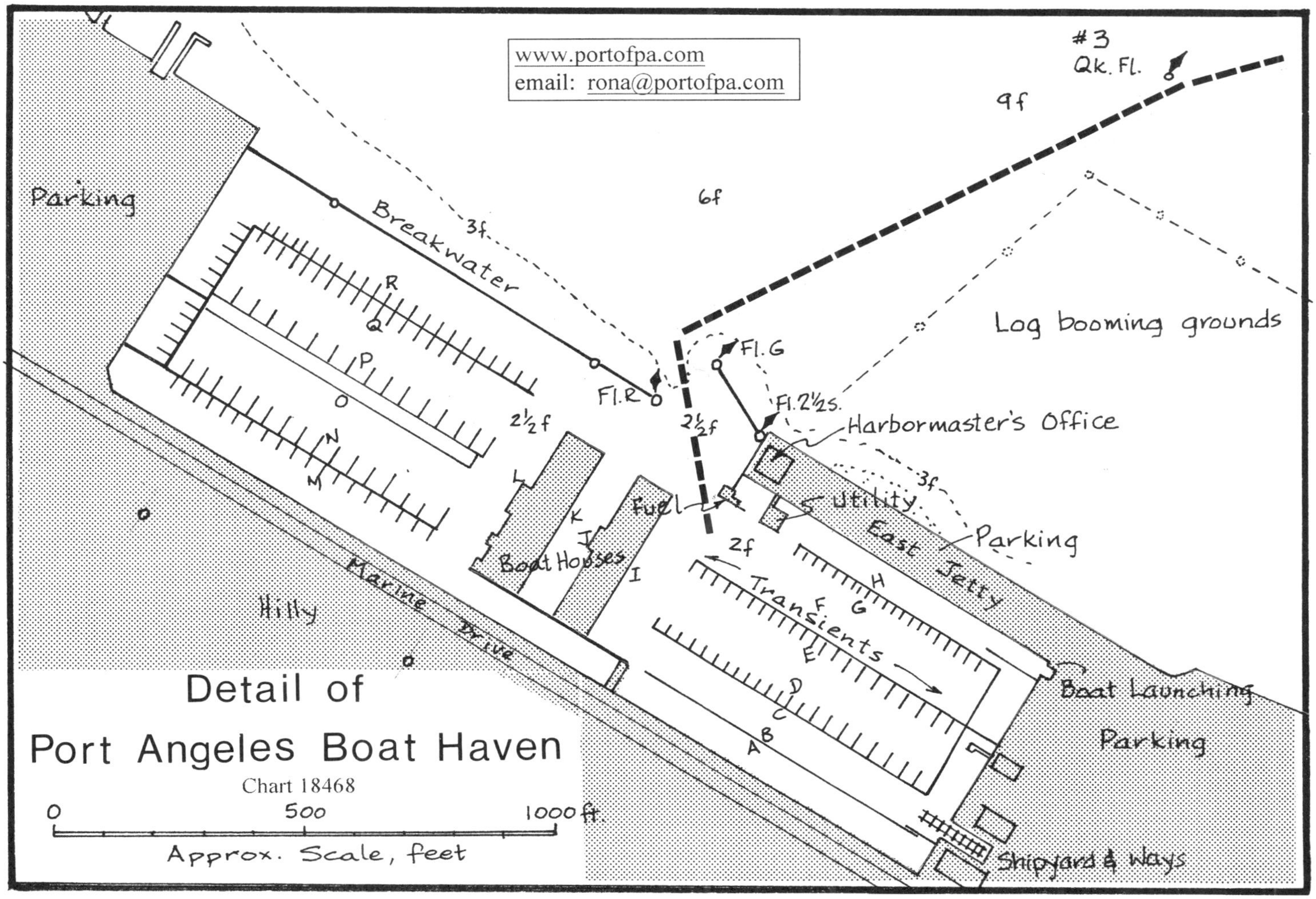

www.portofpa.com
email: rona@portofpa.com
#3 Qk. Fl.
9f
6f
3f
Parking
Breakwater
R Q P O N M
Fl.R
2½f
Fl.G
Fl.2½s.
Log booming grounds
Harbormaster's Office
3f
Fuel
Utility
East Jetty
Parking
K J I
Boat Houses
2f
Transients
H G F E D C B A
Hilly
Marine Drive
Boat Launching
Parking
Shipyard & Ways
Detail of
Port Angeles Boat Haven
Chart 18468
0 500 1000 ft.
Approx. Scale, feet

NEAH BAY

This harbor is located close to Cape Flattery, about 6 miles east of the cape itself. It is a convenient stop-over for cruising vessels entering or leaving the Strait of Juan de Fuca. A long, low, rocky breakwater connecting the west side of the bay to Waadah Island provides protection from the prevailing northwesterly winds. Waadah Island is .5 mile long, with extensive shoal areas around it and rocky ribs extending northwesterly from the rocky island. Similar rock ribs run out from the mainland shore.

Entry into the bay is made via a buoyed channel between Waadah Island and Baadah Point. The least depth is about 4.5m (15 feet) through the channel. Just within the entrance a T-shaped pier lies off the Makah Tribal Office and Museum, with lights at each end. Further along the U.S. Coast Guard pier is located. Although Neah Bay is officially listed as a **Port of Entry**, it is not a full-time Customs Reporting Station. To report in at any time call the toll free number 1-800-562-5943 to obtain clearance. Paper work should be completed at Port Angeles where there is a full-time Customs Office.

A breakwater protects the modern Makah Marina which has 200 slips and can accommodate vessels up to 30m (100'). For moorage contact Big Salmon Resort at (360) 645-2374. Facilities include restrooms, a dump station and waste pump-out and a grocery store and showers are within easy walking distance. The harbor office monitors VHF Ch 66 and 16 and is adjacent to an interesting structure featuring colorful panels of stylized native Indian figures. Anchor in the bay in 4 to 6 fathoms, sand; avoid the western and northern sides and unmarked wrecks. Diesel is available year round, gasoline is seasonal. Fuel is available from the fishing plants with large piers. Call Makah Fuel Co. on VHF Ch 66.

Neah Bay is within the Makah Indian Reservation and is the home port for the local fishing fleet. Indian arts and crafts are available in one or two shops on the main street. Basic food and supplies are available from the store on the main street behind the harbor. Since this is a dry reservation alcohol is not available. For more information call (360) 645-3015 or fax (360) 645-3016. Guided tours from the museum take people on the 7-mile drive to the trail head followed by a .5-mile hike to Cape Flattery on a well maintained easy trail.

Sekiu is a resort and sport fishing town on the W end of Clallam Bay which is a wide bight about 15 miles SE of Neah Bay and 20 miles northwest of Port Angeles. Give safe clearance to the reef marked by a light off Slip Point at the E end of the bay. Anchorage in 3 – 4 fathoms can be taken SE of the sport breakwater protected sport fishing docks where there is some protection from northwesterly winds. Shoals extend for some distance from the shore. Diesel, gasoline, launch ramp, showers, restrooms, laundromat, telephone, fax and E-mail service are available at Olson's Resort. At low water there is 1 fathom of water at the fuel dock. The resort monitors VHF Ch 21 or can be reached at (360) 963-2311, Fax (360) 963-2928.

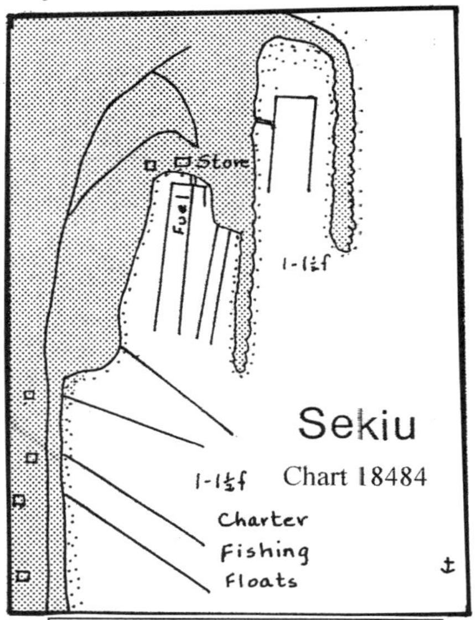

www.olsonsresort.com
E-mail: olsons@olypen.com

Big Salmon Fishing Resort is the tenant of the port. They handle commercial bookings through the summer. Call (360) 645-2374, 1-866-787-1900 or Fax (360) 645-0772. E-mail: bigsalmonresort.com

NEAH BAY, WA

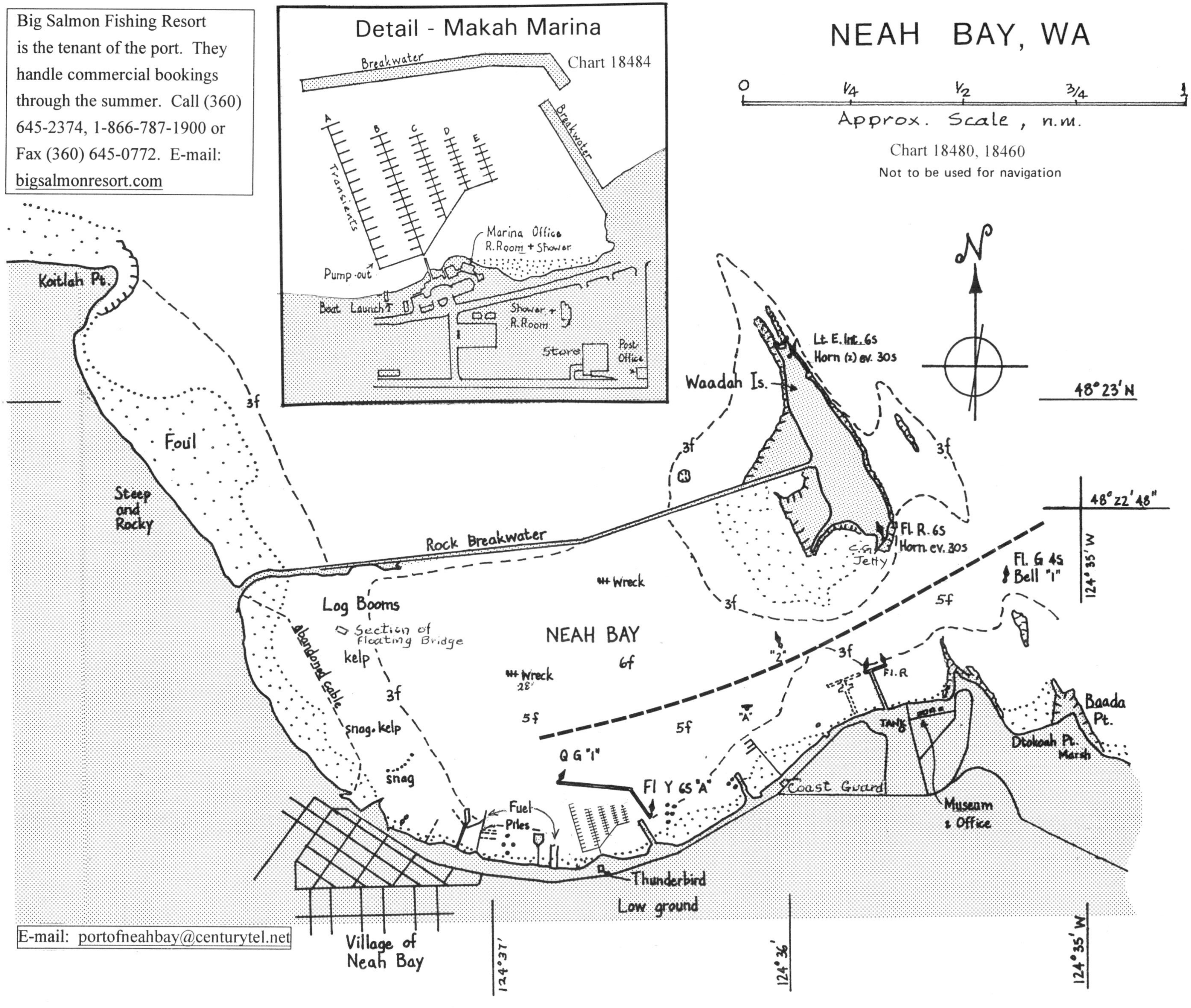

E-mail: portofneahbay@centurytel.net

CAPE FLATTERY and the STRAIT OF JUAN DE FUCA

This famous landmark marks the western gateway to the Strait of Juan de Fuca and thus is a major course change for vessels traveling the west coast of the continent. For sailors making their first offshore voyage from Alaska and the Pacific Northwest the passing of this area marks a commitment to the first important leg of a trip down the coast. Similarly, for returning sailors Cape Flattery means that the last leg of the journey will now be completed in the increasingly protected waters leading off the Strait of Juan de Fuca.

Cape Flattery is a steep, 37m (120') cliff, rising behind to a rocky peak some 2 miles from the cape. Atop this peak a large radar dome is a prominent feature and a good landmark even in fog when it tends to stick out above the lower clouds. Tatoosh Island is a flat-topped, steep-sided island about .3 miles northwest of the Cape. Two rocks awash lie in the center of this passage, often called Hole-in-the-Wall. Though a vessel can cross through this passage holding closer to the Tatoosh side, cross currents and dense fog can make this route hazardous.

In traveling northward around Cape Flattery, stand well out from the land (at least 3 miles) to clear the buoy beyond Duncan and Duntze Rocks, the two main dangers lying about 1.25 miles further NNW of Tatoosh Island. A combination of fog and strong currents requires this much clearance. In good weather it is possible to sail close past Tatoosh Island. However when sailing against a current in fog it is very difficult to determine when one is clear of the dangers here, and it would be wise to power well past this threshold to the open sea. Though the use of radar or GPS can assist in setting a course to clear hazards in the vicinity of the Cape, give them a safe berth to allow for engine failure or other unexpected loss of power.

Looking northwest from Tatoosh Island

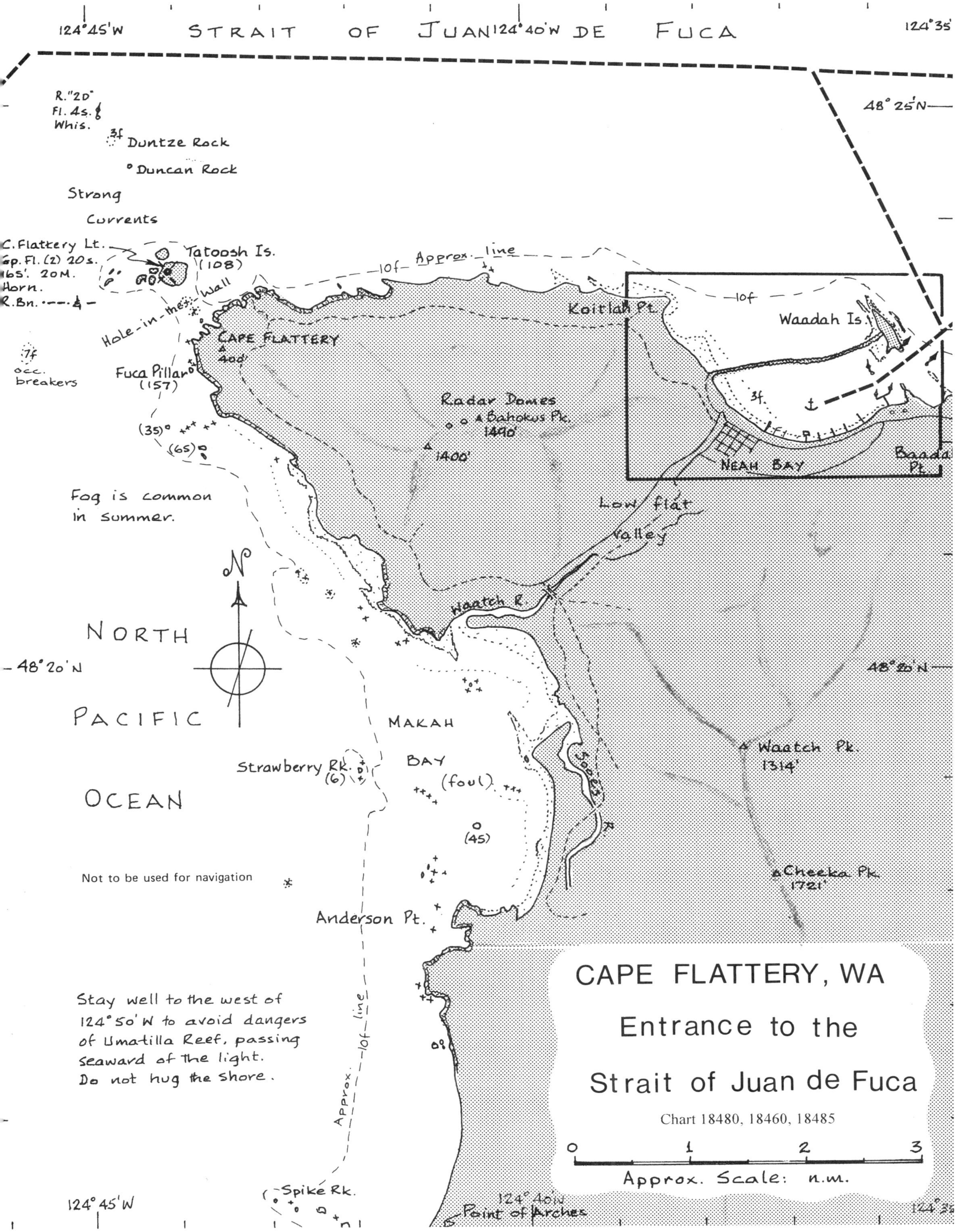

CAPE FLATTERY, WA

Entrance to the Strait of Juan de Fuca

Chart 18480, 18460, 18485

0 1 2 3

Approx. Scale: n.m.

LA PUSH

This small Indian fishing village is about 35 miles south of Cape Flattery. The harbor entrance is between James Island on the west and a jetty extending southwestward from the north end of a long curving beach. Good clearance must be given to the outer end of the jetty as about 7.5m (25 ft.) is awash at high water. An approaching vessel should pick up the lighted whistle buoy that is almost 2 miles southwest of the end of the jetty. A directional light indicates the proper course until past the end of the jetty, after which you should favor the south side of the channel. A light and seasonal buoys mark the channel, which is supposed to be about 18m (60 ft.) wide but seems narrower. Keep a sharp eye for Indian fishnets after entering the river.

Though the controlling depth in the entrance channel is 1.5 fathoms, constant shoaling can reduce this to about 1 fathom in the shallowest spots before semi-annual dredging occurs. Check with the Coast Guard to obtain the current minimum depth before entering. Do NOT attempt to enter at night even though seasonal floodlights illuminate the entrance channel between James Island and the jetty. An overhead cable (clearance of 30m/100 ft.) crosses the entrance channel. Southbound vessels will pass the entrance before it becomes obvious.

In heavy southerly weather the channel is dangerous. Fog often forms at night and may be a problem until burning off around noon. The Coast Guard maintains a rough bar advisory sign. If in doubt, ask the Coast Guard for a report on current bar conditions. On request, they will guide a vessel into the boat harbor if it is feasible. Re-read the section on crossing bars found in the introduction (p. 9 - 10).

The Quileute Marina is seldom crowded. During the summer months there is increased activity in the area from visiting sports fishermen. Pick a berth, then check in at the harbormaster's office. Since he is not always in the office you may have to ask some of the locals for his whereabouts. Diesel, gasoline, limited marine supplies, groceries and fishing tackle are available at the fuel dock. About .5 mile south of the village is Shoreline Resort and Lonesome Creek store where showers, laundry and the Post Office are located.

Destruction Island, which lies 15 miles SSE of La Push and 20 miles NNW of Cape Elizabeth, is a flat-topped, brush-covered island with rocks and ledges extending about 1 mile seaward from the cliffs on the north end. A light and fog signal are on the island. Anchorage in 10 fathoms, sand may be taken off the SE face of the island during settled northwesterly winds with the light bearing 293° to 315°. This isn't a particularly good anchorage but it is the only one available in the area. If the wind hauls west or south you must leave immediately.

LA PUSH, WA

QUILLAYUTE RIVER

Chart 18480

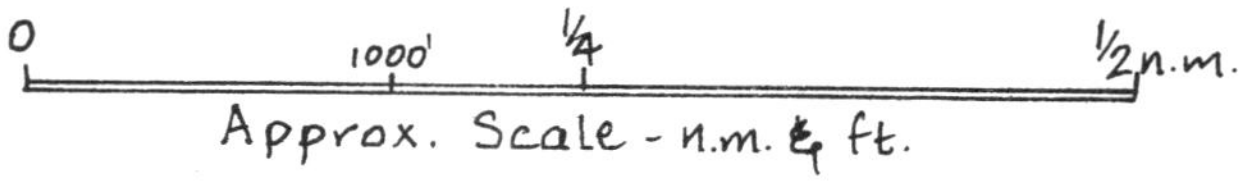

Not to be used for navigation

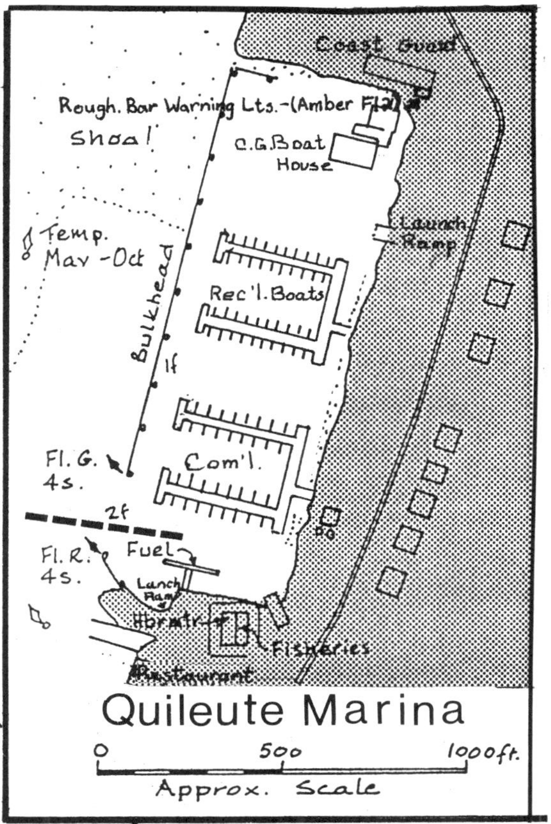

E-mail: qmarina@reachone.com

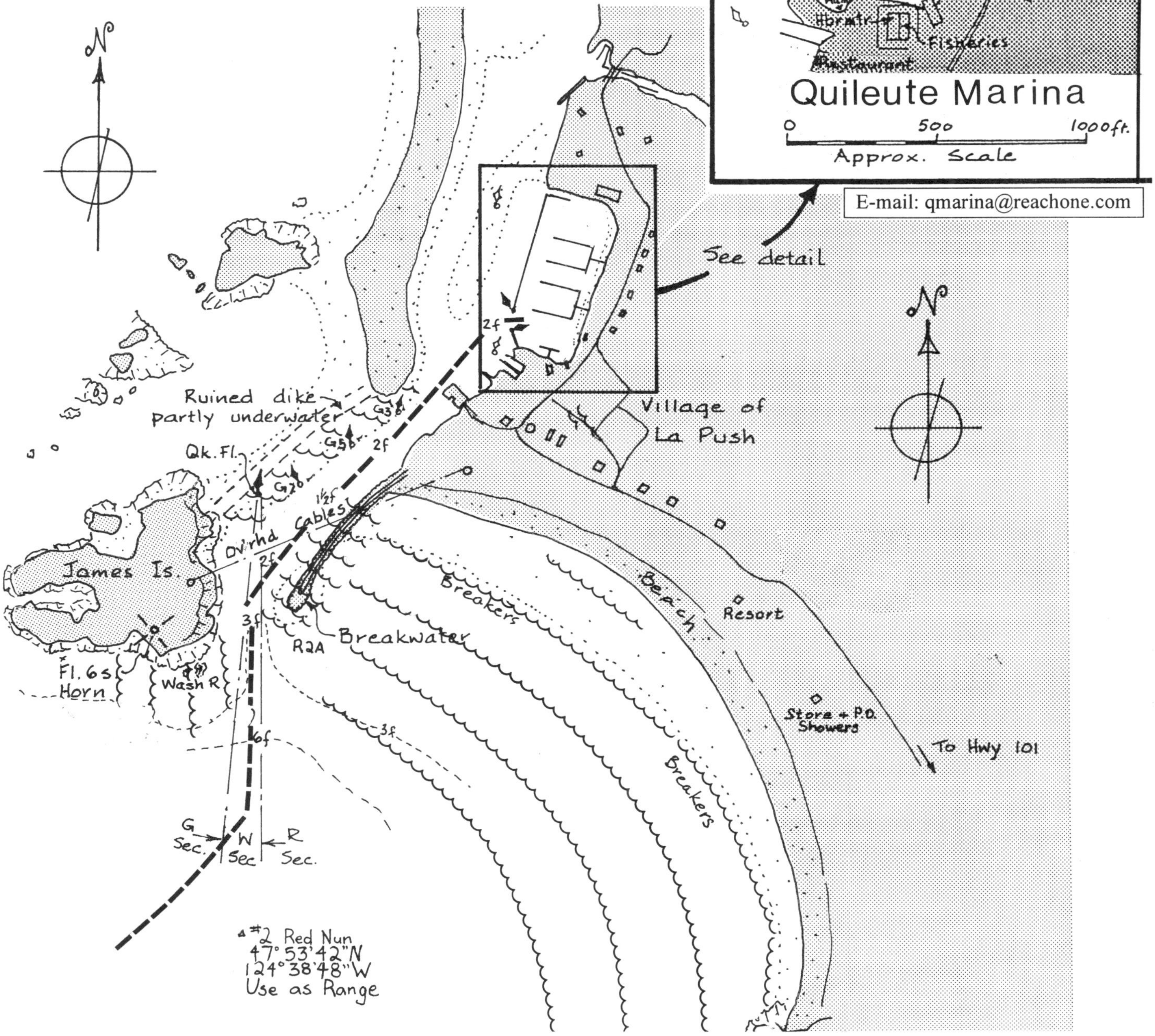

GRAYS HARBOR

The old Coast Guard station

The entrance to this large bay and harbor lies about 93 miles south of Cape Flattery and about 40 miles north of the Columbia River entrance. It is an important lumber port and deep sea charter fishing center, but the approach and entry into the harbor still require a great deal of caution and deliberation. Willapa Bay is about 15 miles south of Grays Harbor but this large, shoal bay is dangerous with its changing bar channel should be avoided.

Two extensive rock breakwater jetties extend out from both entrance points to Grays Harbor. The South Jetty from Point Chehalis extends well out to sea and is submerged at its end. The North Jetty, off Point Brown is shorter. Between these jetties the entrance is about 2 miles wide. A bar formed by shoals extends south and west from near Point Brown, and together with the shoals off Point Chehalis, causes rough water and breakers to occur, especially when the ebb meets the prevailing northwesterly winds. A severe storm in the early part of 1994 eroded and broke through the sandy neck of land connecting the South Jetty to the peninsula on which Grays Harbor is situated.

The channel is subject to change, but a range is established leading in from the outer buoy. As shown by the sketch, a westerly swell curves around and runs right into the bay. Though you should favor the south side of the entrance, the prevailing northwesterlies put a vessel dangerously close to a lee shore and great care is needed to stay out of trouble. A rough bar warning sign is shown on the north side of Westhaven Cove within the harbor. In addition, two heavy weather flags may be shown, one on the Coast Guard lookout tower 70 ft. above the water on the south side of Point Chelalis, the other on the northwest side of the Coast Guard station boathouse 50 ft. above the water. Of additional concern in these waters is the presence of (Dungeness) crab traps, which provide a major source income for this area. Though most are removed by mid-April, a few can be found throughout the summer.

Westport Marina in Westhaven Cove is on the inner side of Point Chehalis, on the south side of Grays Harbor. Breakwaters, jetties and reclaimed land areas form an extensive small boat basin, operated by the Port of Grays Harbor. There are a few transient berths. The harbormaster's office is on Lamb Street and it monitors VHF Ch 71. For information call (360) 268-9665 or Fax (360) 268-9413. Fuel is available from 8 am to 5 p.m., Monday to Saturday. Restrooms, showers and laundry facilities are available at Totem RV. Bus service links the town to Aberdeen. The nearest haul-out facilities are in Astoria, OR. Computer modem access is available at the marina office and an internet café is at Jetty Java.

Many sports fishing charter businesses line the main street, as well as restaurants and gift shops. Grocery stores, Englund Marine and Harbor Marine and Supply are close to the harbor. A visit to the museum in the old Coast Guard Station behind the waterfront is a must for all sailors. A boardwalk leads to the tallest lighthouse in Washington, built in 1898.

A small craft basin also lies behind Point Brown and Ocean Shores on the north shore. It is accessible via a line of dolphins (some of which are lit), leading from the inner channels of Grays Harbor. Shoals all through this area do not recommend it to a stranger. Ferry service joins Ocean Shores and its seaside cabin areas to Westport.

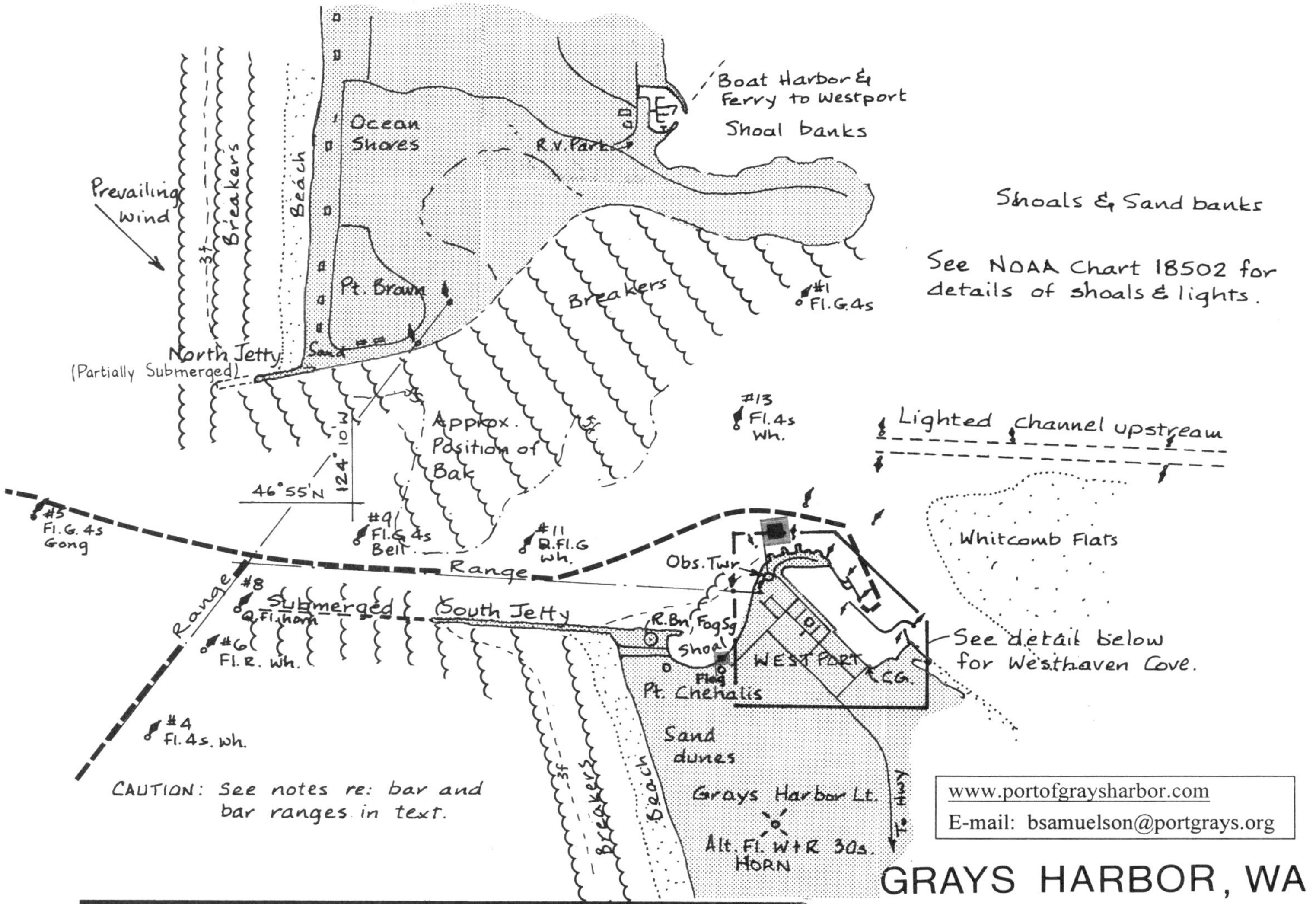

www.portofgraysharbor.com
E-mail: bsamuelson@portgrays.org

GRAYS HARBOR, WA

Chart 18500

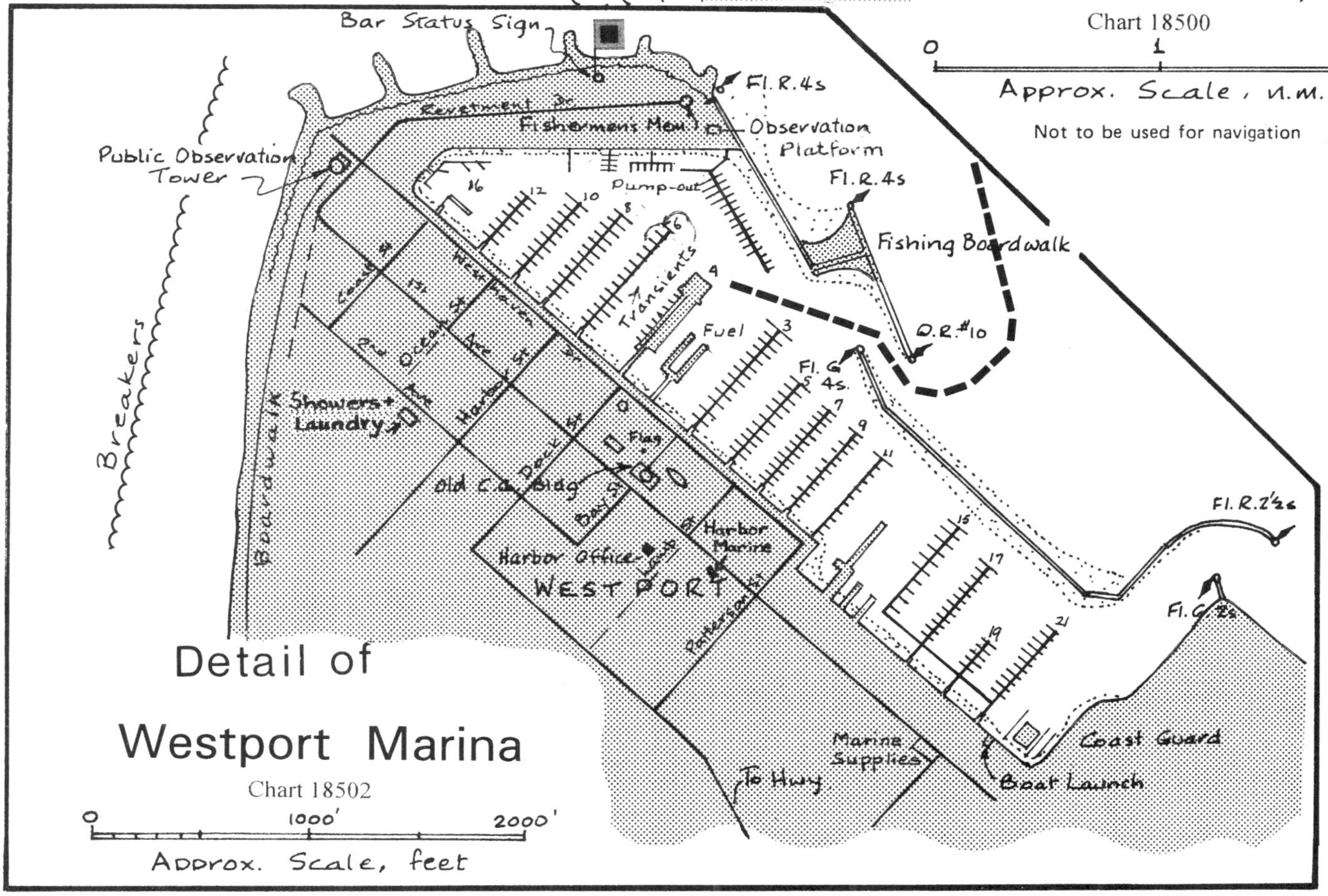

Detail of Westport Marina

Chart 18502

COLUMBIA RIVER ENTRANCE

Cruisers transiting the coast should approach this entrance only during calm conditions, with a rising tide and after contacting the Coast Guard for a bar report. The entrance is 145 miles south of Cape Flattery and 550 miles north of San Francisco. This is the most infamous of west coast bars because of the 2,000 vessels which have foundered while attempting to make or leave port. To reach a moorage you must travel a considerable distance up a channel bordered by hazardous shoals. But when the weather is calm—just the time when one wants to make good mileage up or down the coast—and it is a flood tide, there are no major problems in transiting the well-marked channel.

Willapa Bay is just 20 miles north of the Columbia River entrance but vessels transiting the coast should avoid this dangerous entrance. The large, shoal bay with its long, frequently changing bar channel is not an area to be visited by cruisers.

Cape Disappointment is a high, rocky headland with steep cliffs on the seaward side. Though the Cape Disappointment Light is 67m (220 ft.) above sea level, it is not always visible from the north. The western point of Cape Disappointment is North Head, an unmistakable steep-sided cliff with a heavily wooded area behind. North Head Light is shown from a white conical tower almost 61m (200 ft.) above the water.

When approaching the Columbia River entrance pick up the offshore buoy "CR" which is about 5.5 miles southwest of the channel entrance. A fog signal and radar transponder are at the 13m (42 ft.) buoy. About 2.75 miles northeast of the Columbia River Approach Buoy is whistle buoy #2 that further identifies the beginning of the channel, while bell buoy #1 is about 2 miles west of the channel. On a clear day Mount St. Helens (2,590m/8,500 ft.) is visible 75 miles to the east.

The Columbia River entrance lies between Clatsop Spit on the south and Cape Disappointment on the north. Clatsop Spit is a low sandy beach extending 2.5 miles northwest from Point Adams. The South Jetty extends almost 3 miles seaward from the end of Clatsop Spit. Take care not to confuse the low sand beach north of Cape Disappointment for the one south of Point Adams.

The entrance is between the South Jetty (extending westerly from Clatsop Spit) and the North Jetty that extends 732m (800 yards) in a southwesterly direction from the northern shore. The outermost mile of the South Jetty is submerged and the channel is clearly marked by buoys, lights, daybeacons and lighted ranges. Great care must be exercised in identifying and following the various navigational aids in their proper order. Bar condition reports are broadcast 15 minutes before and after the hour on KVAS (1230 kHz) and KAST (1370 kHz) or you may call the Coast Guard for a report.

Peacock Spit is the area north and west of the North Jetty. This is the most dangerous area and has claimed the majority of the ships lost while attempting the entrance. Breakers extending 4 miles to seaward are heavy in any type of current and must be given more than .5 mile clearance even during calm conditions. "Sneaker waves" are common up to .5 mile outside the breakers off the end of the North Jetty. If power is lost in the channel the vessel can be drawn onto Peacock Spit as the ebb tide runs across its tip.

Photograph courtesy of the U.S. Coast Guard

Coast Guard vessels crossing the Columbia River Bar, Oregon. The nearest vessel is a 40-footer. (Page 44)

Entering the narrow gorge at Depoe Bay is always exciting. (Page 52)

The morning mist soon disappears from Charleston Boat Basin in Coos Bay, Ore. (Page 60)

It can be sunny and clear along the Oregon coast while the offshore fog hovers in the distance.

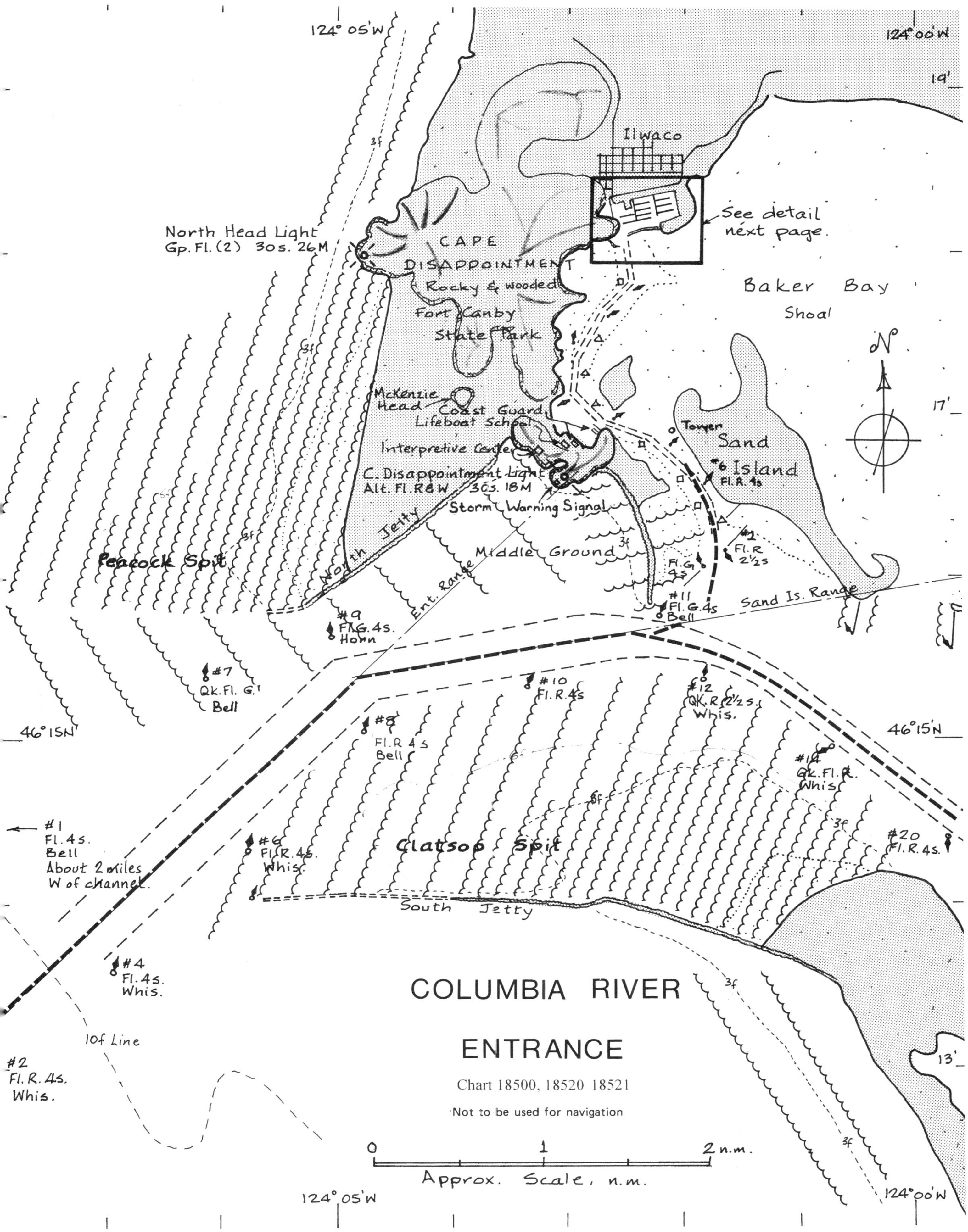

124° 05'W
124° 00'W
19'
Ilwaco
See detail next page.
North Head Light
Gp. Fl. (2) 30s. 26M
CAPE DISAPPOINTMENT
Rocky & wooded
Fort Canby
State Park
Baker Bay
Shoal
McKenzie Head
Coast Guard Lifeboat School
Interpretive Center
Tower
Sand Island
#6
Fl. R. 4s
17'
C. Disappointment Light
Alt. Fl. R&W 30s. 18M
Storm Warning Signal
#2
Fl. R
2½s
Peacock Spit
North Jetty
Ent. Range
Middle Ground
Fl. G
4s
#11
Fl. G. 4s
Bell
Sand Is. Range
#9
Fl. G. 4s.
Horn
#7
Qk. Fl. G.
Bell
#10
Fl. R. 4s
#12
Qk. R. 2½s.
Whis.
46° 15'N
46° 15'N
#8
Fl. R 4s
Bell
#14
Qk. Fl. R.
Whis.
#1
Fl. 4s.
Bell
About 2 miles
W of channel.
#6
Fl. R. 4s.
Whis.
Clatsop Spit
#20
Fl. R. 4s.
South Jetty
#4
Fl. 4s.
Whis.
COLUMBIA RIVER
ENTRANCE
Chart 18500, 18520 18521
Not to be used for navigation
10f Line
#2
Fl. R. 4s.
Whis.
13'
0
1
2 n.m.
Approx. Scale, n.m.
124° 05'W
124° 00'W

To the south is Clatsop Spit which is dangerous for a number of reasons. During flood or slack currents it may be comparatively calm but within a few minutes, when the current starts to ebb, treacherous breakers suddenly extend far out towards the channel. It is essential to maintain a course north of the red buoys. Another danger of this area is that if a vessel is set back after rounding the curve of the channel (resulting from engine failure, lack of fuel etc.) it is set onto the breakers of Clatsop Spit. During poor visibility care must be taken not to miss buoy #8 by cutting across from buoy #6 to buoy #10.

Middle Ground is southwest of Cape Disappointment Light as well as the adjoining area south of Peacock Spit. This shallow area has breakers even when only a 1.2m (4 ft.) swell is running. Conditions can change drastically when the current changes to ebb.

Though the normal river current is 1.5 to 2 knots, the velocity due to tidal outflow may reach as high as 9 knots. It is essential that the vessel is well found and the crew is prudent in negotiating this long channel while using the latest official NOAA chart for navigation.

The first place where moorage can be obtained is at **Ilwaco**, a town located northwest of Sand Island on the shore of Baker Bay. A dredged 3-mile channel marked by lights and daybeacons leads north from the Columbia River east of Buoy #11. The entrance to the harbor is between two jetties with lights at the ends. Depths in the mooring basin vary from 2m to 4.8m (7 - 16 ft.) in the western part to less than 1.2m (4 ft.) in the eastern part. The entrance is subject to constant change as a result of shoaling which fans out from the southern jetty. The channel was dredged in 2003 but it is still important to stay in the channel as the water is very shallow alongside of it.

Contact the harbormaster for dock assignment; VHF Ch 16 is monitored, working channel is 69. Office hours are 8 a.m. to 4:30 p.m. during the summer. A pump-out station is located on the dock just outside the Port Office. The docks can accommodate boats up to 30m (100 ft.) and marine supplies are available in the harbor area as indicated. The port of Ilwaco operates the boatyard and it is the only self-service facility of its kind on the lower Columbia River. Very reasonable rates are charged for use of the 50-ton Travelift. For information call the port office at (360) 642-3143 or Fax (642-3148. Several boat repair and electronic shops operate in the town as well as dive services and marine construction. It is only two blocks to the town center where groceries, laundry services and propane can be obtained. An internet café is located on the waterfront at Festivals Coffee Net.

The town has undergone major revitalization and has attracted gourmet restaurants, art galleries, gift and book shops along with a large fleet of charter boats. It is well worth visiting local points of interest such as Cape Disappointment Lighthouse built in 1856 and North Head Lighthouse constructed in 1898 and the Lewis and Clark Interpretive Center at Fort Canby. Views overlooking the Columbia River and bar can make one glad to be ashore when the tide is ebbing or strong winds cause breakers to form far into the distance.

The Coast Guard Station at Cape Disappointment is home to the Coast Guard Lifeboat School. This location was deliberately chosen since very rough water for "practice" rescues is almost always available.

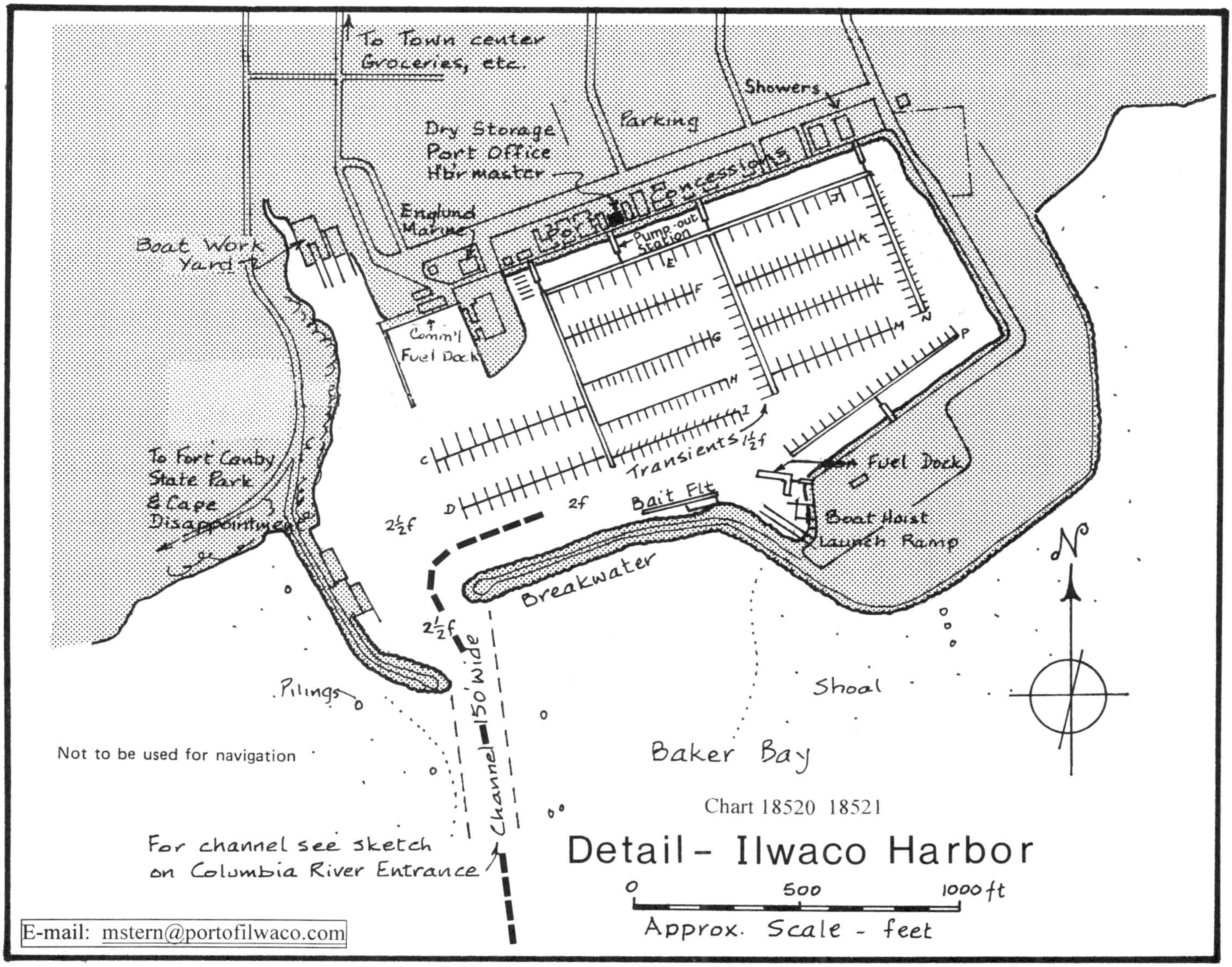
To Town center
Groceries, etc.
Showers
Parking
Dry Storage
Port Office
Hbr master
Concessions
Englund
Marine
Pump-out
Station
Boat Work
Yard
Comm'l
Fuel Dock
Transients
$1\frac{1}{2}$f
Fuel Dock
Bait Flt
Boat Hoist
Launch Ramp
To Fort Canby
State Park
& Cape
Disappointment
$2\frac{1}{2}$f
2f
Breakwater
$2\frac{1}{2}$f
Channel 150 wide
Pilings
Shoal
Baker Bay
Not to be used for navigation
Chart 18520 18521
For channel see sketch
on Columbia River Entrance
Detail- Ilwaco Harbor
0
500
1000 ft
Approx. Scale - feet
E-mail: mstern@portofilwaco.com

ASTORIA

A **Port of Entry**, this historic city is located on the Oregon side of the Columbia River. From Mile 0 (see detail drawing) Astoria is located at about mile 13. To reach it, follow the well marked channel passing through Lower and Upper Desdemona Shoal before following the Tansy Range. The impressive 4-mile long Astoria bridge which crosses the Columbia River and links Washington to Astoria, clearly pinpoints its location. Having a clearance of about 60m (200') at the center, it was designed to withstand gusts of wind of 150 miles per hour and still have a safety factor.

Astoria had its beginnings shortly after the Lewis and Clark Expedition ended its trans-continental journey in Clatsop County in 1805. The area is still settled mainly by people of Scandinavian heritage, who are proud of their past as evidenced by beautifully restored Victorian homes and several carefully maintained museums.

The Port of Astoria a wood and seafood processing center as well as a transportation terminal for rail, highway, air traffic and commercial deep-sea port operations. The Columbia River Channel is maintained by the U.S. Army Corps of Engineers and extends 100 miles upriver from Astoria to Portland, Oregon.

Of the two small craft facilities, the **West End Mooring Basin** is more convenient for transient traffic. It is located .25 mile west of the Astoria Bridge. The entrance is between two breakwaters, with a light at the outer end of each one. Transients may tie to the first six fingers between docks B and C. The harbormaster monitors VHF 16 with working channel 74 or telephone (503) 325-8279 (Fax (503) 325-4525) for dock assignment. Plans call for the marina office to be moved to Pier 1. Office hours are 8 a.m. to 5 p.m. A hoist is located at Pier three; for information call (503) 325-4521. Pump-out facilities are at the fuel dock. Fuel is not available on Sundays from October to May.

The nearest grocery store is by the East Basin, about 3 miles distant. Marine supplies are available at Englund Marine at the foot of 15th Street on Marine Drive. Propane is available a block away at the Texaco station on Highway 101 South. Englund Marine Supply is at the foot of 15th Street on Marine Drive. (Plans call for a possible move to Pier 3.) Computer modem access can be obtained at Gunderson's Sunnyside Cyber Cafe about a mile away.

The **East End Mooring Basin,** 2 miles east of the Astoria Bridge, has a launching ramp, restrooms and shower. Both pleasure and fishing vessels moor here and there are also a few slips for transients.

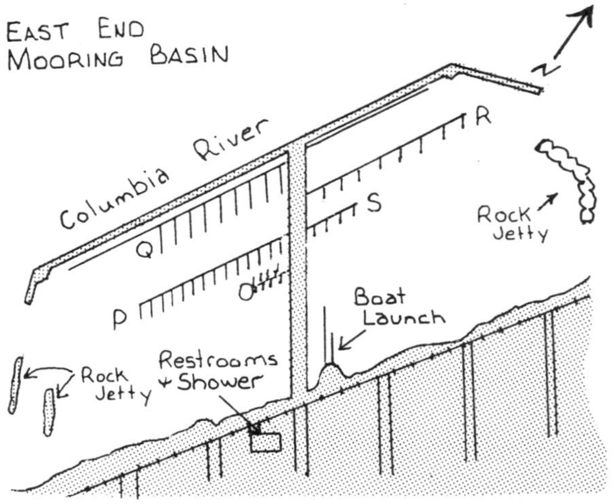

An appreciation of the colorful past of this area can be found by visiting some of the local points of interest, including the Columbia River Maritime Museum, Flavel House Museum, Fort Clatsop National Memorial. A climb to the top of the Astoria Column is well worth the effort for the magnificent view.

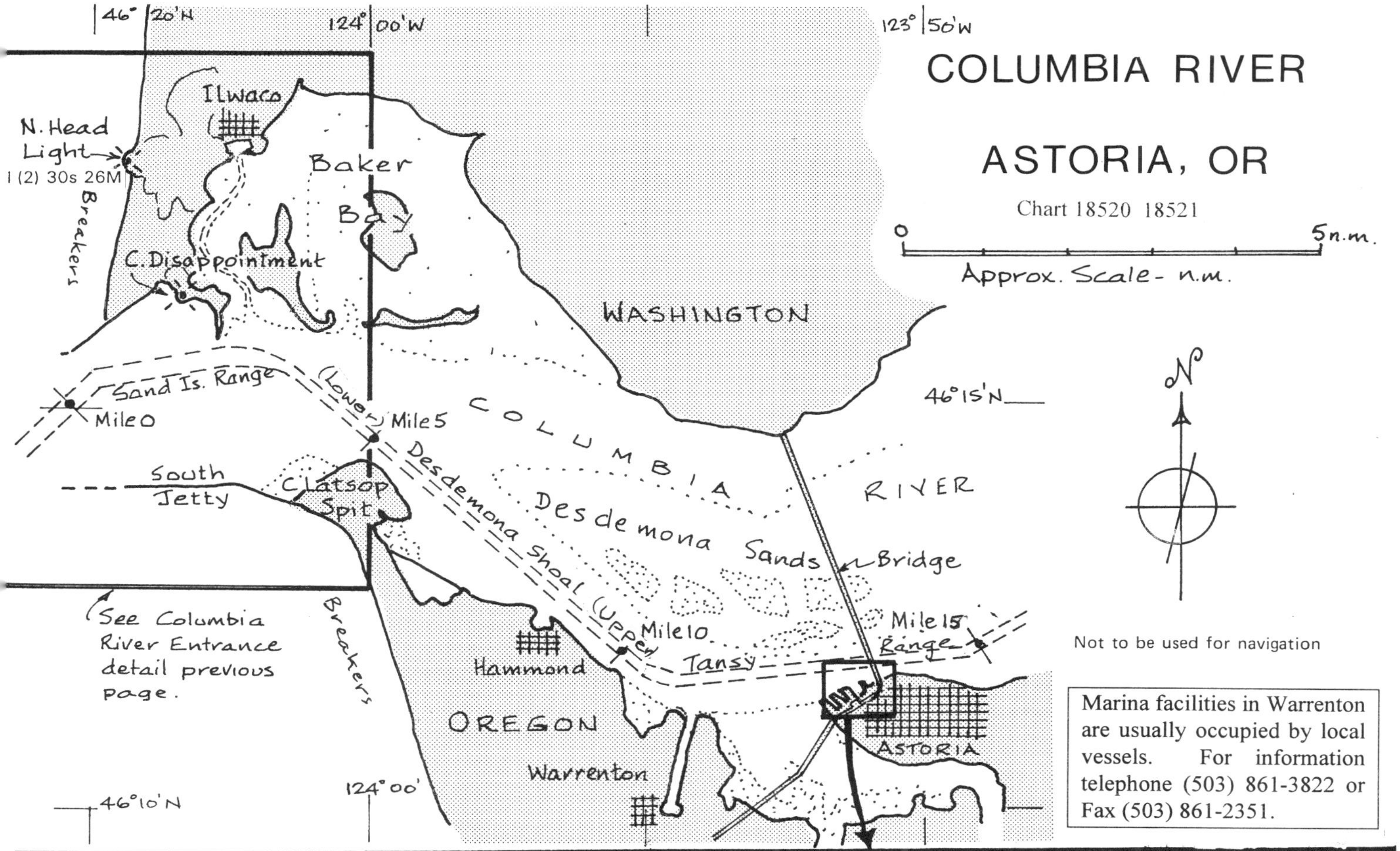
COLUMBIA RIVER
ASTORIA, OR
Chart 18520 18521
0
5 n.m.
Approx. Scale - n.m.
46° 20'N
124° 00'W
123° 50'W
46° 15'N
46° 10'N
124° 00'
Ilwaco
N. Head Light
I (2) 30s 26M
Baker Bay
Breakers
C. Disappointment
WASHINGTON
Sand Is. Range
Mile 0
(Lower)
Mile 5
COLUMBIA
RIVER
South Jetty
Clatsop Spit
Desdemona Shoal
Desdemona Sands
Bridge
(Upper)
Mile 10
Mile 15
Range
Tansy
Hammond
OREGON
Warrenton
ASTORIA
N
See Columbia River Entrance detail previous page.
Not to be used for navigation
Marina facilities in Warrenton are usually occupied by local vessels. For information telephone (503) 861-3822 or Fax (503) 861-2351.

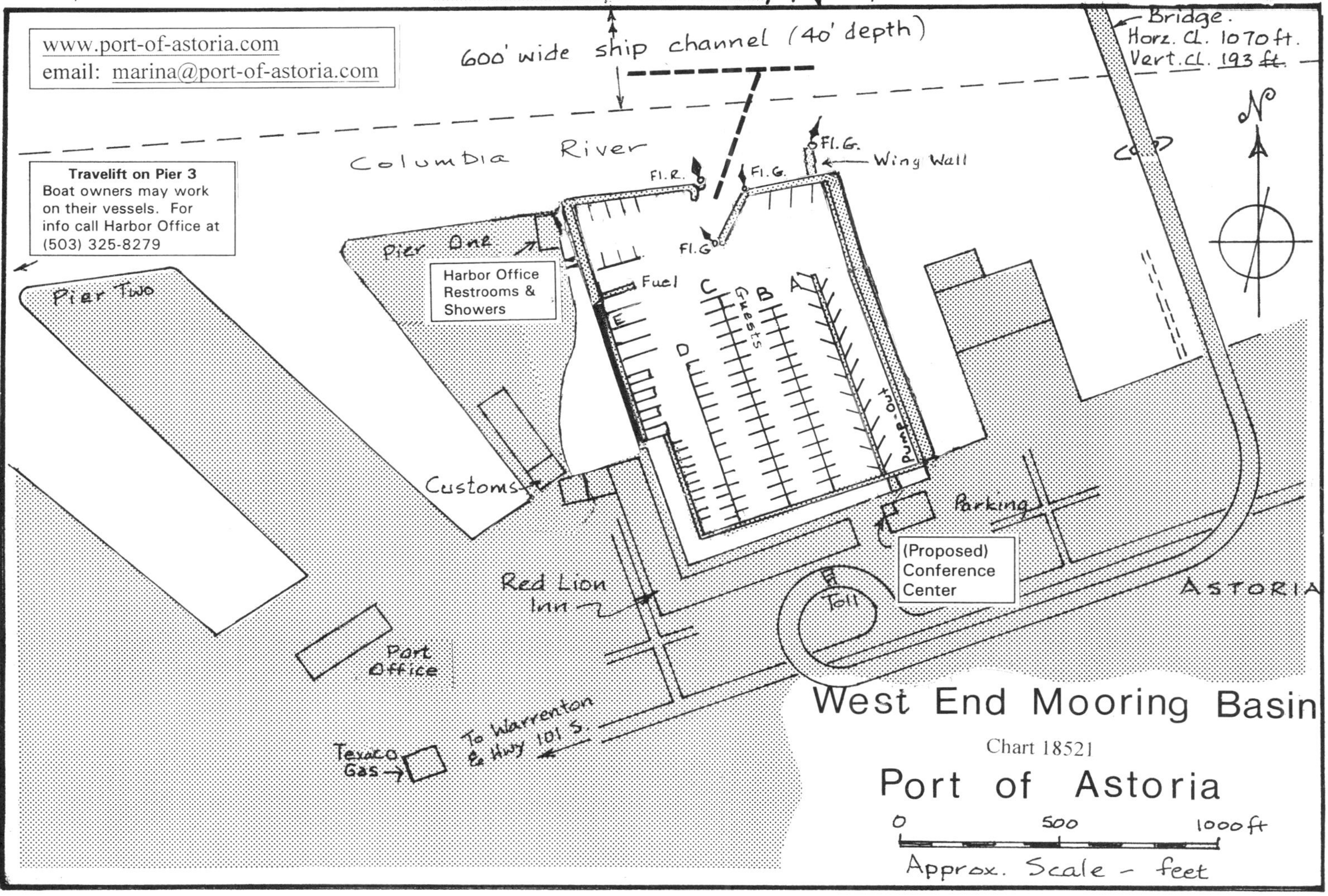
www.port-of-astoria.com
email: marina@port-of-astoria.com
600' wide ship channel (40' depth)
Bridge
Horz. Cl. 1070 ft.
Vert. Cl. 193 ft.
Columbia River
Fl. R.
Fl. G.
Fl. G.
Wing Wall
Fl. G.
N
Travelift on Pier 3
Boat owners may work on their vessels. For info call Harbor Office at (503) 325-8279
Pier One
Pier Two
Harbor Office Restrooms & Showers
Fuel
C
B
A
E
D
Guests
Pump-out
Customs
Parking
(Proposed) Conference Center
Red Lion Inn
Toll
ASTORIA
Port Office
Texaco Gas
To Warrenton & Hwy 101 S.
West End Mooring Basin
Chart 18521
Port of Astoria
0
500
1000 ft
Approx. Scale - feet

TILLAMOOK BAY (GARIBALDI)

About 12 miles north of the entrance to Tillamook Bay is Cape Falcon. This rocky cape jutting out about 2 miles from the general trend of the coast is easily identified by the 61m (200 ft.) vertical cliffs on the southwestern tip. In settled northwesterly weather anchorage may be taken in Smuggler Cove (a small cove tucked in, just south of Cape Falcon). When approaching, avoid Falcon Rock which is located about .75 mile west of the Cape. Anchorage should be taken close to the northern shore of Smuggler Cove in 4 to 5 fathoms. **Caution**: there are two isolated rocks visible only at low low water about 137m (150 yards) from the north shore in otherwise deep water.

A tough entrance makes Tillamook Bay one to avoid, unless you have a very good reason for stopping here. At best, the entrance is not easy; at other times, it is impossible. Located 42 miles south of the Columbia River and 5 miles north of Cape Meares, Tillamook Bay is adjacent to a rich agricultural area which is famous for its cheese.

Offshore whistle buoy "T" is located about 2.25 miles from the west end of the North Jetty, while bell buoy #1 is a little over a mile closer in. Two jetties protect the entrance—the North Jetty extending about .5 mile westward from the shore while the South Jetty reaches out about 1.3 miles from the sand dunes of Kincheloe Point. Lights, a seasonal fog signal and horn are at the end of the North Jetty and a Coast Guard lookout tower is adjacent. Note that Kincheloe Point appears to be a low island when seen at some distance from the north. The channel is marked by a sector light, lights, buoys and a daybeacon. A large "G" is seen on the hillside overlooking the community.

This is a dangerous channel which must be navigated only with extreme caution and local knowledge. Since the channel changes constantly because of natural silting and scouring you must check with the Coast Guard as to the accuracy of the range markers marking the channel. A Rough Bar Advisory sign is on the Coast Guard Station boathouse. Bar condition reports are also broadcast on KTIL (1590 kHz) once in the morning, 30 – 45 minutes after sunrise. For current bar conditions call (503) 322-3242 or contact the Coast Guard on VHF Ch. 16; switch to VHF Ch. 22A.

Enter only on a rising tide. Once the vessel has started to cross do not attempt to turn around if the bar is breaking. When the tide is ebbing, the current can run up to 6 knots, creating heavy breakers which extend over a mile offshore. On the north side of the dredged channel there are a number of visible and covered rocks. Note the location of Sow and Pigs ledge, which is almost 457m (500 yards) off the north shore, in the vicinity of buoy #11. **Caution**: On an ebb tide the current sets towards this dangerous rocky ledge. Having passed this area of danger, the vessel may proceed to the boat harbor as indicated on the sketch.

The Port of Garibaldi docks are for commercial fish boats and local vessels. The office is open 9 a.m. to 5 p.m., Monday to Friday. A dock for transient use is located on the NE side of the basin. Registration forms are available at the head of the dock. Both diesel and gasoline can be obtained a the fuel dock in the northern cove. Garibaldi Marina has gasoline for small craft, and groceries, laundry facilities and some supplies can be obtained in the town, two blocks away. There is a launch ramp and Garibaldi Dry Dock has haul-out facilities.

Garibaldi is quiet, laid back town where transient boaters are welcome. A museum at the old Coast Guard facility is worth visiting.

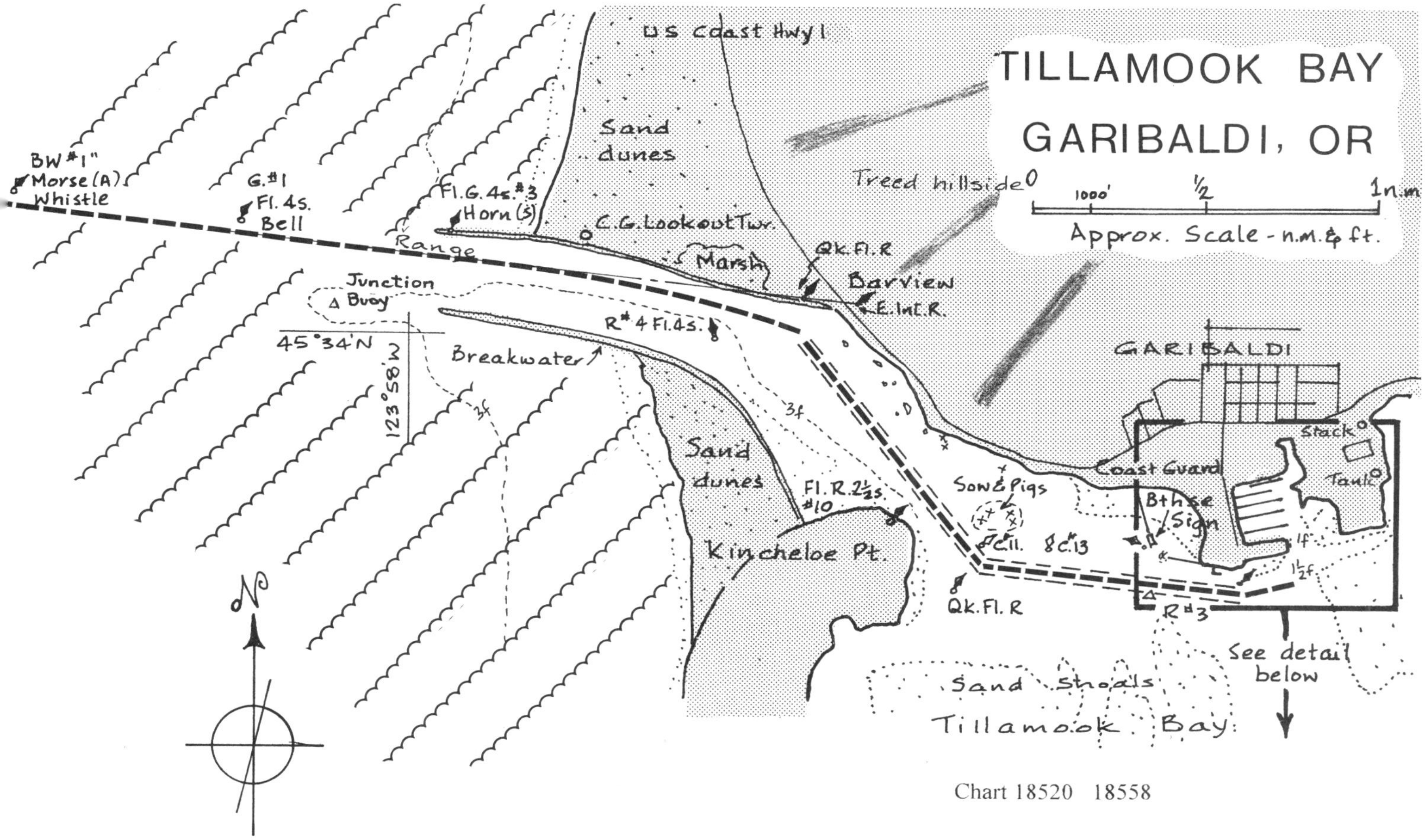

Chart 18520 18558

Not to be used for navigation

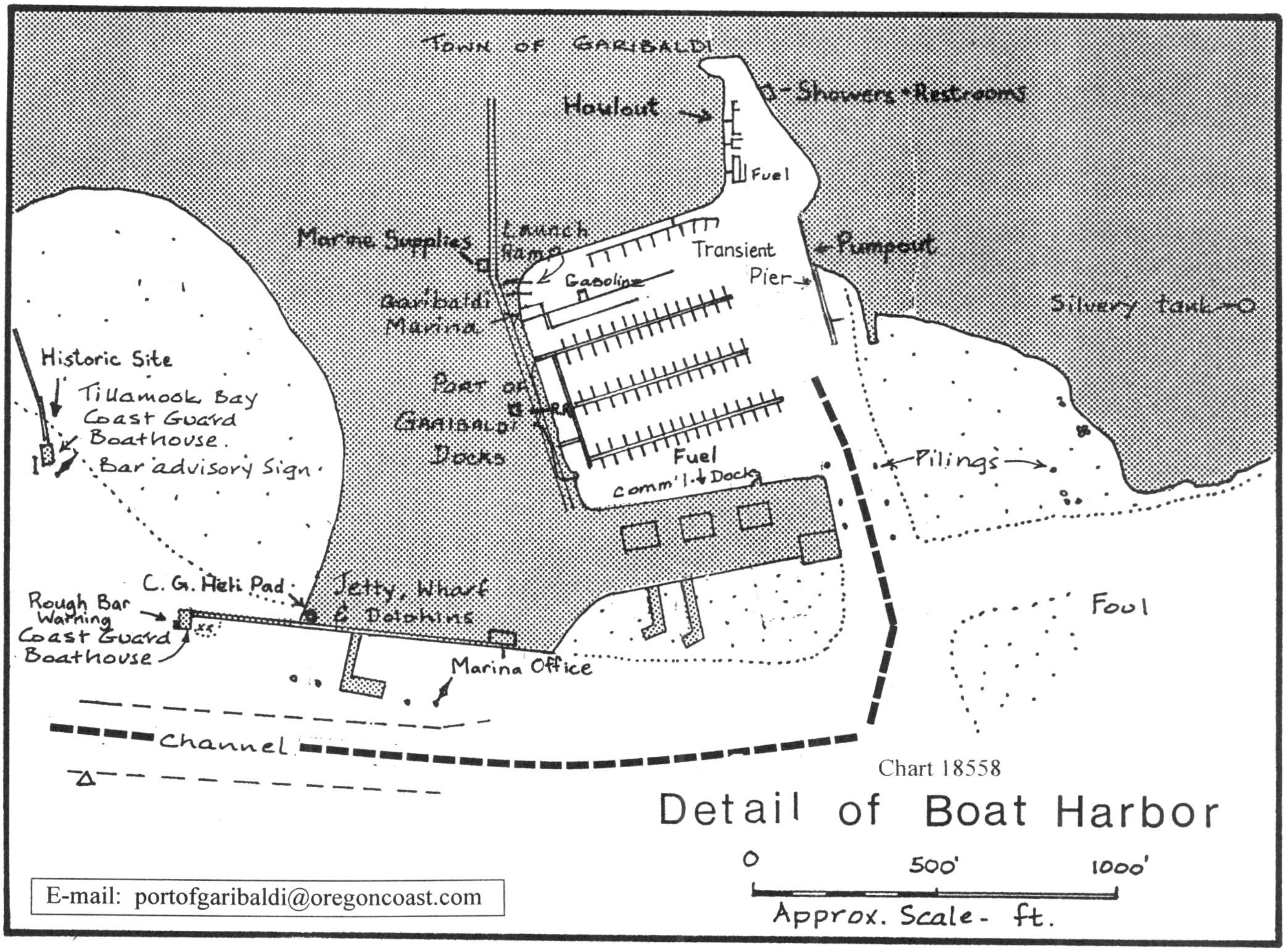

E-mail: portofgaribaldi@oregoncoast.com

DEPOE BAY

Having an area of only 6 acres, Depoe Bay has the distinction of being the world's smallest natural navigable harbor. Often referred to as "Hole-in-the-wall," it also has a well-deserved reputation for having one of the most intimidating entrances of any west coast harbor. It is about 100 miles south of Cape Disappointment and 12 miles north of Yaquina Bay and Newport. If you manage to negotiate its short but difficult entrance successfully, the reward is an excellent small boat haven.

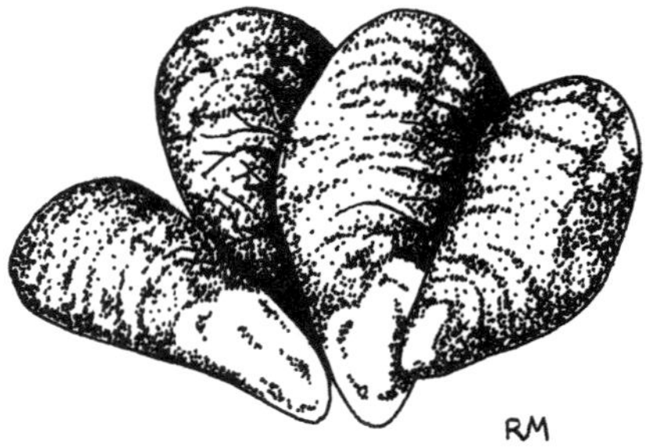

Edible mussels (Mytilus edulis)

The offshore lighted whistle buoy, "DB," is a little over a mile off the entrance to the bay, while bell buoy #2 is about .5 mile closer in. Since North Reef reaches out to within 91m (100 yards) of the bell buoy the approach must be made from the offshore buoy. The arched highway bridge over the entrance has a vertical clearance of 13m (42 ft.) which prohibits its use by many sailing vessels. The depth of the channel is 2.5m (8 feet) and within the basin depths range from 1.8m to 2.5m (6 to 8 ft.) with lesser depths along the edges. Foul ground and breakers extend well offshore on each side of the entrance and at times breaking seas occur clear across it.

Vessels over 15m (50 ft.) require prior permission from the harbormaster before entering the harbor, and then only at high water. Unless you are familiar with the entrance do not enter in rough weather or at night even though floodlights illuminate the passage to the basin. There is a dangerous surge in the narrow, curved 9.1m (30 ft.) wide entrance. The vessel must wait for a period of smoother waves when entering to avoid being pushed on to the rocks on the south side. Then in order to pass under the bridge into the harbor power must be applied to overcome effects of the surge. Some vessels report touching bottom under the bridge at low water where there is little sand over a rock bottom.

A rough bar advisory sign is located 7.5m (25 ft.) above the water on a building on the north side of the entrance channel. In addition a heavy weather flag (a square red flag with a square black center) may be displayed on a pole located about 48m (55 yds.) north of the bridge on the west side of Hwy. 101. Call the Coast Guard on VHF Ch 16 or at (541) 765-2123 for an up-to-date report on bar conditions. Think carefully about what you are taking on before entering, for once you have commenced this short entrance there is absolutely **NO** turning back.

If you are unable to contact the harbormaster prior to entering, tie up at the first dock and call City Hall at (503) 765-2361 to arrange for moorage. The facility is managed from Newport. Restrooms are located adjacent to the store on the east side of the basin. Marine supplies and a launching ramp are handy and a small shipyard is available for repairs. Grocery stores and a laundromat are a short walk up the hill to the town bordering the highway. Many fishing and whale-watching charter boats operate out of Depoe Bay. Two blocks from the marina is Java Bean Espresso which offers internet connections.

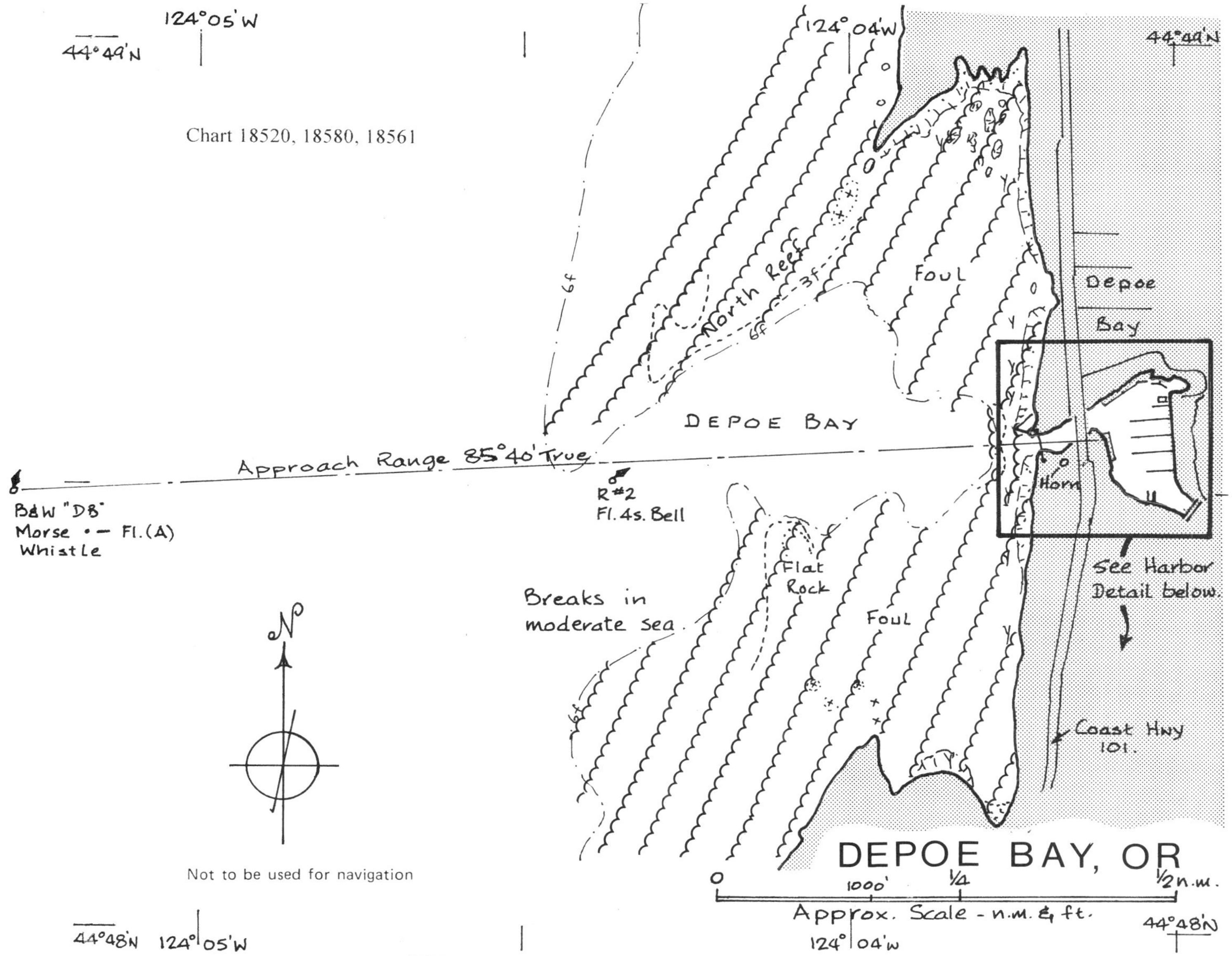

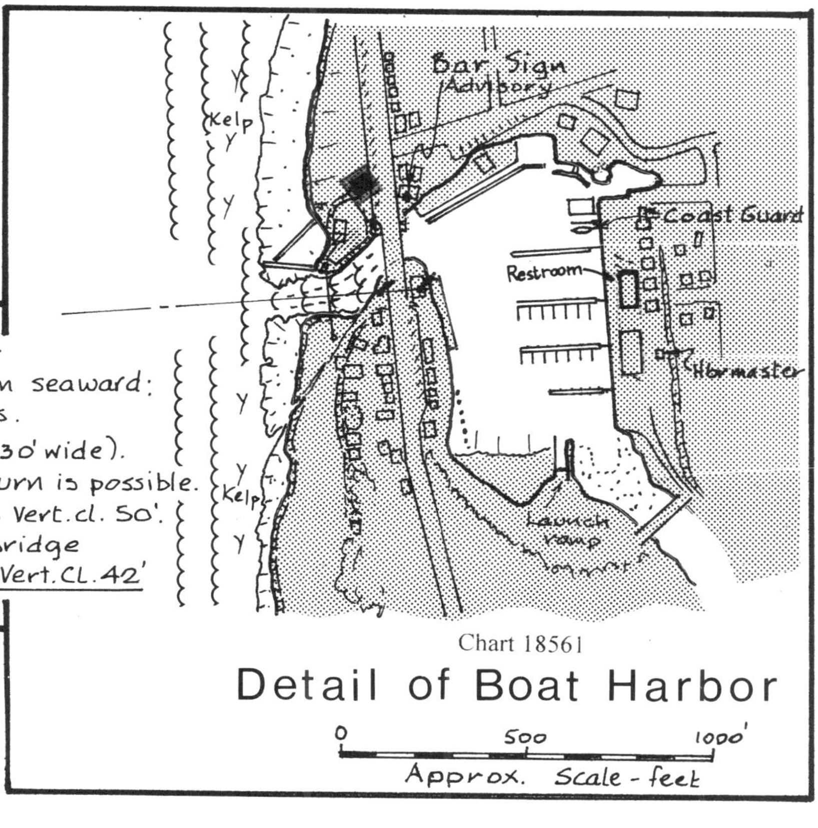

— ATTENTION —

At the Entrance, from seaward:

- Surge is dangerous.
- Channel is narrow (30' wide).
- Once started in no turn is possible.
- Overhead power line, Vert. cl. 50'.
- Conc. Arch Highway Bridge
  Horz. Cl. 30' wide for Vert. CL. 42'
- THINK!!!

Note that the entrance to Yaquina Bay and Newport is 12 n.miles South of Depoe Bay.

YAQUINA BAY (NEWPORT)

A favorable entrance distinguishes Yaquina Bay as the first place south of the Colombia River with a relatively trouble-free entrance. It is 12 miles south of Depoe Bay and 45 miles north of Florence. Inside the northern entrance point of Yaquina Bay, Newport (famous for Dungeness crabs) is the largest city on the central Oregon coast.

When approaching, pick up offshore lighted whistle buoy "Y," 1.5 miles southwest of the entrance, followed by 2 buoys leading to the entrance. An unmistakable landmark is the 40m (129 ft.) high Newport Bridge over the Yaquina River less than 1.5 miles upstream. An abandoned lighthouse and Coast Guard tower are on the sandy bluff on the northern side of the entrance. The channel is protected by two jetties, the .75-mile long North Jetty being slightly longer than the South Jetty. The channel is well marked with lighted ranges, lights and buoys but the Pilot cautions that the buoys cannot be relied upon to indicate the best water, and in the river, depths are subject to frequent change.

Cruisers are advised to contact the Coast Guard on VHF Ch 16 or telephone (541) 265-5382 to obtain the latest bar report. A rough bar advisory sign at the C.G. station, 25 ft. above the water is visible from the channel when looking seaward. In addition, heavy weather flag may be displayed on a pole on the western corner of the C.G. station. When leaving port, clear bell buoy #1 before turning north or south.

On the north shore is a large facility patronized by fish and charter boats: the **Port of Newport Commercial Boat Moorage.** Nearby are ship chandlers, mechanical and electronic repair shops as well as numerous restaurants, bars and souvenir stores. To the east is **Embarcadero Resort and Marina**, a full service facility that welcomes transient vessels up to 18m (60 ft.). For information call (541) 265-8521. On the south shore the **Port of Newport Marina** is a full-service facility catering to pleasure craft. Easy entrance can be made from the main channel. To arrange moorage call (541) 867-3321 or Fax (541) 867-3352. Haul-out facilities are at the river bend, about 6 miles upstream. Anchorage may be taken up the river, but watch the depth sounder carefully.

Interesting exhibits may be found in the Mark O. Hatfield Marine Science Center, the Oregon Coast Aquarium and the Lincoln County Historical Museum.

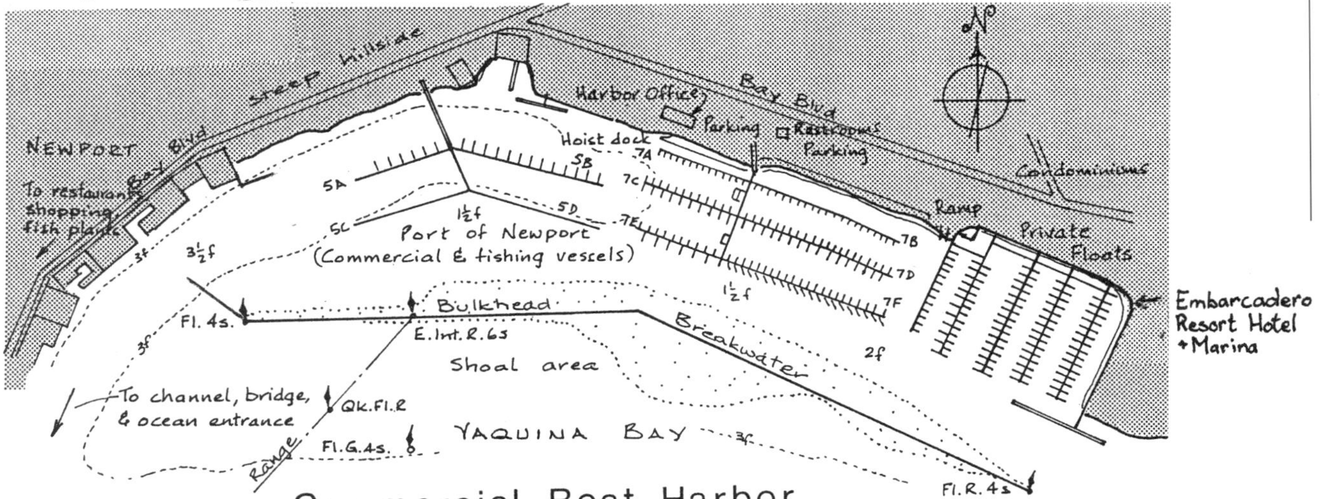

Commercial Boat Harbor

Port of Newport

Chart 18581

Not to be used for navigation

0 500' 1000'

Approx. Scale - feet

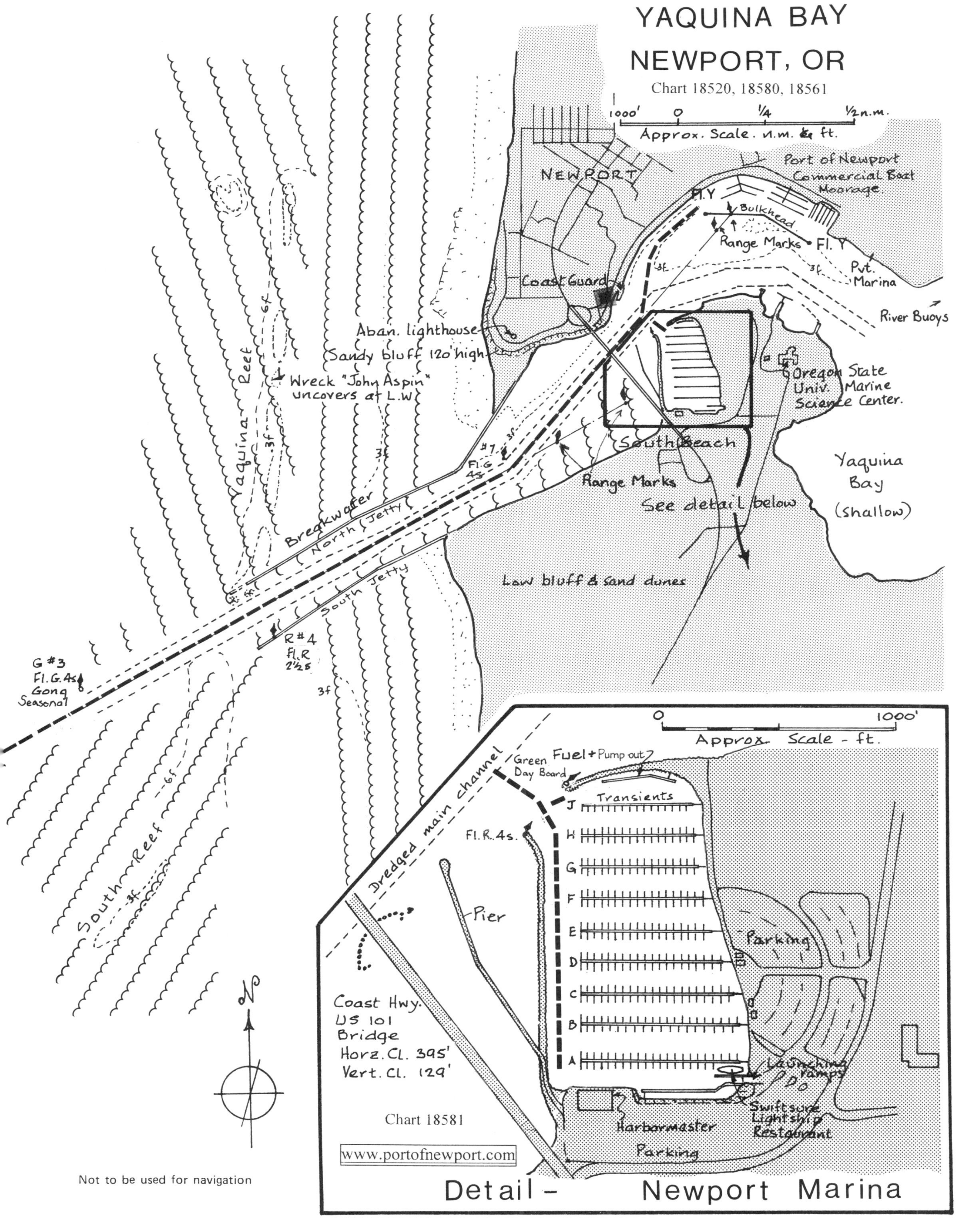

Not to be used for navigation

SIUSLAW RIVER (FLORENCE)

It is recommended that entry be made here only when there is little or not swell and the sea is calm. Caution is required because depths over the narrow channel constantly change as a result of shoaling and action of freshets, and the bar becomes impassable even during a moderate swell. This area is a transition point where coastal scenery abruptly changes from heavily vegetated mountains to a long chain of sandy beaches and dunes. The Siuslaw River entrance is about 40 miles south of Newport and 21 miles north of the mouth of the Umpqua River. Florence is located 12 miles south of the lighthouse at Heceta Head.

The offshore whistle buoy "S" is about .75 mile west of the end of the jetties protecting the entrance. Closer in, but NOT in line with the entrance is gong buoy #1, followed by seasonal buoy #3 (generally in place from May 1 to October 15) on the north jetty. The channel is between two Y-shaped jetties projecting westward from the river mouth. A Coast Guard tower is located on the North Jetty which is about .5 mile long and has a light and fog signal. The South Jetty is slightly longer, though the seaward ends of both jetties are even. Cannery Hill 45m (150 ft.) is a wooded hill on the east side of the river almost 1.5 miles south of the entrance. The C. G. Station is located south of this landmark.

When seas are from the west or southwest, breakers extend off the ends of the jetties as far as the gong buoy. Breakers are usually experienced off the ends of both jetties, especially on an ebb tide or when seas are 1m (3 ft.) or more. A large area of shoal water is on the south side of the channel well in from the bar where breaking seas are common. Another area of shoal water is on the northeast side of the channel where depths are less than 1m (3 ft.) at high tide. In response to the constantly changing channel, uncharted buoys are frequently shifted to mark the best water. The controlling depths are 3.6m (12 ft.) at mid-channel and 3.35m (11 ft.) along the waterfront at Florence.

The Coast Guard operates a rough bar advisory sign on an 11m (37 ft.) tower situated on the North jetty. In addition, a heavy weather flag (a square red flag with a square black center) may be shown on the southwest corner of the Coast Guard Station. Mariners are advised to contact Siuslaw River Coast Guard Station on Ch 16 or 22A before attempting to cross the bar. They can also be reached by telephone at (541) 997-2486 or (541) 997-3631. A recorded bar report can be heard by calling (541) 997-8303.

Florence is about 4.5 miles upriver from the entrance and a walk through carefully preserved Old Town is certainly worthwhile. A bascule bridge with clearance of 5m (17 ft.) connects Florence to a small settlement on the south shore. This bridge is unique because of its large Gothic columns and decorative concrete etchings on each arch. Call (503) 378-2299 to arrange for the bridge to be opened; 2 hours notice is required.

The **Port of Siuslaw Sport Marina** is located east of the bridge at mile 7.5. For dock assignment call (541) 997-3254. Diesel is available during daylight hours at the floating fuel dock on the east side of the launch ramp by calling (503) 999-1256. The marina office is across the parking lot from the launch ramp. An internet café and computer modem access available at Old Towne Coffee, just behind the marina office. It is only a short walk to town for groceries, postal services and restaurants. The small craft marina at mile 6 (just before the bridge) may resume operations in the future.

North of Florence are the Sea Lion Caves which are a rookery for Steller's Sea Lions. To the south, the Oregon Dunes National Recreation Area extends for 47 miles.

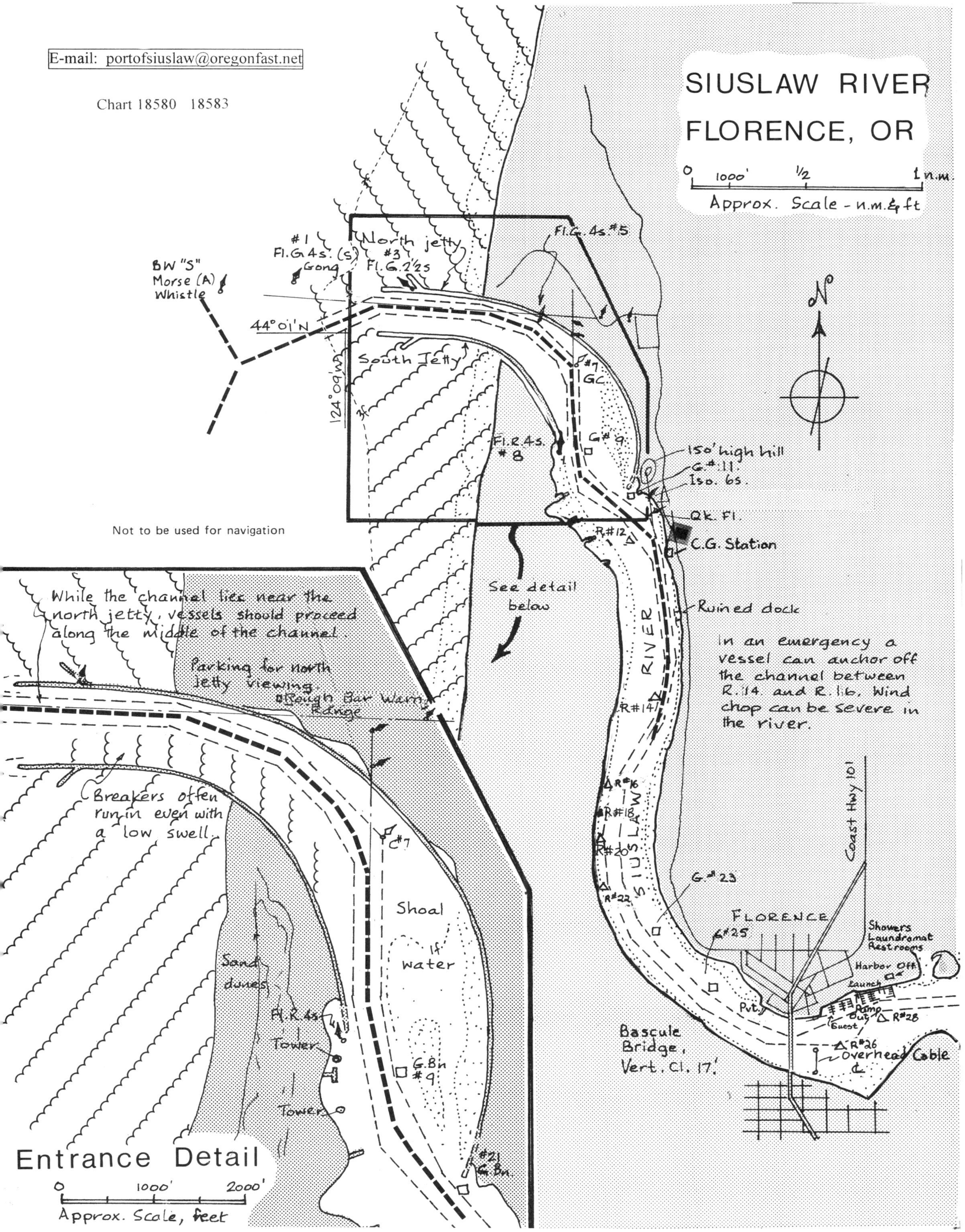
E-mail: portofsiuslaw@oregonfast.net
Chart 18580 18583
SIUSLAW RIVER
FLORENCE, OR
0 1000' 1/2 1 n.m.
Approx. Scale - n.m. & ft
N
BW "S"
Morse (A)
Whistle
#1
Fl.G.4s. (S)
Gong
North jetty
#3
Fl.G.2½s
Fl.G.4s.#5
44°01'N
124°09'W
South Jetty
#7
GC
Fl.R.4s.
#8
G#9
150' high hill
G#11
Iso. 6s.
Qk. Fl.
R#12
C.G. Station
Not to be used for navigation
See detail below
Ruined dock
In an emergency a vessel can anchor off the channel between R.14 and R.16. Wind chop can be severe in the river.
R#14
R#16
R#18
R#20
R#22
SIUSLAW RIVER
G#23
G#25
FLORENCE
Coast Hwy 101
Showers
Laundromat
Restrooms
Harbor Off.
Launch
Pump Out
Guest
R#28
R#26
Pvt.
Overhead Cable
Bascule Bridge,
Vert. Cl. 17'
While the channel lies near the north jetty, vessels should proceed along the middle of the channel.
Parking for north jetty viewing.
Rough Bar Warn. Range
Breakers often run in even with a low swell.
C#7
Shoal
1f water
Sand dunes
Fl.R.4s.
Tower
G.Bn
#9
Tower
#21
G.Bn.
Entrance Detail
0 1000' 2000'
Approx. Scale, feet

UMPQUA RIVER (SALMON HARBOR at Winchester Bay)

The mouth of the Umpqua River is 21 miles south of Siuslaw River and about 20 miles north of Coos Bay. The coast is a succession of sand dunes behind and on which the trees' dark foliage is visible. Following a period of calm weather with little or no swell running this entrance can be transited during a flood tide. In rough sea conditions the entrance becomes very dangerous and should be by-passed.

When nearing the entrance approach from the lighted offshore whistle buoy "U" which is almost a mile west of the South Jetty light. About .2 of a mile off the end of the jetty is gong buoy #2. The Umpqua River Light is south of the river's mouth in a treed area.

Two jetties form the entrance to the river: the .6-mile long North jetty extends seaward from sand dunes; the slightly longer South Jetty has a light and fog signal at its tip. About 145m (160 yards) of the outer end of the South Jetty is submerged. Connected to the seaward end of the South Jetty and bordering the channel some distance upriver is another breakwater sometimes referred to as Middle Jetty. The channel is marked by lights, daybeacons and lighted ranges. A television monitor visible from the channel is on a pedestal next to the entrance front range light. When operating, it shows river traffic which may include barges hauling sand, fuel, or lumber. On Winchester Point is a Coast Guard rough bar advisory sign. Contact the Coast Guard on VHF Ch 16 to check bar conditions.

Caution is advised as there is considerable variation in the channel and basin depths, depending on how much time has elapsed since the last dredging took place. Dangerous breaking seas occur seaward of the North Jetty even when a low swell is running. These breakers may cover half of the distance between the jetties and often continue well around the curve of the North Jetty until past buoy "6A". Breakers may also be found between the two sections of Salmon Harbor. Prior to leaving the harbor some local boaters drive out to what is known as "Chicken Point," on the North Jetty, in order to check conditions before venturing out.

Salmon Harbor Marina is located about 1.5 miles beyond the entrance and consists of two sections. A dredged channel leads in from the main river channel to a turning basin almost .5 mile from the harbor's entrance. Depths in the entrances to the basins are generally maintained at two fathoms. The marina office is between the two basins at their southern end; office hours are 8 a.m. to 4:30 p.m. during the summer. and since they do not monitor VHF channels at the present time, transients must tie to an empty slip and then register in the office. For reservations call (541) 271-3407 or Fax (541) 271-2060. Several restrooms are located throughout the facility; the only one with a shower is at the head of the crabbers' dock. The fuel dock and pump-out station operate Monday to Friday. Docks 1 to 8 are used primarily by small craft. Docks F, G, and H are seasonal docks.

The second basin is just to the east and two lights mark the entrance—the westernmost has a fog signal. There is usually space available for transients on Dock A, often called, "Sailboat Dock." It is about .75 miles to the village of Winchester Bay for shopping, laundry facilities or a restaurant meal. A source for propane is about 1.5 miles distant. Internet services and computer modem access are available at Village Roastery.

The village of Winchester Bay is a resort known for its excellent steelhead fishing. The river is a busy one, with fuel oil barges serving the large paper and lumber mills upriver at Gardiner and lumber barges working out of Reedsport, 10 miles from the entrance.

UMPQUA RIVER
WINCHESTER BAY, OR

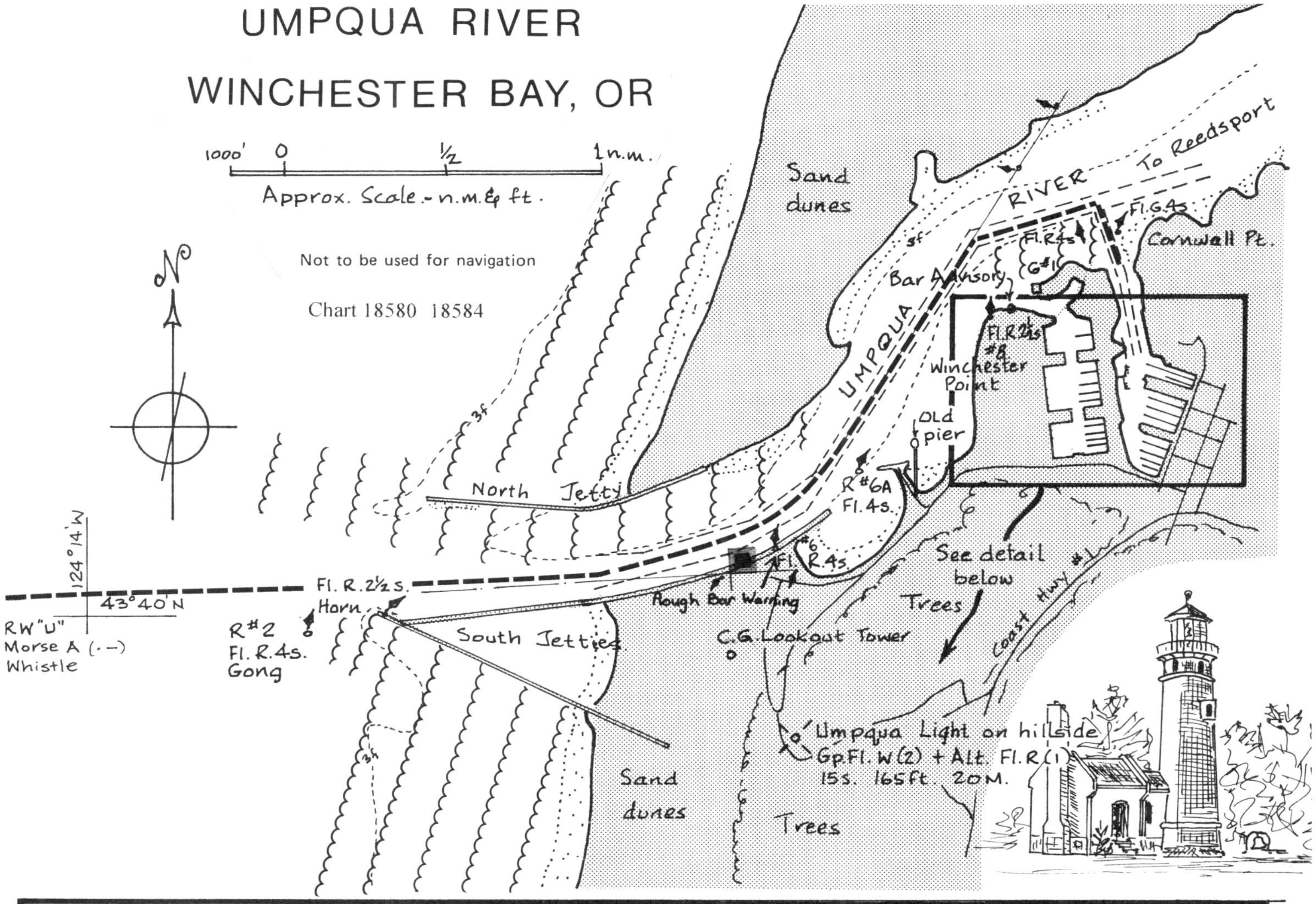

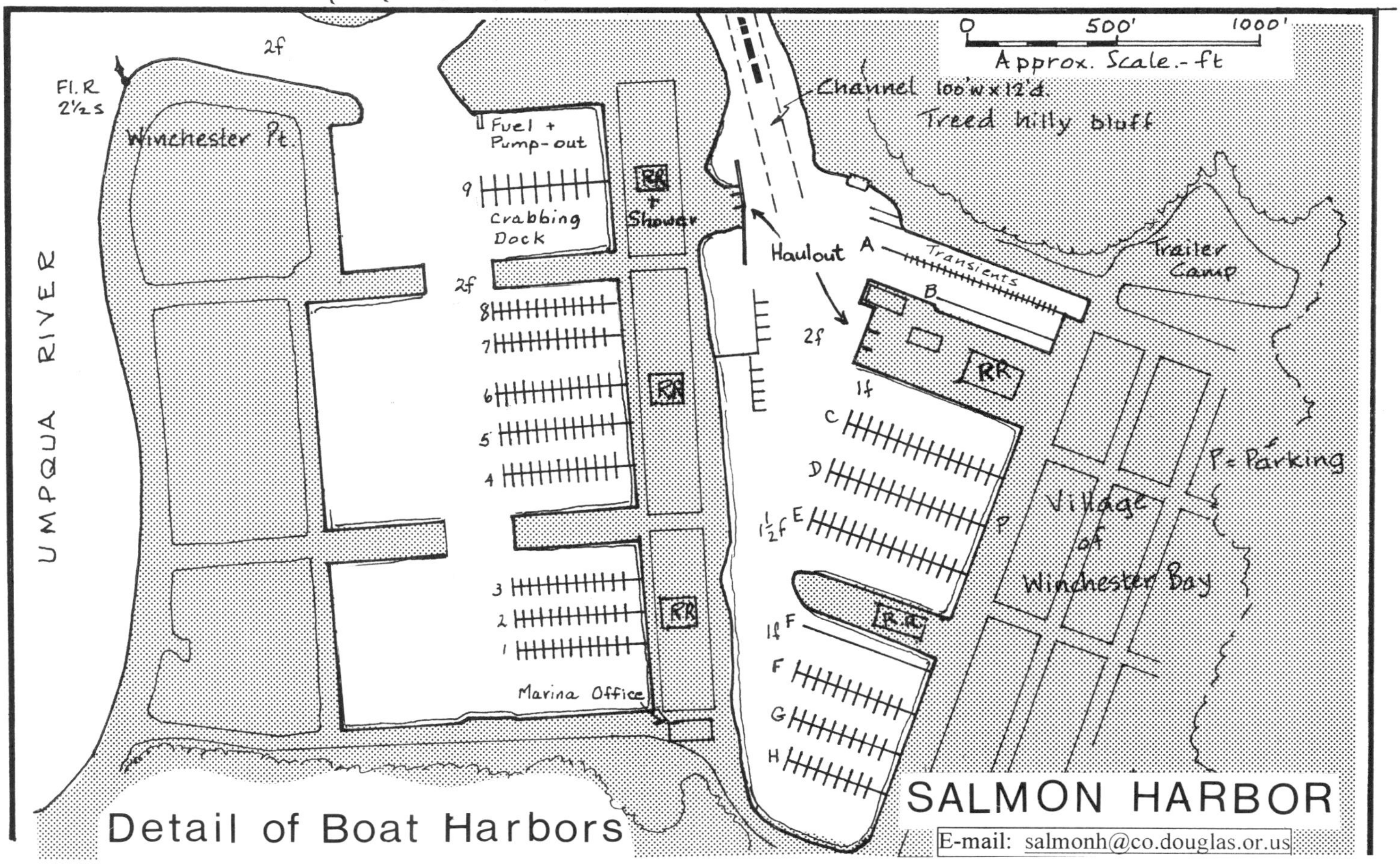

COOS BAY

Coos Bay is the most important harbor between San Francisco and the Columbia River, not only because it is a **harbor of refuge** but also because it is one of the largest wood products ports in the world. It is almost 20 miles south of Umpqua River and 33 miles north of Cape Blanco. It is easily identified by Cape Arago Light as well as Coos Head, a high, steep-sided, rocky head on the south side of the entrance. The U.S. Naval buildings are conspicuous on the bluffs south of Coos Head.

The offshore lighted whistle buoy "K" is almost 2 miles WNW of the entrance. Bell buoy #1 is 1 mile closer in and then comes gong buoy #3 a short distance further. Bar reports are given each hour on KBBR (1340 kHz) or one may contact the Coast Guard on VHF Ch. 12 or 16 or listen to a recorded update by calling (541) 888 - 3102. The Coast Guard maintains a rough bar advisory sign on the east side of the breakwater near the fuel dock. In addition, a heavy weather flag (a square red flag with a square black center) may be shown on a pole on the north side of the Coast Guard lookout tower at the Umpqua River entrance. Two jetties form the entrance to the bay, the North Jetty having a light and fog signal at its seaward end. Both jetties are submerged at their tips—the North Jetty for 274m (300 yards), the South Jetty for 91m (100 yards). Even moderate swells break heavily on the ends of the jetties, those on the south side of the channel rolling in for a considerable distance along the edge. Buoys, lighted ranges, lights and daybeacons mark the channels.

Enter only on a flood tide, preferably toward the end of the flood current. **Caution**: just off the ends of the jetties there is a current of variable strength sweeping either to the north or south, a southerly current being more common during summer months. The impressive size of the swells, coupled with breakers thundering on the two jetties, can make this an awesome entrance even during favorable conditions.

Heavy breakers occur in a wide arc off North Spit in the vicinity of bell buoy #5 and rogue waves are common in this area. Guano Rock (uncovered at low water) is on the south side of the channel about 256m (280 yards) northwest of Coos Head. Another danger is a submerged jetty which reaches out .3 miles off the east shore of Coos Bay. The sharp curve in the channel must be negotiated carefully in this area. Stay southwest of lighted buoy "1" when approaching or leaving the Charleston Boat Harbor.

Charleston Marina, with 700 slips, is easily entered and the people are very friendly. Call the harbormaster for dock assignment at (541) 888-2548 or Fax (541) 888-1611. Fuel, propane (10% valve only), marine supplies, fishing tackle, laundry and washroom facilities are handy and groceries may be obtained from a store near the cafe. Excellent repair services of every kind are available at reasonable prices and there is a three-lane launch ramp. The nearest marine supply store is Englund Marine, a comfortable one-mile walk away. As this is the sports fishing headquarters for the Bay area many charter and fishing boats are in evidence.

Anchorage in Coos Bay can be taken outside of the dredged channels but it is a long trip by dinghy to pick up supplies. Alternatively, anchorage may be taken upriver from Charleston opposite Dairy Queen; going ashore is easy any place on the beach.

A few miles north of the entrance lie the rusting, remains of half of the cargo ship, *New Carissa.* In 1999 it landed on the beach near Coos Bay. Salvage crews towed, floated, pumped, shot, torpedoed and burned the vessel but it refused to sink. The vessel broke in two and half of it drifted ashore where it is partially buried.

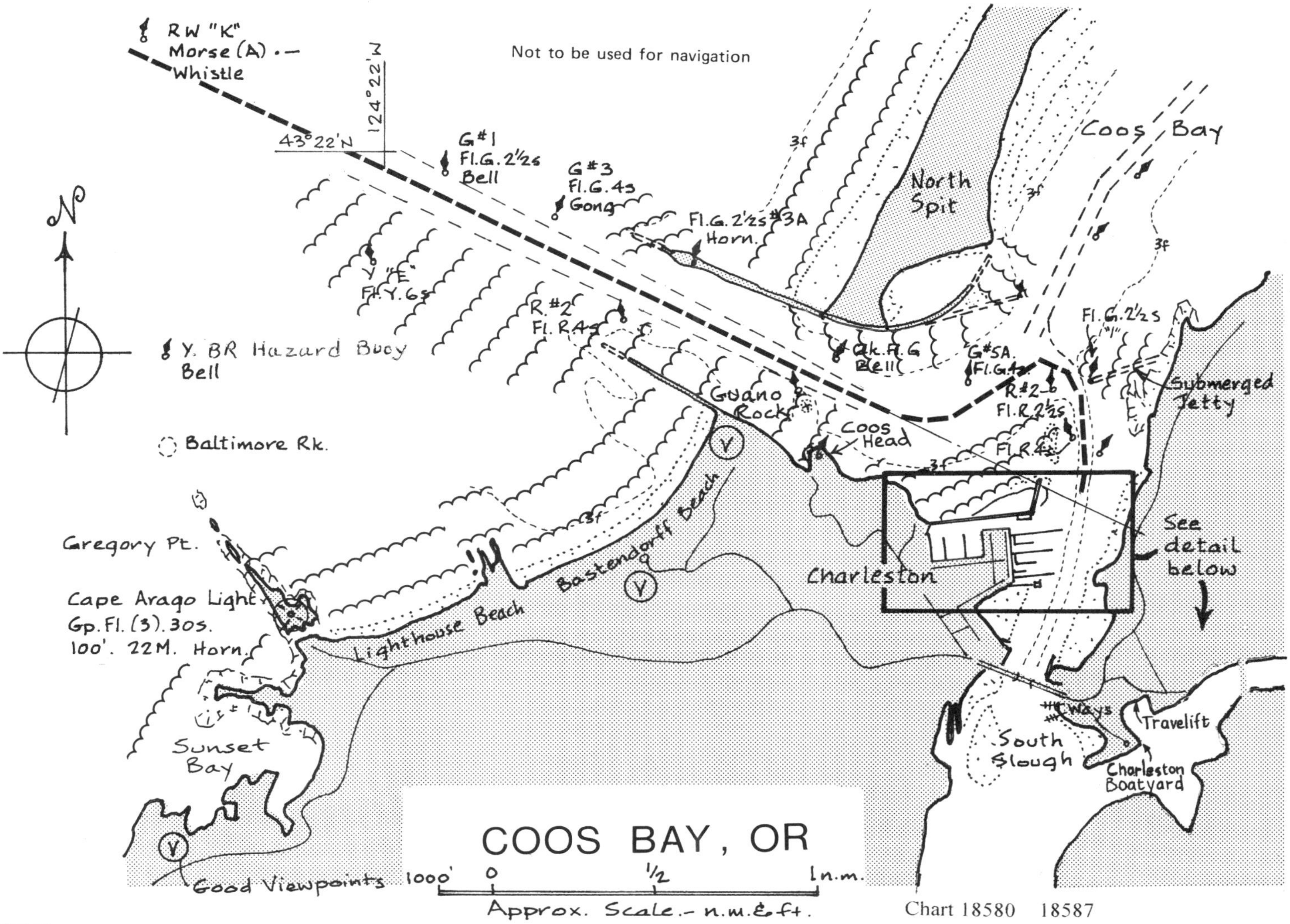

Chart 18580 18587

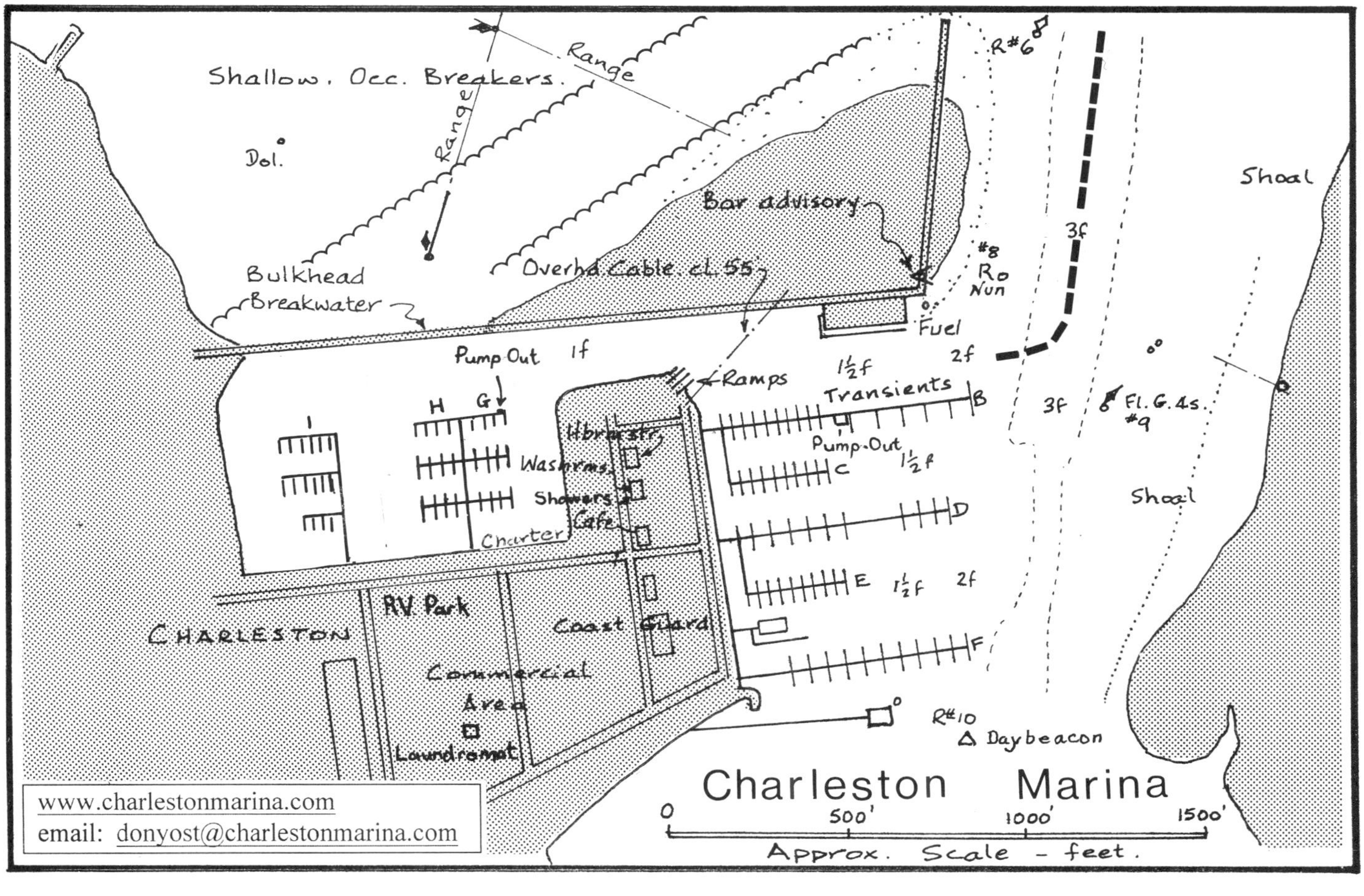

COQUILLE RIVER (BANDON)

Bandon calls itself the "Storm Watching Capital of the World." That in itself should tell you something!

The Coquille River entrance is about 15 miles south of Coos Bay and 30 miles north of Port Orford. The area's economy relies mainly on tourism, though cheese and cranberry industries make a contribution also.

The offshore lighted whistle buoy "2" is about 2 miles northwest of the entrance to the river. About .25 mile SSE of this buoy is Coquille Rock where breakers occur in heavy weather. Approximately .6 mile closer to the entrance is bell buoy "4" followed in another .6 mile by whistle buoy "6", which is in turn about .6 mile from the ends of the jetties. The entrance is about .6 mile northeast of Coquille Point. The countless upstanding rocks, ledges and reefs which form a huge fan-shaped area of spume and breakers off Coquille Point are a great encouragement to stay on course and not to take short-cuts. Breakers over a mile offshore are seen in this area.

When traveling south of Bandon on the coastal route take care to give at least a safe clearance to Cape Blanco, for Blanco Reef extends almost two miles off the outer extremity of the Cape and the extensive dangers of Orford Reef just to the south protrude an additional two miles. Pay particular attention to the weather forecast, since waters off Cape Blanco have a reputation for boisterous conditions at times.

The entrance is protected by two jetties on which the sea breaks heavily. A horn (operating June 1 - September 15) and light are on the South Jetty. Breakers roll up the southern shore of the channel where submerged rocks are evident. **Caution**: a current in the channel sets a vessel towards the south. On the North Jetty, just inland from the shore, is an abandoned lighthouse; partially submerged rocks extend from it to the end of the jetty. The channel is marked by a lighted range, lights and buoys. Local knowledge is advised since the channel changes frequently and so the range does not always indicate the deepest water.

From May 15 to Memorial Day or Labor Day the Coast Guard operates a Boat Patrol and can be contacted for information on bar conditions at (541) 347-2038. Out of season, call the Charleston Coast Guard at (541) 888-3267 for information. A rough bar advisory sign is located as shown on the sketch.

The **small craft basin** is tucked behind a high breakwater with a boardwalk on top. To arrange moorage call (541) 347-3206 or Fax (541) 347-4645. The Port Office is in the old Coast Guard office .5 miles to the west; office hours are 7:30 a.m. to 3:30 p.m., Monday to Friday. Fuel is available from Port O'Call which monitors VHF Ch 19 or phone (541) 347-2875. Limited marine supplies and fishing tackle are available from a shop on the waterfront. Many restaurants and shops selling sundry items can be found in the town. Propane, groceries and laundry facilities are about a mile distant.

A walk through restored Old Town Bandon (near the waterfront) is a pleasant way to work off one's sea legs, or you may prefer a hike along the beach to better enjoy the safe side of the spectacular offshore rocks. Unique to Bandon is a metaphysical exhibit known as the Continuum Center. The Coquille River Museum is located in an historic Coast Guard building across the river at Bullards Beach State Park.

Rocks south of the Coquille River entrance to Bandon, Oregon, make a spectacular scene. (Page 62)

The beautiful coast south of Port Orford, California beckons. (Page 64)

Pleasure craft and fishing vessels find protection in Shelter Cove on California's "Lost Coast." (Page 76)

Dredging in a narrow channel can make passage very tight as at Noyo River. (Page 78)

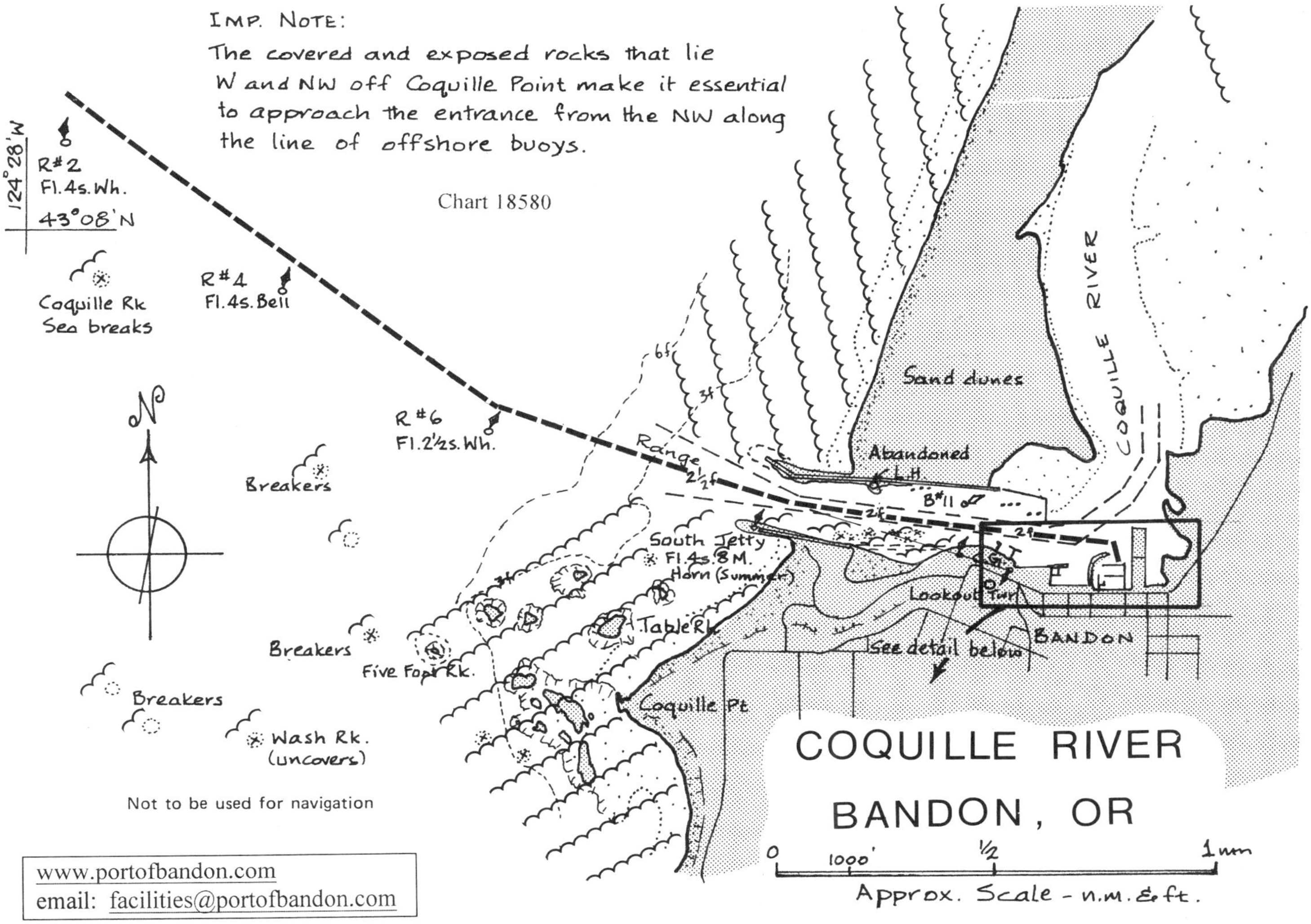

www.portofbandon.com
email: facilities@portofbandon.com

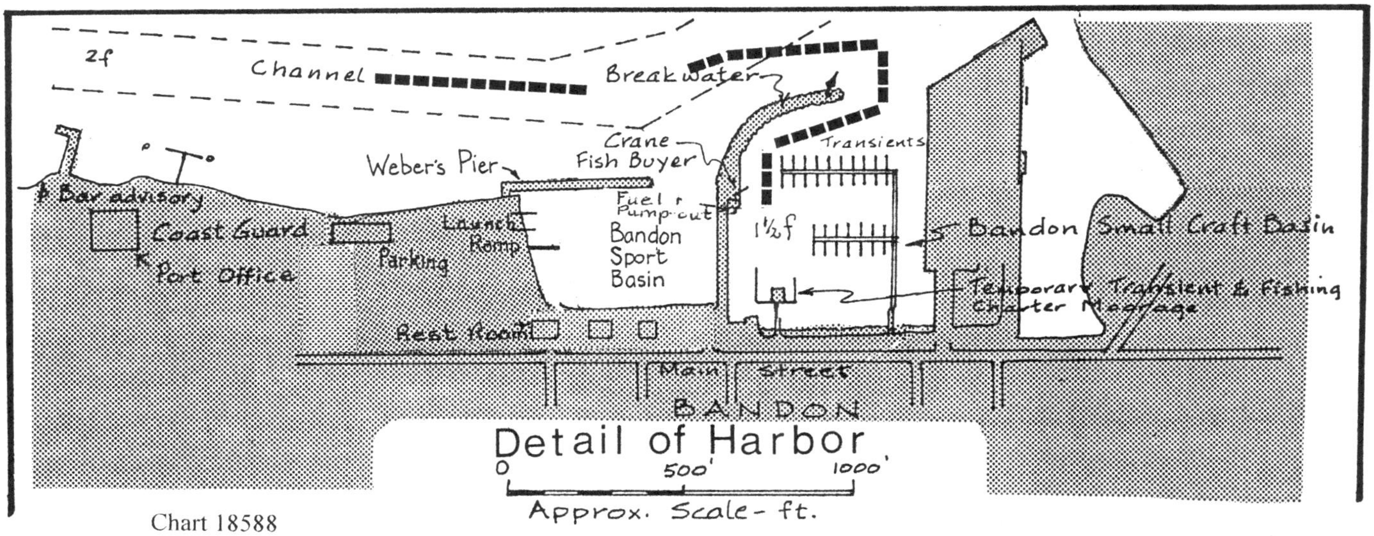

Chart 18588

LOOKING E INTO THE COQUILLE RIVER FROM THE ENTRANCE

PORT ORFORD

A very good anchorage in northwesterly weather, Port Orford is 30 miles south of Coquille River and 19 miles north of Rogue River. Except in southerly weather this is the best anchorage north of Point Reyes. There is no bar to cross, no narrow channel between jetties through which one must thread one's, way and the coastal area is beautiful.

When approaching from the north, maintain a course well clear of the extensive areas of danger west and southwest of Cape Blanco. Blanco Reef and various other rocks are scattered in a large area west and southwest of the Cape. This hazard is followed by Orford Reef which covers an area 2 to 5 miles offshore between the Heads and Cape Blanco. Do not pass between these reefs or take a route between Orford Reef and the coast unless you have local knowledge. Similarly when southbound from Port Orford following the coastal route, give a generous clearance to Rogue River Reef which extends about four miles off the coast prior to reaching the offshore buoy off Rogue River.

Identification of Port Orford when one is approaching from the north is achieved by pinpointing Cape Blanco (Oregon's westernmost headland); from the south, Humbug Mountain is a reliable landmark. The anchorage is 6.5 miles south of the former and about 4 miles north of the latter. Offshore whistle buoy "20R" is about 7 miles northwest of bell buoy "1" which is a mile from the anchorage.

When approaching from the south you should also maintain an offshore course to be well clear of Island Rock (a little over a mile southwest of Humbug Mountain) and Redfish Rocks (2 miles north of Island Rock and 1 mile offshore). Rocks border the cove up to .5 mile offshore, while a 168m (550 ft.) breakwater extends southeasterly from Graveyard Point. The cove is tucked behind The Heads which from the south appear to be a triple-lobed ridge. Don't mistake the white tank on the top of The Heads for Cape Blanco lighthouse which is located 6 miles to the north.

Anchorage may be taken in 6 - 10 fathoms, sand. Vessels may anchor the lee of The Heads if strong northwesterlies are funneling across the draw and sweeping the inner area. During settled weather vessels can anchor east of the wharf, but a 2.5 fathoms shoal .2 miles east of the south end of the breakwater is reported to be troublesome for fouling anchors.

Fuel is available at the wharf daily from 5 a.m. to 12 p.m. In addition water and showers are available and there is a hoist capable of lifting boats up to 13m (42 ft.). The loading crane is near a ladder that gives access to a dinghy landing below. Located on the wharf, the port office is open from 8:30 a.m. to 5:00 p.m. For information call (541) 332-7121. A pleasant hike takes you to the community of Port Orford for groceries and other supplies.

Featured in local shops is a wide variety of goods made from beautifully patterned myrtle wood. This wood is also known by the following names: California laurel, California sassafras, California olive, myrtle, spice tree, pepperwood, bay, bay laurel and yellow, black, or white myrtle. Port Orford yellow cedar is unique to this area and is also highly valued.

An interesting historical fact is that Battle Rock is the site of the first European settlement in this area in 1851. Two attacks by local Indians were repelled before the settlers' enthusiasm and ammunition ran low and they escaped to a settlement on the Umpqua River.

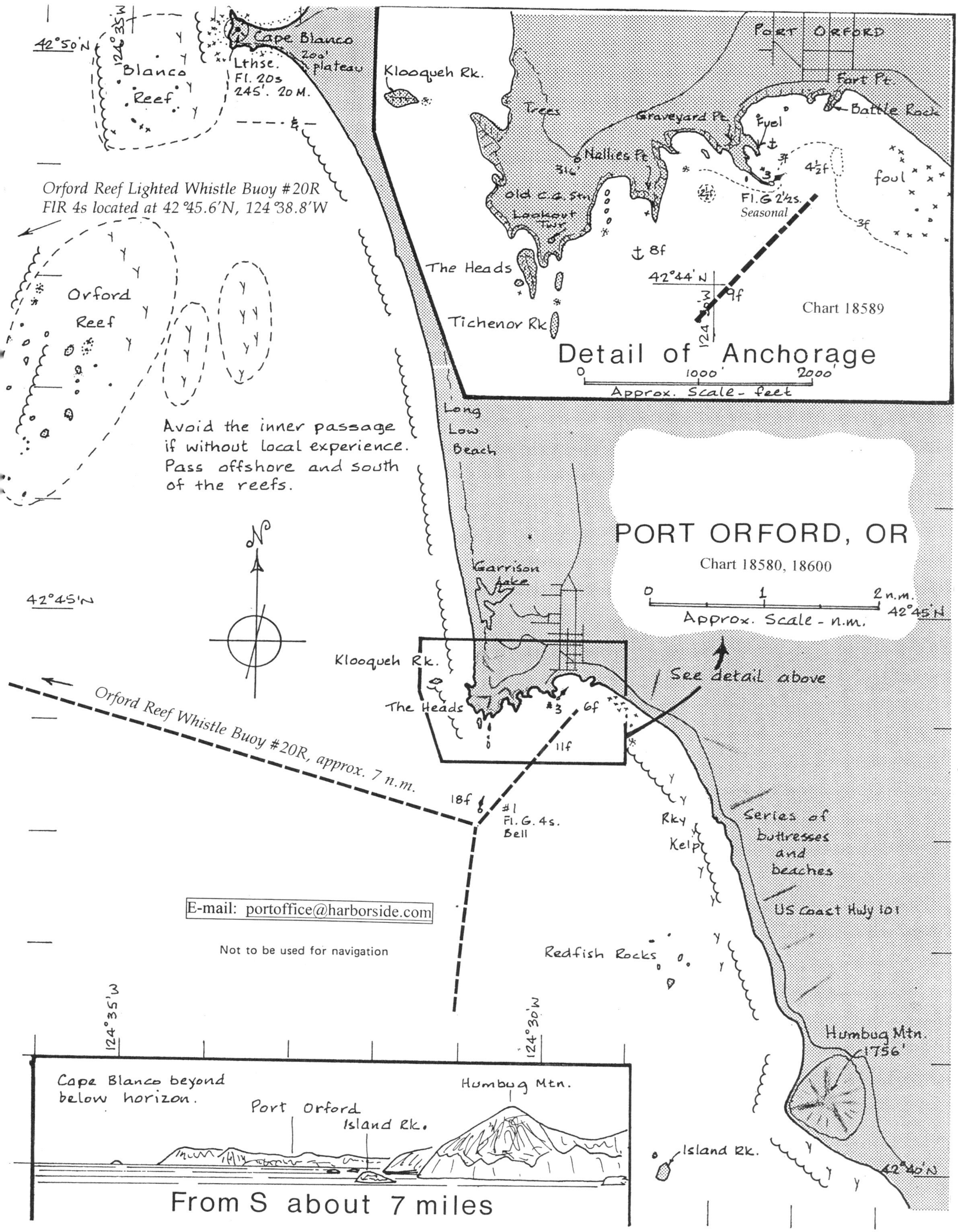

PORT ORFORD, OR
Chart 18580, 18600
0 1 2 n.m.
Approx. Scale - n.m.
42°45'N
Detail of Anchorage
Chart 18589
0 1000' 2000'
Approx. Scale - feet
Port Orford
Fort Pt.
Battle Rock
Graveyard Pt.
Fuel
Nellies Pt.
Klooqueh Rk.
Trees
Old C.G. Stn.
Lookout Twr
The Heads
Tichenor Rk
8f
9f
3f
4½f
foul
Fl. G 2½s.
Seasonal
42°44'N
124°30'W
Cape Blanco
200' plateau
Lthse.
Fl. 20s
245'. 20 M.
42°50'N
124°35'W
Blanco Reef
Orford Reef Lighted Whistle Buoy #20R
FlR 4s located at 42°45.6'N, 124°38.8'W
Orford Reef
Avoid the inner passage if without local experience. Pass offshore and south of the reefs.
N
42°45'N
Long Low Beach
Garrison Lake
Klooqueh Rk.
The Heads
#3
6f
11f
See detail above
Orford Reef Whistle Buoy #20R, approx. 7 n.m.
18f
#1
Fl. G. 4s.
Bell
Rky
Kelp
Series of buttresses and beaches
US Coast Hwy 101
E-mail: portoffice@harborside.com
Not to be used for navigation
Redfish Rocks
Humbug Mtn.
1756'
Island Rk.
42°40'N
124°35'W
124°30'W
Cape Blanco beyond below horizon.
Humbug Mtn.
Port Orford
Island Rk.
From S about 7 miles

ROGUE RIVER (GOLD BEACH)

The Rogue River entrance is about 22 miles south of Port Orford and 27 miles north of Brookings. This is not a recommended stop-over unless a vessel draws less than 1.8m (4 ft.). Rogue River is known not only for its excellent sport fishing but also for its popular white-water rafting and jet boat trips.

The offshore whistle buoy "R" is approximately 3 miles southwest of the channel entrance, and closer in is bell buoy "1". Two jetties extending in a southwesterly direction protect the entrance; a seasonal light and fog signal are at the seaward end of the northerly one. Breakers are often present at the outer ends of the jetties, being especially dangerous when the sea is running from the west or southwest. When conditions warrant a warning, a heavy weather flag (a square red flag with a square black center) is shown on a pole located near the south side of the Coast Guard lifeboat station. Rogue waves may be experienced up to .25 miles off the ends of the jetties even on a calm day.

Prior to approaching the entrance it is essential that cruisers contact the Coast Guard on VHF Ch 16 to obtain a current report on sea conditions at the entrance as well as depths in the channel and within the basin. When approaching from the north the course should be at least 5 miles off the coast to clear Rogue River Reef which extends 4 miles northwest of the entrance. The outermost visible rock of the reef is Northwest Rock and a rock covered by 2.5 fathoms is .3 miles west of Northwest Rock.

The entrance should be attempted only at the latter part of a rising tide or at high water. When entering the channel the vessel must wait for a period of lower seas and time the passage carefully to get past the breakers. At times, small boat traffic can be congested just inside the bar where trolling vessels congregate when the fish are running.

Depths over the bar are constantly changing and are often less than the 4m (13 ft.) projected depth. Water depths are dependent on the amount of run-off in the river; in dry years or winters with little snowfall the snowmelt will be less, and above normal shoaling will occur. The depth across the bar can be less than 1.8m (6 ft.) and within the boat basin it may be less than 1.2m (4 ft.) with shoaling to bare near the entrance and within. Shoal water and a large gravel bar extend about three-quarters of the way across the channel making it essential to favor the north side of the entrance channel. The Coast Guard is stationed here from May to September. For information call the Brookings office at (541) 469-2242. A rough bar advisory sign is about .6 miles upstream of the entrance.

A buoyed channel leads directly into the Gold Beach small craft harbor where fuel, water, fishing tackle and limited marine supplies are available from 8 a.m. to 4:30 p.m. The marina monitors VHF Ch 16 and office hours are 8 a.m. to 5 p.m. daily. For information call (541) 247-6269 or Fax (541) 247-6268. Some restaurants and gift shops are adjacent to the waterfront but the main part of the town is up the hill to the south. Propane is available at New Hope Plumbing.

LOOKING S FROM ENTRANCE TO ROGUE RIVER

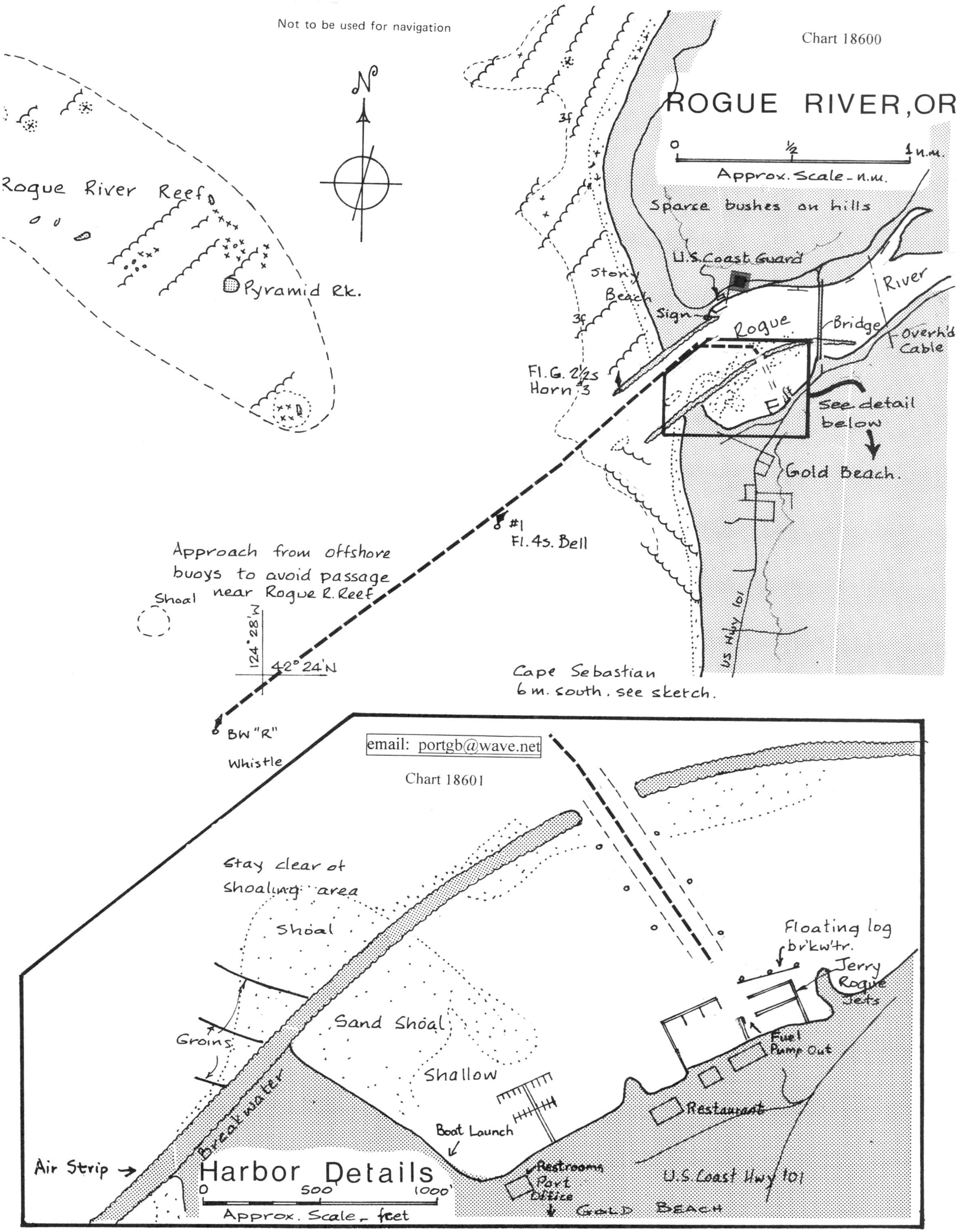
Not to be used for navigation
Chart 18600
ROGUE RIVER, OR
0 1/2 1 n.m.
Approx. Scale - n.m.
Sparse bushes on hills
U.S. Coast Guard
Stony Beach
Sign
Rogue
River
Bridge
Overh'd Cable
Rogue River Reef
Pyramid Rk.
3f
3f
Fl. G. 2½s Horn "3"
See detail below
Gold Beach
#1 Fl. 4s. Bell
Approach from offshore buoys to avoid passage near Rogue R. Reef
Shoal
124° 28' W
42° 24' N
US Hwy 101
Cape Sebastian 6 m. south. see sketch.
BW "R" Whistle
email: portgb@wave.net
Chart 18601
Stay clear of shoaling area
Shoal
Sand Shoal
Shallow
Groins
Breakwater
Floating log br'kw'tr.
Jerry Rogue Jets
Fuel Pump Out
Restaurant
Boat Launch
Air Strip
Harbor Details
0 500 1000'
Approx. Scale - feet
Restrooms
Port Office
U.S. Coast Hwy 101
GOLD BEACH

CHETCO RIVER (BROOKINGS)

Cruisers following a harbor-hopping route should stand well off the coast to avoid scattered rocks about a mile offshore north of Point St. George and must allow extra clearance to avoid St. George Reef extending 6.5 miles northwest and west from Point St. George. This hazardous area is particularly dangerous since it rises abruptly from deep water. St. George Reef Lighted Horn Buoy #2SG (41°50.1'N, 124°23.2'W) is about .5 miles west of Northwest Seal Rock and Little Black Rock, the westernmost rocks of St. George Reef. The red buoy shows a light 4.5m (15 ft.) above sea level and has a fog signal. During good visibility and settled conditions it is safe to transit St. George Channel which is over a mile wide and clear between visible rocks fringing Pt. St. George and the eastern rocks of St. George Reef. The St. George Reef Lighthouse is no longer maintained or recognized by the Coast Guard but is a privately maintained navigational aid that was relit in 2002.

The offshore whistle buoy "CR" is about 1 mile southwest of the entrance. A half mile closer in is bell buoy "2", followed by gong buoy "3". The smokestack of a plywood plant in Brookings is noticeable from some distance offshore.

The entrance to the Chetco River is about 26 miles south of Rogue River and 18 miles north of Crescent City, California. This part of the coast is composed of rocky cliffs bordered by numerous spectacular rocks and backed by high tree-covered hills. Chetco Point marks a change in the generally NW/SE trend of the coast, for here the land recedes abruptly to the east for a mile, forming Chetco Cove. This gives Brookings protection from northwesterly winds when it is considered a "harbor of refuge" but makes use of the channel very difficult in southerly weather when the entrance should not be attempted. Brookings is a commercial fishing port and sports fishing center for deep-sea and freshwater fishing. The surrounding area is the largest center in the nation for the cultivation of Easter lilies.

The entrance is protected by two jetties. The West Jetty is the longer of the two and has a light and fog signal at the outer end. A lighted range marks the narrow (18m/60 ft.) channel. Caution: depths in the channel are only 3m (9ft.); it is essential to stay within the channel because of the shoals and numerous rocks that are just alongside. The channel leads directly to a turning basin which is protected on the west by a 550m (1,800 ft.) dike.

Bar reports are given every hour during daylight hours on KURY (910 kHz). A rough bar advisory sign facing northwest is located on the Coast Guard fuel dock and a Heavy Weather Flag is shown on a pole at the north end of the Coast Guard station. Call the Coast Guard on VHF Ch 16 or (541) 469-2242 or (541) 469-3885 for current bar conditions.

There are two mooring basins with a capacity of 678 berths, over 65 slips for visitors and a 150m (485 ft.) transient dock. The southern basin is for commercial boats, the northern for pleasure craft. Call the harbormaster on VHF Ch. 12 or at (541) 469-2218 or Fax (541) 469-0672 before proceeding beyond the commercial docks unless tie-up in the commercial basin is preferred. The port offices are open 8:00 to noon and 1:00 p.m. to 5 p.m. Monday to Saturday during the summer. Showers are available at the RV park.

Fuel and water are available. There is a 60-ton travel lift and work yard for vessel repair; marine supplies can be obtained at Chetco Chandlery. Numerous shops and restaurants line the shore of the northern basin. A short walk up the road leading to the city level takes you to a shopping mall. On the south side of the river are several large shopping centers and a choice of many shops and restaurants.

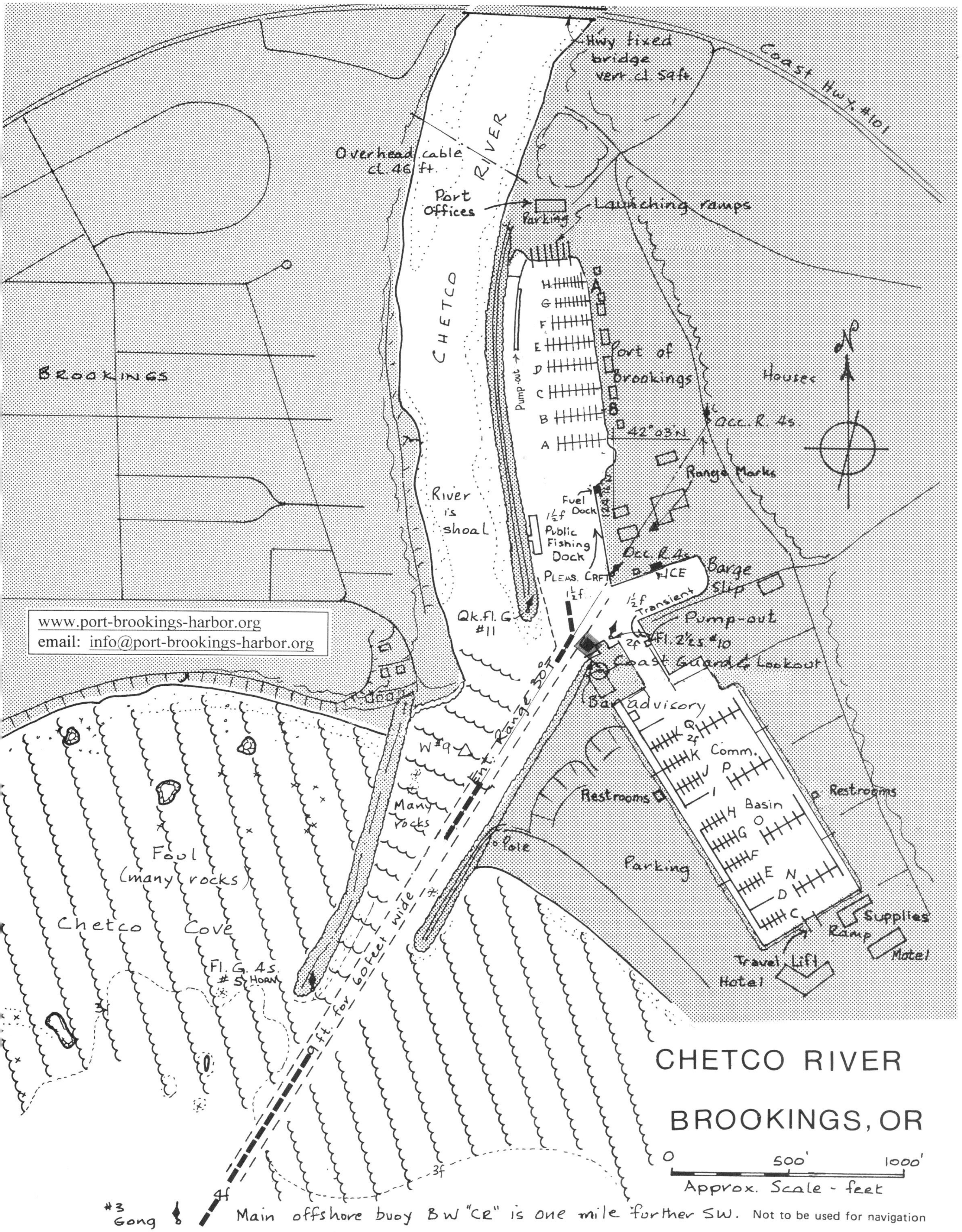

CHETCO RIVER
BROOKINGS, OR
Hwy fixed bridge vert. cl. 54 ft.
Coast Hwy. #101
Overhead cable cl. 46 ft.
CHETCO RIVER
Port Offices
Parking
Launching ramps
BROOKINGS
Pump-out
H
G
F
E
D
C
B
A
Port of Brookings
Houses
Occ. R. 4s
42°03'N
Range Marks
River is shoal
Fuel Dock
Public Fishing Dock
Occ. R. 4s
PLEAS. CRFT
Barge Slip
Transient
Pump-out
Qk. Fl. G #11
Fl. 2½s #10
Coast Guard & Lookout
Bar advisory
Ent. Range 30°T
W #9
Many rocks
Comm.
Basin
Restrooms
Restrooms
Pole
Parking
Foul (many rocks)
Chetco Cove
Supplies
Ramp
Motel
Travel Lift
Hotel
Fl. G. 4s #5 HORN
9 ft. for 60 feet wide
3f
4f
#3 Gong
0
500'
1000'
Approx. Scale - feet
www.port-brookings-harbor.org
email: info@port-brookings-harbor.org
Main offshore buoy BW "CR" is one mile further SW. Not to be used for navigation

CRESCENT CITY

This **harbor of refuge** is straightforward and easy to enter, though strangers should not attempt an entrance at night.

Just 3.5 miles southeast of Point St. George, Crescent City is 18 miles south of Brookings and 45 miles north of Trinidad Head. For sailors from the north this initial moorage in Californian waters is often marked by the first sighting of pelicans, evidence of progress to warmer, southern waters. It is also the beginning of the famous California Redwood Forest. A trip inland to admire these awe-inspiring giants is a must.

Offshore whistle buoy "2" is a little over a mile WSW of the entrance to the harbor. Closer in is whistle buoy "4" followed in .5 mile by bell buoy "6". Extreme care must be taken when approaching the entrance because of the rocks and shoals that extend about a mile offshore to the northwest.

A breakwater protects the western side of the harbor, at the seaward end of which is Crescent City Entrance Light, a fog signal and radio beacon. A private light is shown from an islet south of Battery Point. The entrance to the harbor lies between the end of the breakwater and bell buoy "6" and it is marked by a lighted range and buoys.

A dredged channel leads into the **Inner Boat Basin**. Contact may be made with the harbormaster for berth allocation on VHF Ch. 9 and 16 from 8 a.m. to 5 p.m. on weekdays, by calling the office at (707) 464-6174 or Fax (707) 465-3535 or by checking in at the office. Transient vessels may tie at dock H after hours or on weekends and security personnel will collect moorage fees.

Moorage for recreational vessels under 7.5m (25 ft.) in length is available in the **Outer Boat Basin**. The Chart Room Marina and Restaurant handles berth allocation and can be reached at (707) 464-5993. A launch ramp is located west of the Chart Room on Anchor Way. Some slips are available year round, with the majority being in the water for summer months and removed during the winter.

The commercial fuel dock adjacent to the channel operates year round selling diesel and gasoline. A recreational fuel dock on the opposite side of Citizens' dock sells only gasoline and operates during summer months. Repair facilities can handle vessels up to 33.5m (110 ft.). Grocery stores and other shops are about a mile away in the downtown area on the north side of the bay. Showers and laundry facilities are next to the Harbor Office. Internet services are available at the public library, about a mile distant. Other facilities in town include an art gallery, beauty shop, fish market, ice plant, Englund Marine Supply, restaurants, scuba gear, a U.S. Coast Guard station and yacht club.

An interesting story lies behind the title "Citizen's Dock." It is aptly named for in 1950 it was built as a community-sponsored project. Because the Federal Government would not fund its construction the people of Crescent City pitched in, raising money and donating all the labor and materials needed for construction. The dock provided a facility vital to the local economy so local products (lumber and fish) could have access to markets elsewhere.

At present there is little evidence of the devastation caused by a seismic sea wave that virtually wiped out the harbor in 1964. Photographs on display illustrate the damage that occurred at that time.

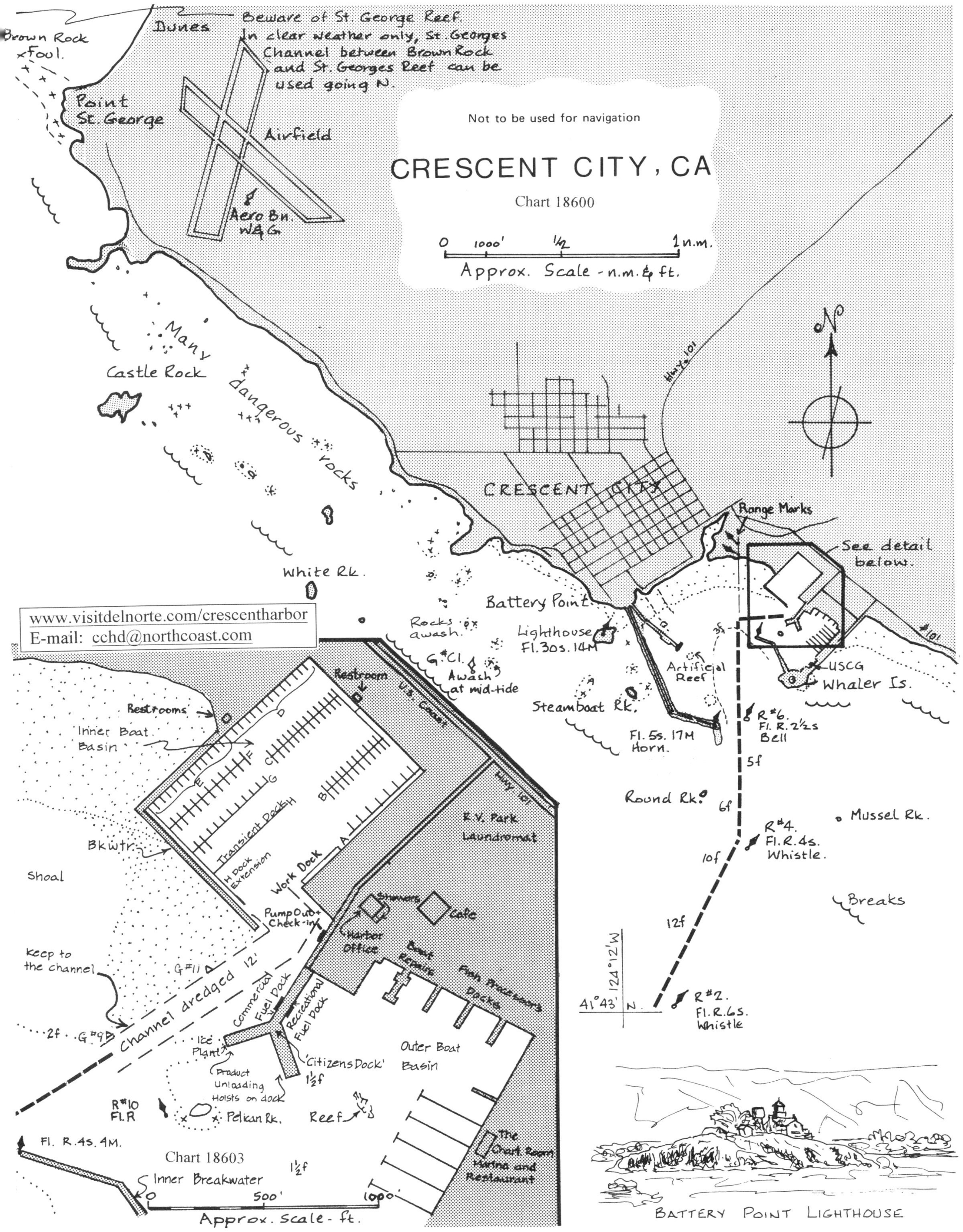
Not to be used for navigation
CRESCENT CITY, CA
Chart 18600
0 1000' 1/2 1 n.m.
Approx. Scale - n.m. & ft.
Brown Rock
Foul.
Dunes
Beware of St. George Reef.
In clear weather only, St. Georges Channel between Brown Rock and St. Georges Reef can be used going N.
Point St. George
Airfield
Aero Bn. W & G
Many dangerous rocks
Castle Rock
White Rk.
CRESCENT CITY
Hwy. 101
N
Range Marks
See detail below.
Battery Point
Lighthouse Fl. 30s. 14M
Rocks awash.
G #C1. Awash at mid-tide
Artificial Reef
USCG
Whaler Is.
Steamboat Rk.
Fl. 5s. 17M Horn.
R #6. Fl. R. 2½s Bell
5f
Round Rk.
6f
Mussel Rk.
R #4. Fl. R. 4s. Whistle.
10f
Breaks
12f
41°43' N.
124°12' W
R #2. Fl. R. 6s. Whistle
www.visitdelnorte.com/crescentharbor
E-mail: cchd@northcoast.com
Restroom
Restrooms
Inner Boat Basin
U.S. Coast
Hwy 101
Transient Dock
H Dock Extension
Work Dock
R.V. Park Laundromat
Bkwtr.
Shoal
Pump Out + Check-in
Showers
Cafe
Harbor Office
Keep to the channel
G #11
Channel dredged 12'
Boat Repairs
Fish Processors Docks
2f
G #9
Commercial Fuel Dock
Recreational Fuel Dock
Ice Plant
'Citizens Dock'
1½f
Outer Boat Basin
Product Unloading Hoists on dock
R #10 Fl. R
Pelican Rk.
Reef
Fl. R. 4s. 4M.
Chart 18603
1½f
The Chart Room Marina and Restaurant
Inner Breakwater
0 500' 1000'
Approx. Scale - ft.
BATTERY POINT LIGHTHOUSE

TRINIDAD HEAD

TRINIDAD HEAD LIGHT

This prominent coastal feature is almost 122m (400 ft.) high, though at a distance from either north or south it resembles a round-topped island. It is 45 miles south of Crescent City and almost 18 miles north of Humbolt Bay. Trinidad Head Light and a fog signal are on the southwestern side. Another landmark is a prominent white cross northeast of the light.

When approaching from the north keep at least 2 miles offshore to clear rocks and reefs north of Trinidad Head. Offshore whistle buoy "26" lies about 1 mile west of Trinidad Head, followed by gong buoy "2" one mile to the southeast. When approaching from the south the coast is clear after passing Humbolt Bay (except for a wreck); the gong buoy is the first navigational aid to be picked up.

The moorage area, to be used only in settled, northwesterly weather, is in a cove east of Trinidad Head. A large number of seasonal mooring buoys are installed for the summer, with rental on a first-come, first-serve basis. Normally, there are plenty left for transient vessels. Arrangements for use of mooring buoys can be made by calling VHF Ch 78 or by tying to the temporary float which juts out on the east side of the pier. The daily fee for use of a mooring buoy is $15 with a similar charge for water taxi service to or from shore. A floating dock for dinghy use is available during summer months. The harbor office operates from May to September from 5 a.m. to 5 p.m. and during the rest of the year from 7:30 a.m. to 4 p.m.

Kelp and rocks extend off the northern shore leaving the only place to anchor about halfway between Prisoner Rock and Trinidad Head in 7 fathoms, mud. (This suggestion is given assuming that mooring buoys do not extend into this area.) There is always some swell in the cove and holding is only fair; nevertheless, many fishing boats use this anchorage during the summer.

The long pier on pilings northeast of Trinidad Head has facilities for offloading fish and a marine ways for launching power boats up to 7m (24 ft.). The wharf houses a recycling facility for sorting garbage, so plan ahead. There is a small but attractive restaurant near the head of the wharf. Propane is available at the gas station in town. Internet services are available in town at Web Cam. Plans call for computer modem access to be available on the pier by 2004.

The lighthouse offers tours during Fish Fry days in the middle of June. Gasoline and groceries are available in the "city" of Trinidad, which now has a population of about 500 but still holds the charter granted in 1852, when it was home to 3,000 people. In its heyday, it was a busy lumbering port and later a whaling station of which it was said, "You can smell it before you can see it."

A good walk can be taken on the trail encircling Trinidad Head and leading to the granite cross. (The view is spectacular—bring a camera.) A stroll into town can take you to the Holy Trinity Catholic Church built in 1873 which is now a US Historical Landmark. Free tours and slide shows are offered at the Humboldt State University Marine Laboratory, open Mon. to Fri. 8 a.m. to 5 p.m. or on weekends from 11 a.m. to 5 p.m

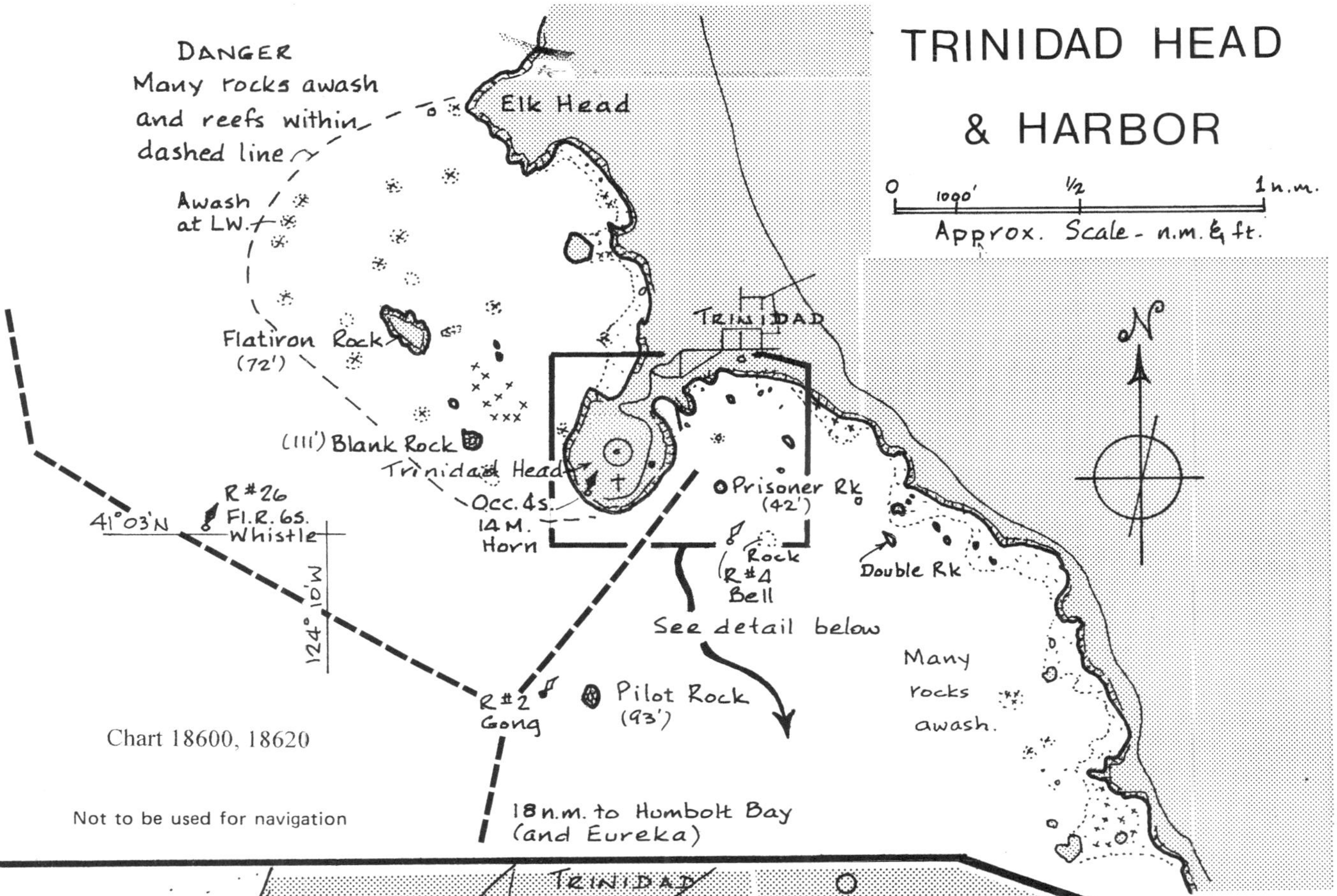
TRINIDAD HEAD
& HARBOR
0
1000'
1/2
1 n.m.
Approx. Scale - n.m. & ft.
DANGER
Many rocks awash
and reefs within
dashed line
Awash
at LW.
Elk Head
Flatiron Rock
(72')
(111') Blank Rock
Trinidad Head
Occ. 4s.
14 M.
Horn
TRINIDAD
Prisoner Rk
(42')
Rock
R #4
Bell
Double Rk
R #26
Fl. R. 6s.
Whistle
41°03'N
124° 10'W
See detail below
Many
rocks
awash.
R #2
Gong
Pilot Rock
(93')
Chart 18600, 18620
Not to be used for navigation
18 n.m. to Hombolt Bay
(and Eureka)

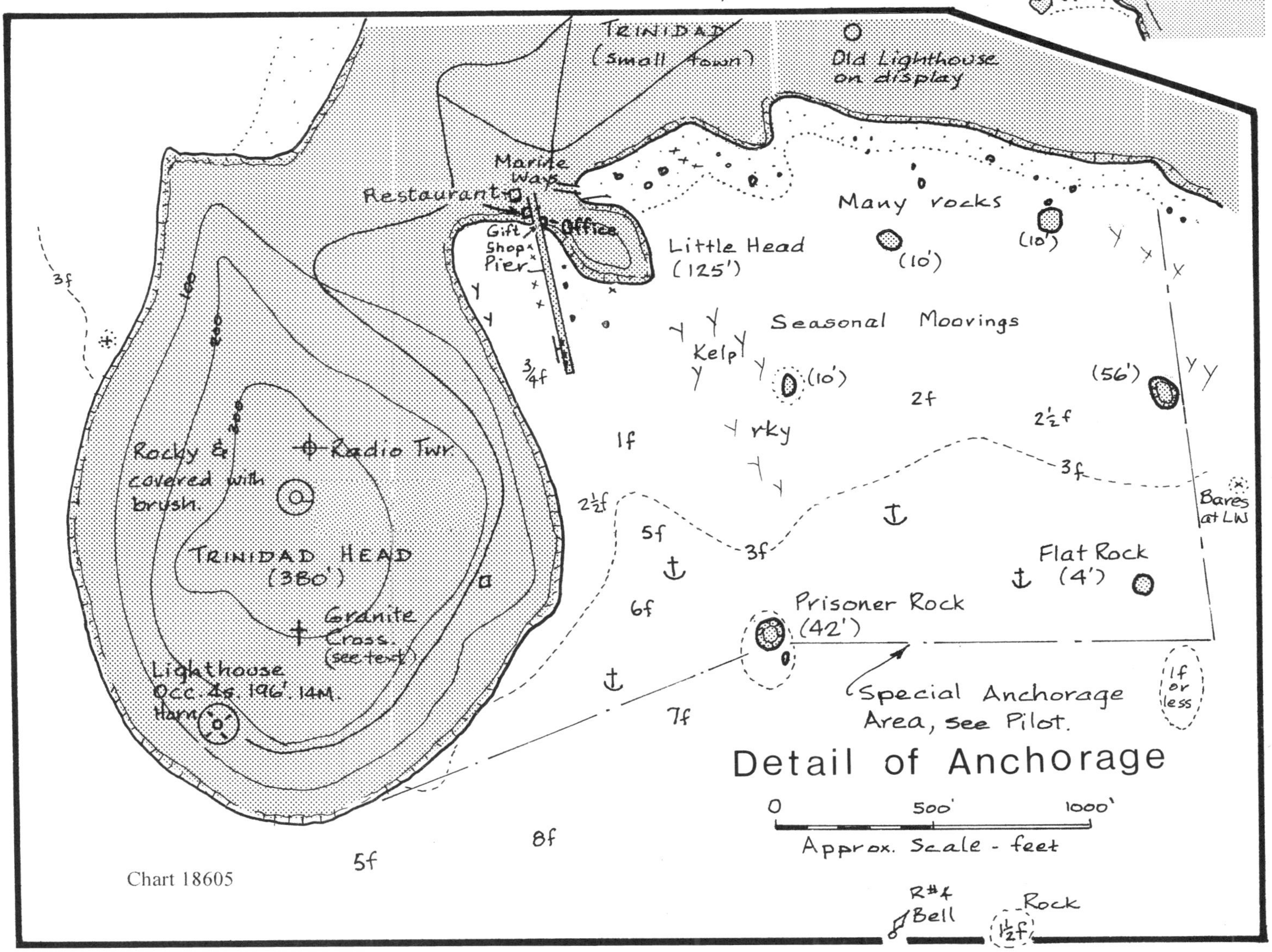
TRINIDAD
(small town)
Old Lighthouse
on display
Marine
Ways
Restaurant
Office
Gift
Shop
Pier
Little Head
(125')
Many rocks
(10')
(10')
Seasonal Moorings
Kelp
(10')
(56')
3/4f
1f
2f
2 1/2f
rky
3f
Bares
at LW
Rocky &
covered with
brush.
Radio Twr
TRINIDAD HEAD
(380')
Granite
Cross
(see text)
Lighthouse
Occ. 4s. 196'. 14M.
Horn
400
300
200
100
3f
2 1/2f
5f
3f
6f
Flat Rock
(4')
Prisoner Rock
(42')
Special Anchorage
Area, see Pilot.
1f
or
less
7f
Detail of Anchorage
0
500'
1000'
Approx. Scale - feet
8f
5f
Chart 18605
R#4
Bell
Rock
1 1/2f

HUMBOLDT BAY (EUREKA)

This historic city is 17 miles south of Trinidad Head and 21 miles north of Cape Mendocino. It is a good place to recover from passing Cape Mendocino or to await the end of a storm prior to making the passage around this important turning point. Eureka is a Customs **Port of Entry** and a harbor of refuge when a storm is predicted, provided the vessel can get past the bar before it becomes impassable. This can be a very dangerous bar as shoaling and strong currents in the channel make it difficult to negotiate the entrance at times. It is advisable to call the Coast Guard on VHF Ch 16 for a bar report.

During good visibility, Trinidad Head and Cape Mendocino are excellent landmarks to identify the general location of Humboldt Bay. Within the bay the stacks and smoke from sawmills pinpoint the city of Eureka. There is an irregular southerly current in this area, so when coming from the north, sailing vessels should approach from off Trinidad Head to get a good slant to the Humboldt Bay entrance. Humboldt Bay is a National Wildlife Refuge.

Offshore lighted whistle buoy "HB" is a little over a mile WNW of the entrance to the bay, followed by bell buoy "2" which is west of the southern jetty. Two jetties, each with lights at their seaward ends, form the channel entrance. Take care not to confuse the radio towers south of the entrance with the lighted range marking the entrance approach from the offshore buoy. Once over the bar another range indicates the entrance to the bay, which is also marked by buoys. The turn is rather sharp when changing from the Approach Range to the Entrance Range—don't carry either one past the beginning of the other when entering or leaving.

Three towns are in the northern part of the bay: Bucksport, 3 miles above the entrance, followed by Fairhaven and Eureka. Very reasonable moorage (25 cents a foot or $5 a day, whichever is greater) in a first-class full-service facility is available at the **Eureka Public Marina.** Daily office hours are 8 a.m. to 5 p.m. with transient slips available on a hot-berth basis. For dock assignment call (707) 441-4203 or Fax (707) 441-4202. The marina is conveniently close to restaurants, shopping and the interesting buildings of Old Town. Facilities include a two-lane boat launch, a handicapped accessible gangway and free use of bicycles for a fascinating tour of this historic city. A short distance to the east is Englund Marine Supplies and fuel dock which operates daily from 8 a.m. to 5:30 p.m. Internet services are available at the Public Library, about 5 blocks away and computer modem access is in the marina office.

Around the corner is another full-service facility at **Woodley Island Marina.** The office monitors VHF Channels 14 and 16 or can be reached at (707) 443-0801 or Fax (707) 443-0800. Arrangements can be made at the office to have a vessel hauled at Field's Landing, about a mile south of the entrance channel in an area known as South Bay. This is not a shipyard—it is a do-it-yourself facility. A good restaurant and bookstore are in the marina complex. Computer modem access is available in the marina office.

Eureka owes its charm and fascination to the many Victorian homes, a legacy of the lumber baron era. If a walking tour of Old Town is not for you then at least have a look at Carson Mansion, the most photographed of all. Celebrating the past are Clarke Museum (historical), Fort Humboldt State Historic Park (logging) and the Maritime Museum.

The fishing fleet based in the bay uses the following system when southbound: aim for the Cape Mendocino Buoy (5 miles offshore) and if it is windy, head for Punta Gorda Buoy (1 mile offshore) to avoid heavy winds and seas. But if there is no wind at Cape Mendocino Buoy carry on to your destination.

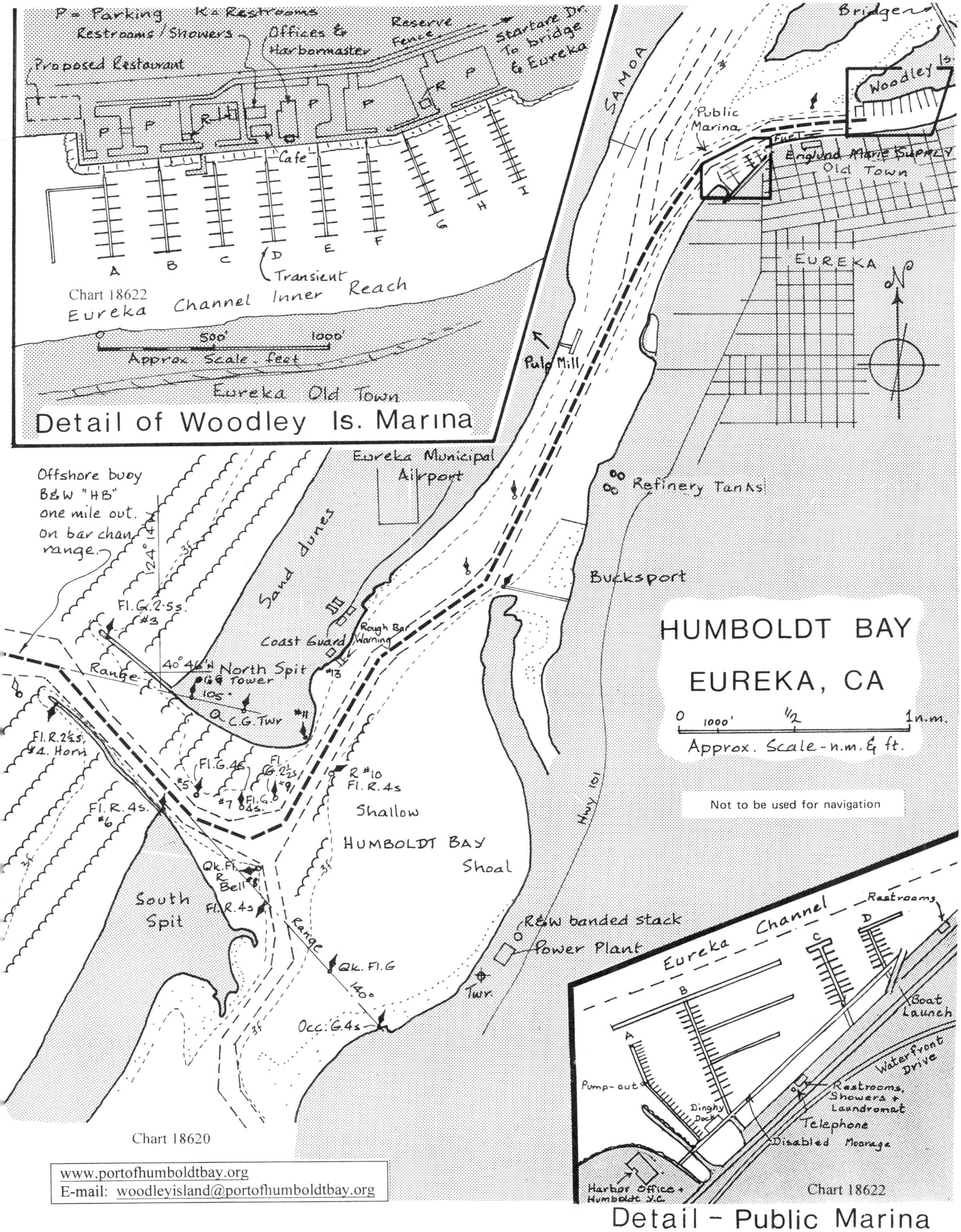

www.portofhumboldtbay.org
E-mail: woodleyisland@portofhumboldtbay.org

Detail - Public Marina

SHELTER COVE

This popular anchorage tucked behind Point Delgada is about 60 miles south of Eureka and 37 miles north of Noyo River. Many small fishing vessels are launched off the beach to enjoy the salmon, ling and rock cod and red and black snapper found in local waters.

Point Delgada is a high plateau bordered by cliffs. It is easily recognized by its proximity to King Peak (over 1,220m/4,000 ft.), which is almost 9 miles to the north. In clear weather it is visible to seaward for 75 miles. The King Range forms a mountainous backdrop for the miles of tidal beach bordering the anchorage.

Offshore whistle buoy "20" is about a mile southwest of Point Delgada. One mile in an ENE'ly direction is bell buoy "1" which marks the southern end of the rocks and kelp extending off the point. Keep to the south of both buoys in order to avoid shoals and rocks in this area. Anchorage may be taken in northwesterly weather, although there is generally some surge in the cove. In south or southeasterly winds the anchorage becomes a dangerous lee shore and vessels must leave.

This area is referred to as "California's Lost Coast" because of the 26-mile steep and narrow road linking it to Highway 101. In spite of its comparative isolation,the community of Shelter Cove has telephone service, restaurants, a gift shop and even a real estate office. Propane, gasoline and diesel are available,but it's a long trip back to the boat. Considerable development has taken place in recent years providing a unique area away from the congested living of many parts of California. The Shelter Cove Port District monitors VHF channel 8 and a local boat launch service monitors VHF channel 68. A pump-out station and dry winter storage are available.

An archeological dig on Point Delgada plateau has traced Indian habitation back several hundred years. A trail leads to King Peak, which is the highest point on the continental U.S. shoreline and offers a great view. Seals and sea lions can be seen on the rocks below the Point as well as migrating gray whales just offshore during their semi-annual transit of the coast.

When northbound, picking up Punta Gorda buoy (1 mile offshore) before proceeding to Cape Mendocino Buoy (5 miles offshore) may enable a vessel to avoid the heavier winds and seas generally found off these points.

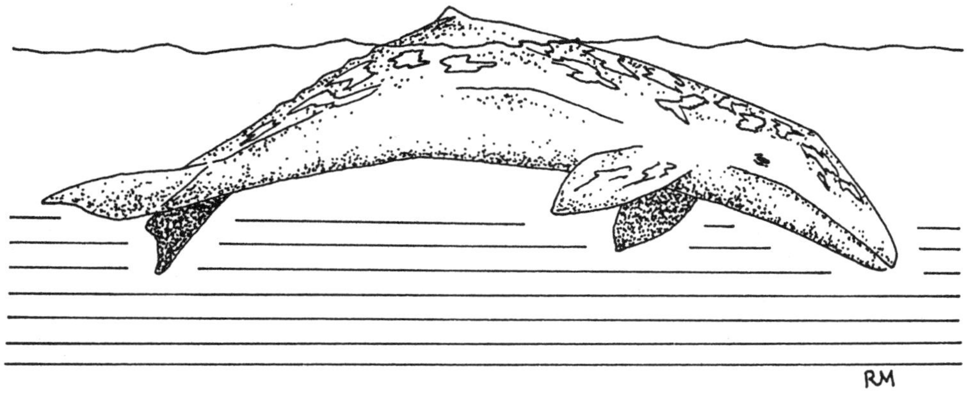

California Gray Whale (Eschrichtius robustus)

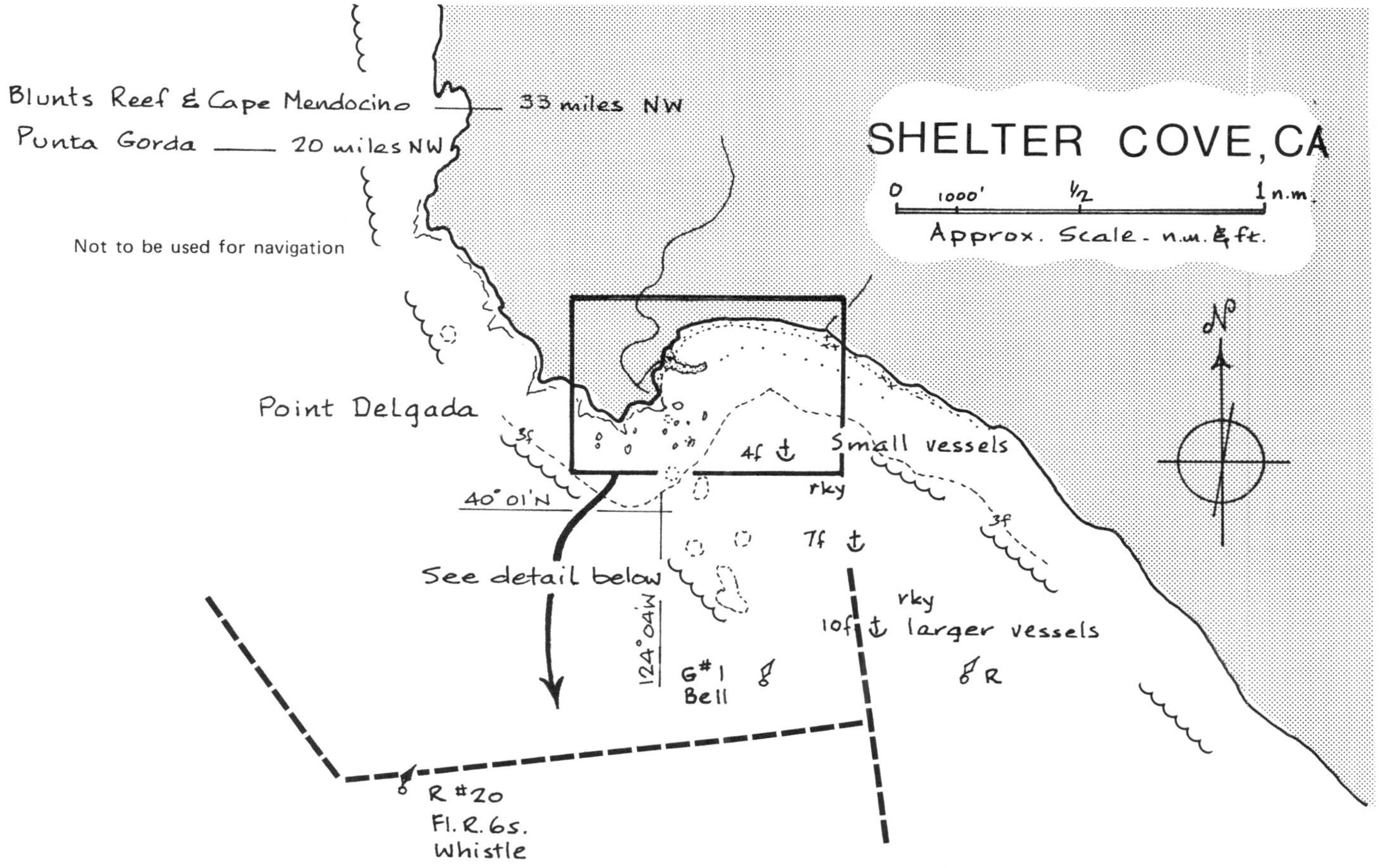

Chart 18620

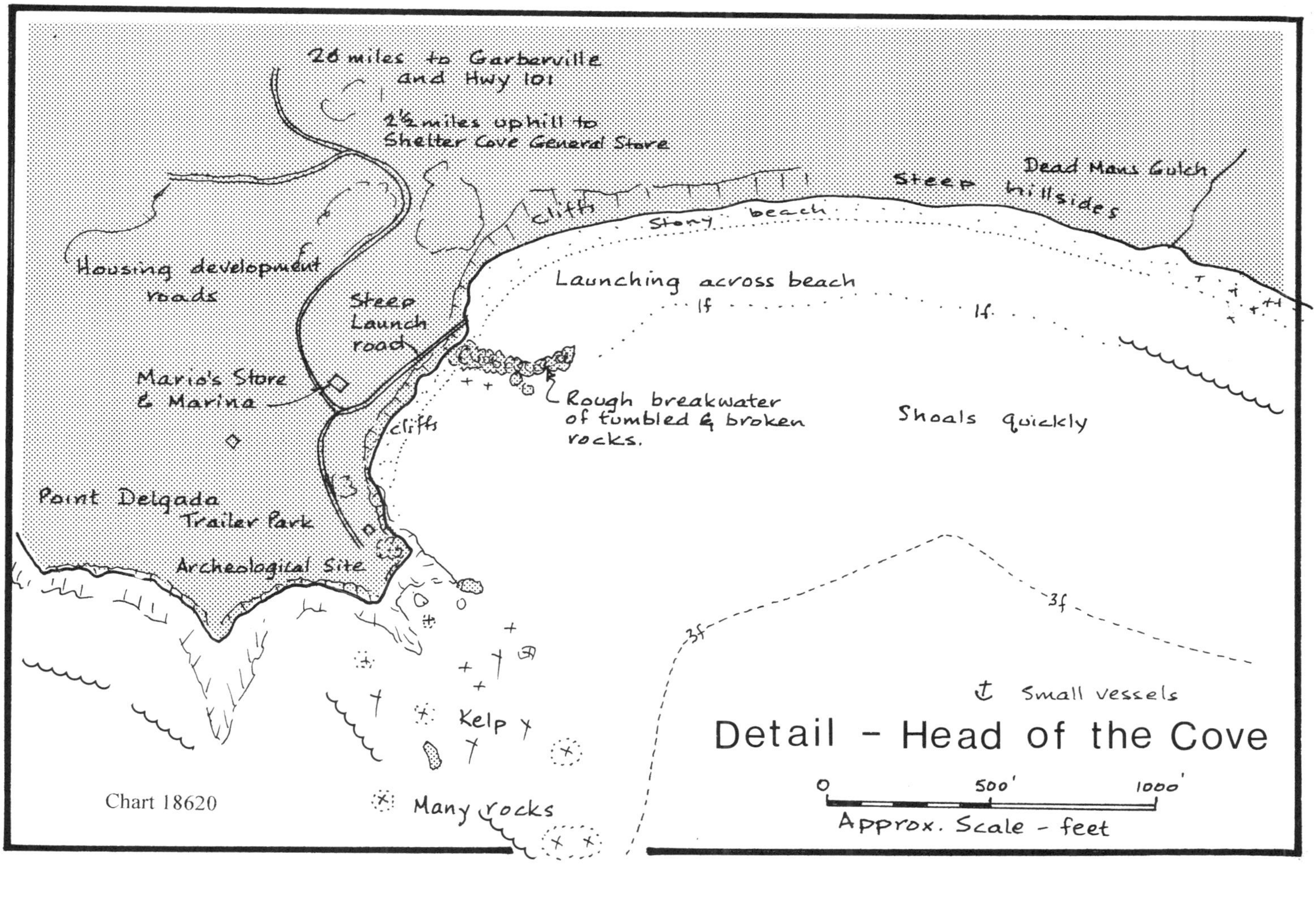

Chart 18620

NOYO ANCHORAGE (FORT BRAGG)

Pass by this stop-over if there is strong westerly weather or poor visibility unless you have local knowledge. Noyo Anchorage is 37 miles south of Shelter Cove and about 80 miles north of Bodega Bay. Noyo Anchorage refers to the bight into which the Noyo River flows; it should not be confused with the Noyo Boat Basin and the small craft anchorage in the Noyo River just off the entrance to the basin. The controlling depth in Boat Basin is about 3m (10 ft.). The fishing community of Fort Bragg borders the shores of the Noyo River.

The offshore whistle buoy "NA" is a mile out from the entrance followed by bell buoy "2". The North Jetty has a light on its outer end and a fog signal closer inshore. The South Jetty has a daybeacon on its seaward end. The highway bridge spanning the river has a clearance of 24m (80 ft.). A dredged channel (2.4m/8 ft.) leads to Noyo Basin, which is protected from surge coming up the river by a breakwater. Dredging is a regular necessity: the river was dredged in 2001, the marina in 2002 and the channel in 2003.

When approaching the channel, take care to clear the many rocks in the southern half of Noyo Anchorage. This is a narrow rocky channel in which fog and shoaling are common problems. Call the Coast Guard on VHF Ch 16or phone (707) 964-6611 for a report on conditions at the entrance if you have concerns. The Coast Guard station is on the south side near the entrance to the boat basin.

Call the **Noyo Basin** harbormaster on VHF Ch 66 or telephone (707) 964-4719 for berth allocation. The office is open on weekdays from 8 a.m. to 5 p.m. Facilities include fuel, water, showers (bring quarters after obtaining a key from the office), marine supplies and a launch ramp. Boats up to 14m (45 ft.) can be hauled for repairs. **Dolphin Isle Marina** is about .5 mile up the river where shoal water limits the use of the facility to shallow draft vessels. Propane is available and a dinghy trip up the river is recommended before committing your cruiser. For information call the marina at (707) 964-4113.

Fuel is available at Fort Bragg Marine, Monday to Saturday. For information call (707) 964-3310. A laundromat and grocery store are in the Boatyard Shopping Center, located about four blocks from the marina. More and larger stores are in the town center; some grocery stores will provide a ride back to the marina. Polly Cleaners Coin Laundry will pick up and deliver if you call them at (707) 964-3293. Tommy's Marine has a good inventory of marine supplies. The Pizza Factory will deliver to the harbor. It is worth a dinghy trip to the other side of the river where a good walk leads to restaurants and markets where fresh and smoked fish are sold.

The people of Fort Bragg have a reputation for being very friendly and willing to help visiting cruisers. What a pleasant change from communities totally oriented towards the tourist dollar!

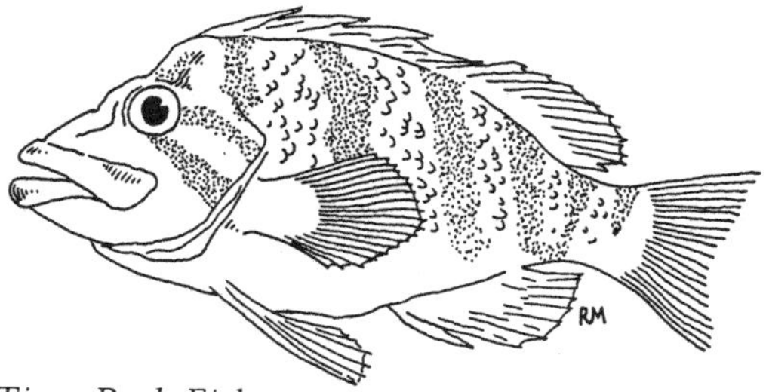

Tiger Rock Fish

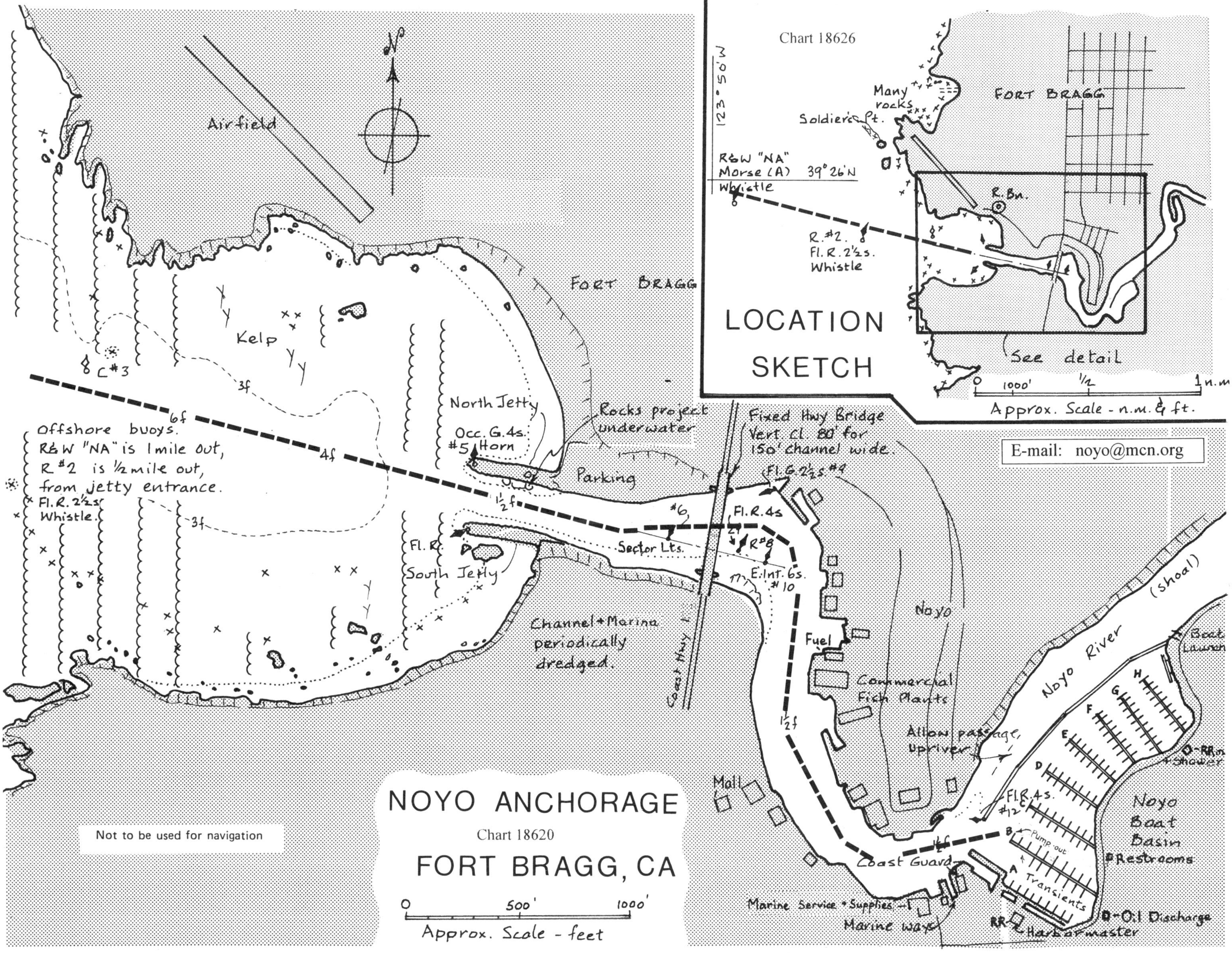
NOYO ANCHORAGE
Chart 18620
FORT BRAGG, CA
0 500' 1000'
Approx. Scale - feet
Not to be used for navigation
E-mail: noyo@mcn.org
Chart 18626
LOCATION SKETCH
FORT BRAGG
Many rocks
Soldiers Pt.
R&W "NA" Morse (A) Whistle
39° 26'N
123° 50'W
R. Bn.
R. #2. Fl. R. 2½s. Whistle
See detail
0 1000' ½ 1 n.m
Approx. Scale - n.m. & ft.
Airfield
Kelp
C #3
3f
6f
4f
½f
Offshore buoys.
R&W "NA" is 1 mile out,
R #2 is ½ mile out,
from jetty entrance.
Fl. R. 2½s
Whistle.
North Jetty
Occ. G. 4s.
#5 Horn
Rocks project underwater
Parking
FORT BRAGG
Fixed Hwy Bridge
Vert. Cl. 80' for
150' channel wide.
Fl. G. 2½s. #4
#6
Fl. R. 4s
R #8
Sector Lts.
E. Int. 6s.
#10
Fl. R.
South Jetty
Channel + Marina periodically dredged.
Coast Hwy 1
Noyo
Fuel
Commercial Fish Plants
Allow passage upriver
Noyo River
(shoal)
Boat Launch
RR + Shower
Fl. R. 4s. #12
Pump-out
Transients
Noyo Boat Basin
Restrooms
Oil Discharge
Harbormaster
RR
Coast Guard
Mall
Marine Service + Supplies
Marine ways
A
B
D
E
F
G
H

BODEGA HARBOR

Located 80 miles south of Noyo River and 25 miles north of Drakes Bay. In clear weather the 60m (200 ft.) cliffs, which run for 2 miles north of Bodega Head, help to identify this grass-covered landmark. Offshore whistle buoy "12" lies a little over a mile southwest of Bodega Head; 3.5 miles slightly north of east is gong buoy "BA". Watch for commercial crab pots in the area from November to June and a horde of small sports fishing boats dashing about from spring to fall.

This is a major fishing and pleasure boating center where many Bay area sailors keep their boats. A wide range of facilities are located in the harbor. This is a beautiful part of the coast and more importantly, provides excellent protection from wind and weather.

When approaching the entrance to Bodega Bay at night it is difficult to identify unlit Bodega Rock. In addition, when approaching from the north the entrance lights are hidden by the rock. Thus it is recommended that your first entrance to Bodega Harbor be made during daylight hours.

When approaching the harbor, a route should be taken as indicated on the sketch. Passage between Bodega Rock and Bodega Head is best accomplished during calm conditions and during a flood tide. If conditions are otherwise it is recommended that you contact the Coast Guard on VHF Ch 16 for the current advisory. Small breakers mark the shoal east of the route between the RW "BA" gong buoy and the entrance channel.

The entrance channel is protected by two jetties. A light and fog horn are at the outer end of the southern jetty, and two lighted ranges lead a vessel through the entrance and beyond. The channel is clearly marked by a buoys, daybeacons and lights leading to a turning basin at the north end of the harbor. Stay in the middle of the channel and take particular care not to hug the northeast side of the passage when making the turns at the southern portion of the entrance channel as shoaling occurs in this area. The sizable fetch in the shallow harbor combined with a strong wind results in a considerable chop developing in the bay. Do not be tempted to tack back and forth around the daybeacons, for outside the channel it IS shallow.

The marked channel leading southeast of Porto Bodega Marina has a mean depth of 2.4m (8 ft.) at mean low water making it important for vessels exploring this part of the bay to stay within the channel.

Bodega Bay marks the northern limit of the Gulf of the Farallones National Marine Sanctuary, which includes the Farallones and offshore waters almost as far south as Muir Beach. Because of a fortuitous mix of salt and fresh water, ocean currents, shifting waters and circulating nutrients, these waters are home to an abundant variety of marine life. Fishing is allowed provided you have a California Fishing License.

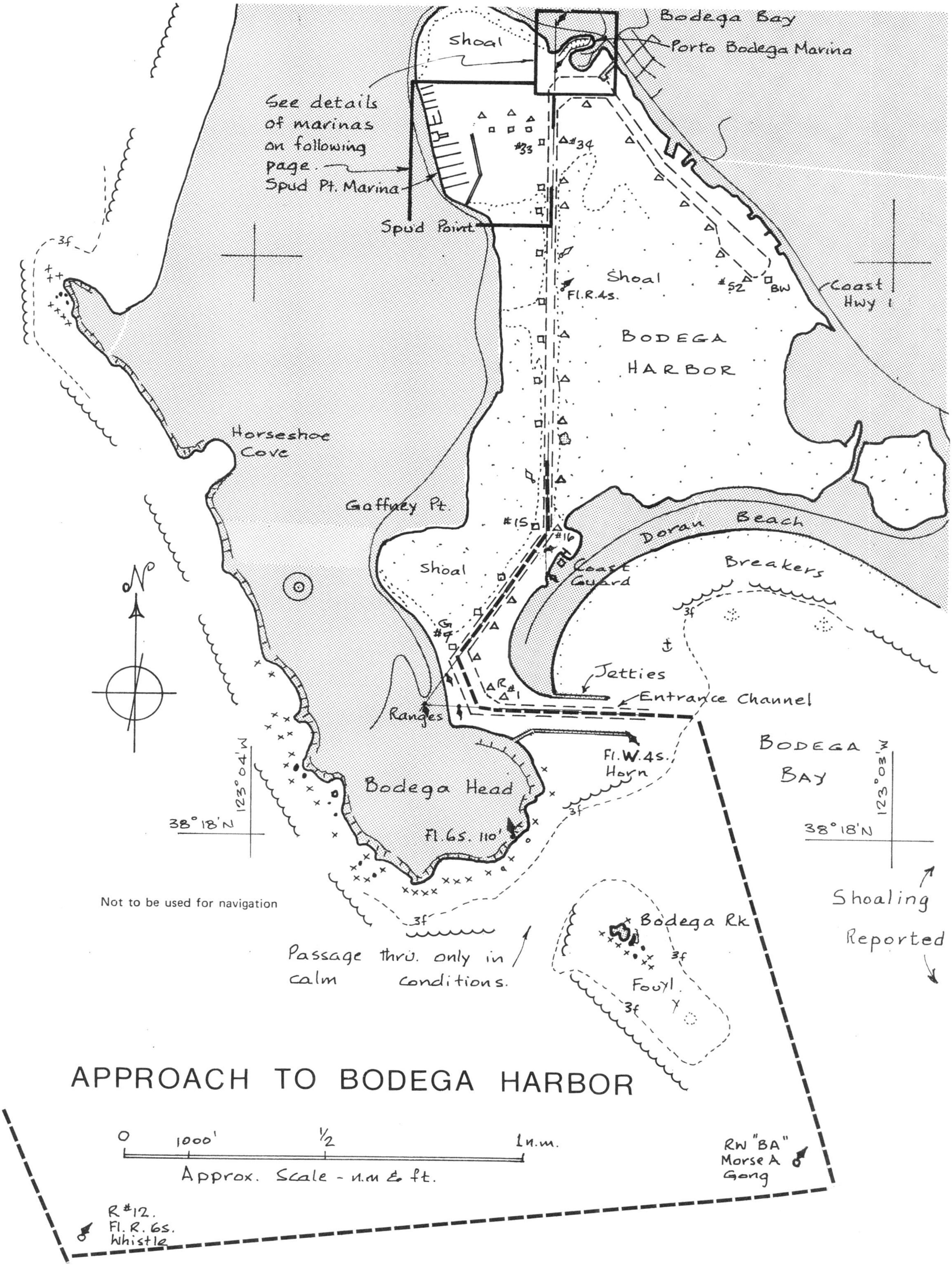

APPROACH TO BODEGA HARBOR

MARINAS IN BODEGA HARBOR

Spud Point Marina is the first facility seen as one proceeds north through the channel. To reach it, turn to port immediately after passing Buoy "33" and follow the marked side channel as shown on the sketch. This 244-berth marina can accommodate vessels up to 24m (80 ft.); dredged to 3.5m (12 ft.), it is well protected by a breakwater.

This is a first-class facility providing all services with marina staff on duty daily for 24 hours. It is operated by the Sonoma County Regional Parks Department. Contact the harbormaster on VHF Ch.16 or by phone (707) 875-3535 or Fax (707) 875-3436 for information on berthing; slips are usually available on a hot-berth basis at a reasonable cost. A $25 fee is charged for failure to register, so don't "forget."

Fuel, pump-out facilities, water, restrooms, free showers, a laundromat, public telephones and vending machines are available. Power is included in the moorage fee. Additional services include a flake ice machine, waste oil disposal, bilge pumps, and welder receptacles. Washroom keys are available to monthly tenants; transients are given access to washrooms by the staff. The marina office is an official NOAA weather display station and it posts messages for yachts in transit; from the number of messages seen, this is a popular spot for a rendezvous.

Internet facilities are available at Business Services Unlimited, telephone (707) 875-2183. Propane is available at Mason's Marina. Bus service around the bay provides access to the convenience store at Porto Bodega Marina where basic groceries and sundry items are sold.

Mason's Marina, just to the north of Spud Point Marina, is used primarily by commercial craft and fish boats. Propane is available at the Shell station located nearby.

Porto Bodega Marina used to be the only place where exhausted cruisers could tie up for a rest or a chance to enjoy terra firma again after grappling with Cape Mendocino. Now it is known as the fishermen's marina. This is a snug little basin that often becomes congested with several vessels in the small berths while tent-trailers, boat-trailers, campers and motor homes crowd into the limited parking spaces.

The marina can only accommodate vessels less than 12m (40 ft.) in length with a maximum draft of 1.5m (5 ft.). Contact the harbormaster on VHF Ch. 16 or 71 or phone (707) 875-3811 to arrange moorage. Moorage fees of $10 per night includes power. Marine supplies, fishing tackle, groceries and almost any basic item can be obtained in the truly "general" general store. At Pelican Plaza, about .5 mile distant, Diekman's grocery has a wider selection of foods. Propane is available at the marina.

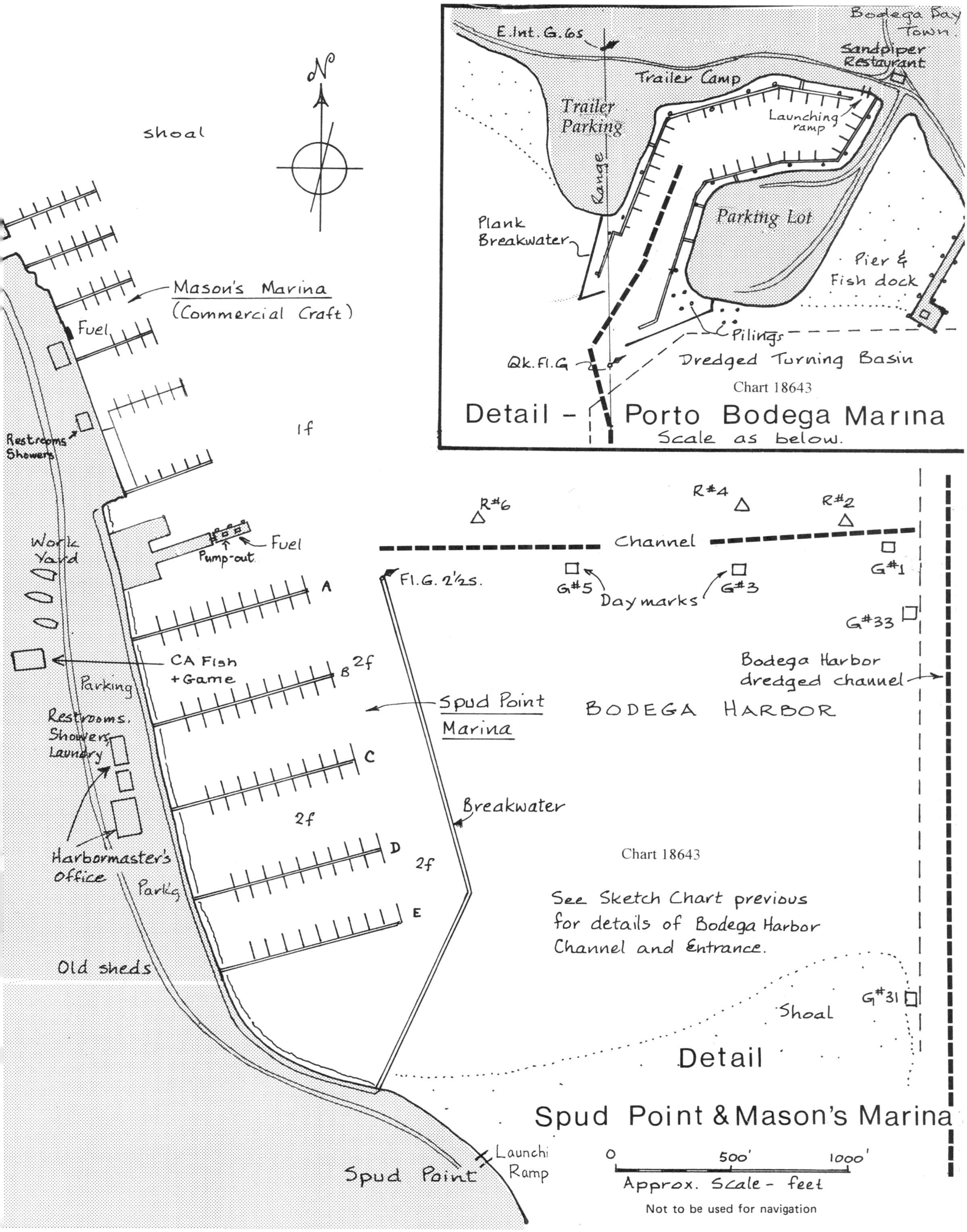
Detail – Porto Bodega Marina
Scale as below.
Chart 18643
Bodega Bay Town
Sandpiper Restaurant
E.Int.G.6s
Trailer Camp
Trailer Parking
Launching ramp
Range
Plank Breakwater
Parking Lot
Pier & Fish dock
Pilings
Qk.Fl.G
Dredged Turning Basin
shoal
N
Mason's Marina
(Commercial Craft)
Fuel
Restrooms Showers
1f
Work Yard
Fuel
Pump-out
Fl.G. 2½s.
R#6
R#4
R#2
Channel
G#1
G#5
G#3
Daymarks
G#33
A
B
2f
C
D
E
CA Fish + Game
Parking
Restrooms, Showers, Laundry
Harbormaster's Office
Park'g
Spud Point Marina
Bodega Harbor dredged channel
BODEGA HARBOR
Breakwater
2f
2f
Chart 18643
See Sketch Chart previous for details of Bodega Harbor Channel and Entrance.
Old sheds
G#31
Shoal
Detail
Spud Point & Mason's Marina
Spud Point
Launchi Ramp
0
500'
1000'
Approx. Scale - feet
Not to be used for navigation

DRAKES BAY

Drakes Bay was first visited by Sir Francis Drake in 1579. Its broad entrance and good protection from northwesterly winds made it an ideal anchorage for a heavy vessel under sail power alone. Nowadays it is a home port to numerous fishing vessels which operate from the bay during the season.

Drakes Bay is 25 miles south of Bodega Bay and 25 miles northwest of the Golden Gate Bridge. This rocky headland marks the northern edge of the Gulf of the Farallons; when seen at a distance in hazy weather it resembles an island. The Point Reyes Light and fog signal is at the end of the point. After proceeding 3 miles east of the lighthouse and rounding lighted whistle buoy "1", you can enter the bay and take good anchorage in 4 - 6 fathoms, sand.

The Farallon Islands (23 miles west of Golden Gate) are a 7-mile chain extending in a northwesterly direction from the largest island, Southeast Farrallon. A lighthouse and radio beacon are located here, while a whistle buoy "NR" is ½ mile west of Noonday Rock (the northern limit of the Farrallons).

APPROACHING SAN FRANCISCO

San Francisco is a normal port of call for vessels transiting this coast. Its large harbor and geographical location have made it a transportation center for sea borne traffic since 1775 when the Spaniards established their first settlement at Fort Point. Sailing your own boat under the Golden Gate Bridge for the first time is a memorable experience for everyone aboard a cruising vessel. But to have this time of exhilaration, there is a price to pay-—and that price is thorough preparation for the entry and good seamanship. Preparation involves a careful study of the weather, tide and current tables, selection of one's entrance route and knowledge of the hazards involved.

The famous San Francisco Bar guards the approach to the Golden Gate Bridge. It covers a large arc extending from 3 miles south of Point Lobos to within .5 mile of Point Bonita, with the outer edge about 5 miles WSW of the Bridge. The northern part of the bar is Fourfathom Bank, part of which is the infamous Potato Patch Shoal with depth of 7m (23ft.). The deepest route over the bar is the Main Ship Channel and this is the path followed by commercial shipping in the designated traffic lanes. It is recommended that pleasure craft avoid the traffic lanes and use Bonita Channel when approaching from the north and South Channel when approaching from the south.

During a heavy swell combined with an ebb tide the bar can be very dangerous, particularly in the vicinity of Potato Patch Shoal south of Bonita Channel. Particularly in heavy fog the main shipping lanes should be avoided even though a vessel is equipped with radar. If weather, tides or visibility make conditions hazardous it is better to wait until seas moderate and the tide is flooding so that a safe passage can be made.

Because of the large volume of commercial shipping the Coast Guard has established a Vessel Traffic Service (VTS) to minimize collisions and aid in the smooth flow of traffic. Pleasure craft are not prohibited from using traffic lanes, but if you choose to use them extreme caution must be practiced and the vessel must follow the established pattern and stay well to the starboard side of the lane. If you must cross a lane, do so at right angles and as quickly as possible. Information on major shipping traffic can be obtained by monitoring VHF Ch.16 or by contacting VTS on VHF Ch 16.

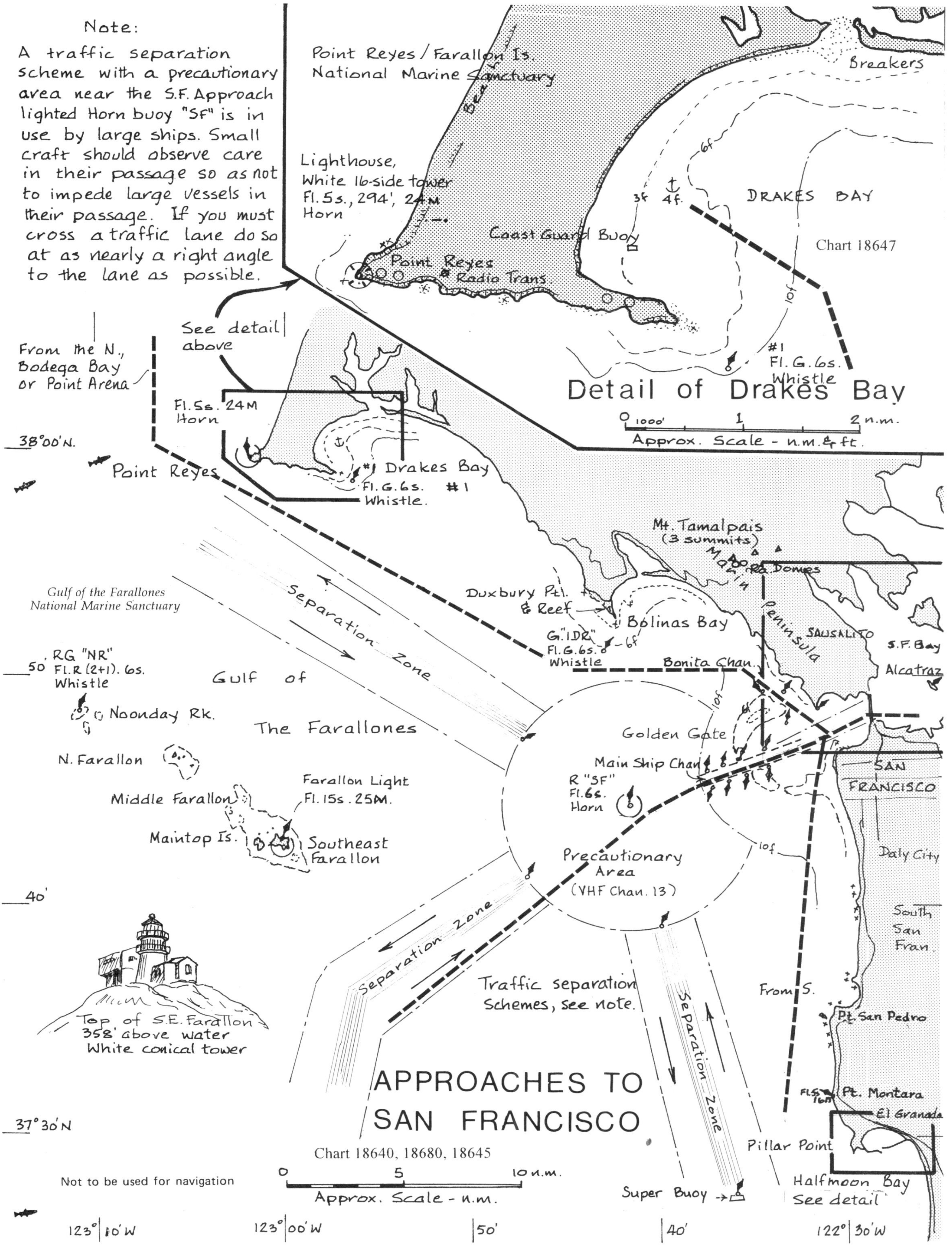

Note:
A traffic separation scheme with a precautionary area near the S.F. Approach lighted Horn buoy "SF" is in use by large ships. Small craft should observe care in their passage so as not to impede large vessels in their passage. If you must cross a traffic lane do so at as nearly a right angle to the lane as possible.
Point Reyes / Farallon Is. National Marine Sanctuary
Breakers
Lighthouse, White 16-side tower Fl.5s., 294', 24M Horn
DRAKES BAY
Chart 18647
Coast Guard Buoy
Point Reyes
Radio Trans
#1 Fl. G. 6s. Whistle
Detail of Drakes Bay
0 1000' 1 2 n.m.
Approx. Scale - n.m. & ft.
See detail above
From the N., Bodega Bay or Point Arena
Fl.5s. 24M Horn
38°00'N.
Point Reyes
#1 Drakes Bay Fl. G. 6s. #1 Whistle.
Mt. Tamalpais (3 summits)
Marin Peninsula
Ra. Domes
Duxbury Pt. & Reef
Bolinas Bay
G."1DR" Fl. G. 6s. Whistle
Gulf of the Farallones National Marine Sanctuary
Separation Zone
SAUSALITO
S.F. Bay
Alcatraz
Bonita Chan.
50'
RG "NR" Fl. R (2+1). 6s. Whistle
Gulf of The Farallones
Noonday Rk.
Golden Gate
N. Farallon
Main Ship Chan
R "SF" Fl. 6s. Horn
SAN FRANCISCO
Farallon Light Fl. 15s. 25M.
Middle Farallon
Maintop Is.
Southeast Farallon
Precautionary Area (VHF Chan. 13)
Daly City
40'
South San Fran.
Separation Zone
Traffic separation Schemes, see note.
From S.
Pt. San Pedro
Top of S.E. Farallon 358' above water White conical tower
Separation Zone
APPROACHES TO SAN FRANCISCO
Fl.5s. 16M Pt. Montara
El Granada
37°30'N
Pillar Point
Chart 18640, 18680, 18645
0 5 10 n.m.
Approx. Scale - n.m.
Not to be used for navigation
Super Buoy
Halfmoon Bay See detail
123° 10'W
123° 00'W
50'
40'
122° 30'W

By monitoring VHF Ch 14 you can hear UTS communications to commercial vessels. The Coast Guard can be reached at (415) 399-3451. If you have an emergency and must talk to the bridge of a commercial vessel use VHF Ch 13.

Hazards particular to the Bay waters include one or all of the following: strong currents, high winds, choppy seas, fog and congested traffic.

Currents are a major consideration since about one sixth of the Bay's entire water volume flows in and out twice a day through Golden Gate. Refer to *Tidal Current Charts, San Francisco Bay* for direction and velocity of daily currents.

The well-known Bay **winds** always live up to their gear- and crew-testing reputation. In summer the normal northwesterlies may blow up to 10 knots in the morning, increasing to 20 to 25 knots in the afternoon, creating choppy waters. In winter, winds are unpredictable, ranging from calms to 75 knots during storms. On average, spring is the windiest season, generally in the 20 to 30 knot range.

Fog is common from July to September; during August fog signals operate 40 - 50% of the time in the Gulf of the Farrallons. Offshore fog usually rolls over Golden Gate by noon and can often be seen boiling over the western hills. Having radar is a distinct advantage, though ghost echoes occurring in the vicinity of the Bridge sometimes make identification of radar targets difficult.

APPROACHES TO SAN FRANCISCO

When approaching from the north, cruising vessels may enter the Bay by rounding Point Reyes and proceeding to whistle buoy "DR" which marks the end of Duxbury Reef before entering via Bonita Channel. A lighted range and buoys mark this channel. Alertness at the helm is essential as there is a dangerous rip off Point Bonita, while on the ebb there is a set towards Potato Patch Shoal that must be countered. Enter only on a flood tide, and if seas are very rough it may be advisable to enter via the main ship channel or await calmer conditions.

Notes on Approaches to San Francisco. The North and South Shipping Lanes have been shifted slightly as of 7/1/00. See Notices to Mariners 49/00 if you have not already made correction on your 24th Edition Chart 18645. The 25th Edition of Chart 18645 provides accurate information on the results of these changes. A hazardous cargo lane has been added to the South Channel as indicated in Notices to Mariners 49/00.

When approaching from offshore, a vessel should identify San Francisco Approach Lighted Horn Buoy "SF," located 14 miles southwest of the Bridge and about 8 miles south of Duxbury Reef whistle buoy "DR". A circular precautionary area, having a radius of 6 miles around the "SF" buoy, should be avoided by small craft. This area may have large ocean-going vessels converging as a result of routing maneuvers, changing from the Main Ship Channel to offshore lanes and/or taking on or dropping off Pilots. Keep clear of traffic while proceeding to the Main Ship Channel, indicated by four pairs of lighted buoys. Keep well to the starboard side of the channel to leave a maximum amount of space for large commercial vessels, which are slow to respond to the helm and need considerable space to decelerate.

Vessels approaching from the south can follow South Channel along the coast but stay about .5 mile offshore to clear gong buoy "1" which marks the end of the dangerous reefs and shoals southwest of Pillar Point. By keeping at least .5 miles offshore you will clear Colorado Reef (south of Montara Point) marked by Buoy ODAS, and the rocks WNW of San Pedro Point. When using the South Channel follow the 7-fathom line to avoid breaking seas off Ocean Beach that extend to Golden Gate Park. During southerly storm conditions breaking waves are found well off the beach due to shallow water. By using unbuoyed South Channel until Seal Rocks have been safely cleared, you may then approach the Golden Gate.

Approaching the Golden Gate Bridge *Margo Wood*

It is important to remember that the VTS traffic patterns continue into the Bay. Extreme caution is essential in the precautionary zone, where large vessels converge to cross and join diverging traffic lanes. Great care must be taken to watch for other vessels of every size and shape and to follow the rules of the road.

To report in, call Customs at (415) 782-9422, 8 a.m. to 4:30 p.m. Vessels making a landfall from a foreign country must also call Immigration at (415) 844-5227 8 a.m. to 4:30 p.m. Monday to Friday or after hours go to the Immigration Office at 444 Washington Street, San Francisco. For general Customs information call (415) 782-9210 or visit their website at www.customs.gov. For marine Customs call (415) 782-9422 or (415) 782-9423. The local Immigration Office may be called at (415) 844-5227 and general immigration information is available by calling 800-375-5283. To contact the San Francisco Coast Guard call (510) 437-3700 and for general Coast Guard information call 800-368-9647.

San Francisco Bay is a large body of water covering some 400 square miles. Cities bordering this extensive bay include San Francisco, Oakland, Emeryville, Alameda, San Leandro, Richmond, Berkeley, Sausilito, Redwood City and South San Francisco. San Francisco/Oakland is a **Port of Entry**.

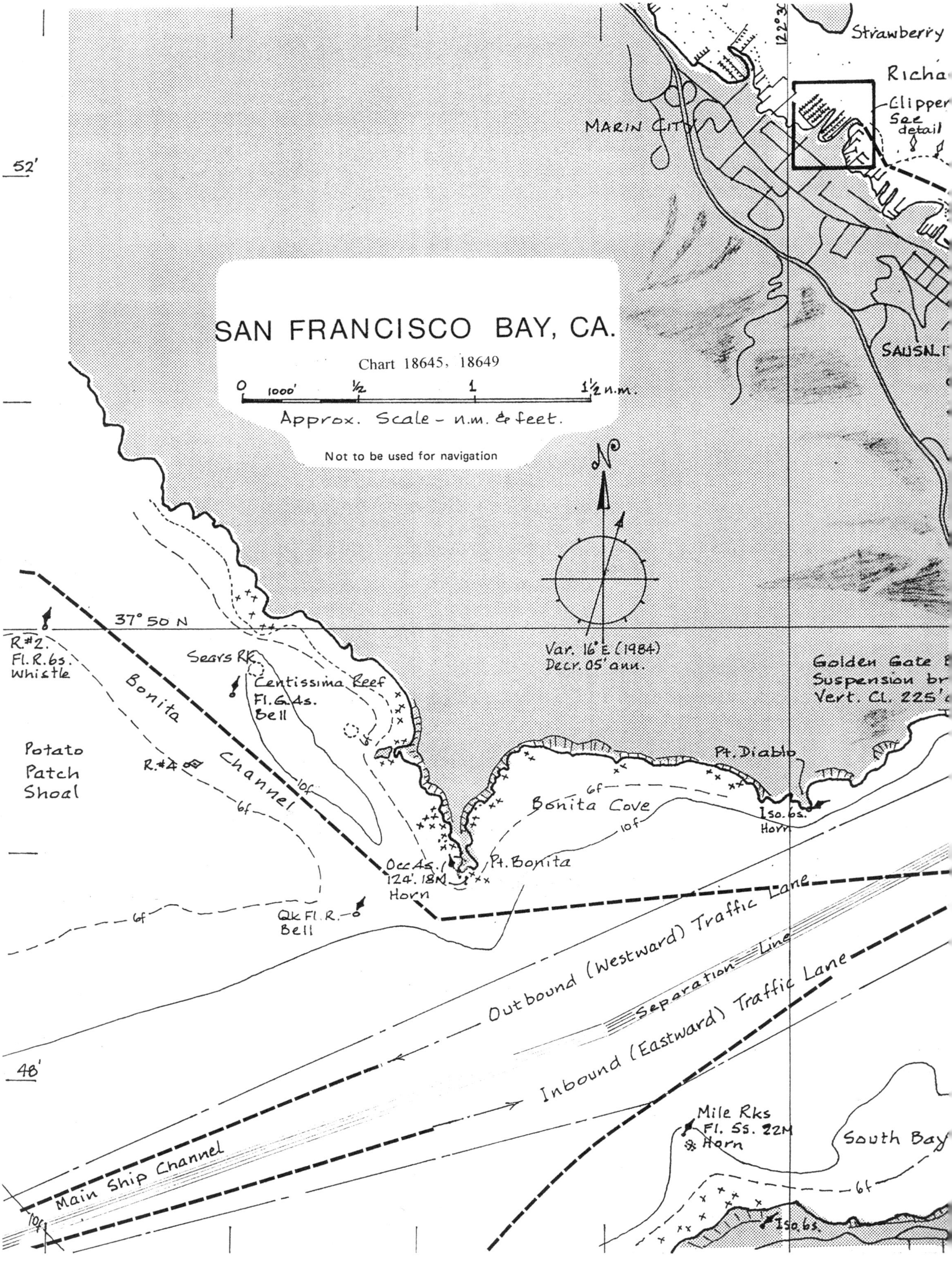
SAN FRANCISCO BAY, CA.
Chart 18645, 18649
0 1000' ½ 1 1½ n.m.
Approx. Scale - n.m. & feet.
Not to be used for navigation
N
Var. 16°E (1984)
Decr. 05' ann.
37° 50 N
52'
48'
Strawberry
Richa
Clipper
See detail
MARIN CITY
SAUSALI
Golden Gate
Suspension br
Vert. Cl. 225'
R.#2.
Fl. R. 6s.
Whistle
Sears Rk.
Centissima Reef
Fl. G. 4s.
Bell
Bonita Channel
Potato Patch Shoal
R.#4
10f
6f
Bonita Cove
Pt. Diablo
Iso. 6s.
Horn
Occ 4s.
124'. 18M
Horn
Pt. Bonita
Qk Fl. R.
Bell
Outbound (Westward) Traffic Lane
Separation Line
Inbound (Eastward) Traffic Lane
Main Ship Channel
Mile Rks
Fl. 5s. 22M
Horn
South Bay
Iso. 6s.

TIBURON PENIN.
122°25
son Bay
Yacht Harbor
Belvedere Cove
S.F.Y.C.
Corinthian Y.C.
V. Shallow
R "6"
Schoonmaker Yt. Hbr
Cone Rk.
Qk. Fl.
Fl. G. 4s #3
Sausalito Yachts
Fl. R. 4s. #2
Buoys
Tide rips
Ayala Cove
Tide rips
52'
ANGEL Is.
(State Park)
Fl. G. 5s.
R#2
Fl. R. 4s.
Recreation Area
Tide rips
G#1
Fl. G. 4s.
10f
Harding Rk
RB "HR"
1 Qk. Fl. R
37°50'N
Sausalito Y.C.
SAN FRANCISCO BAY
GR Bell
Tank
ALCATRAZ Is.
Light tower
Fl. 5s. 214'. 31M.
Tide rips
Separation Line
Traffic Lanes
Tide rips
Recreation Area
S.F. Maritime
State Hist. Park
Fisherman's Wharf
Pier 39
Marina
& Shops
Anita Rk
Qk. Fl
Qk. Fl. R
E. Int. R 6s.
Fl. R. 4s
Coast Guard
See detail
Marina Blvd
Fort Mason
Aquatic Pk.
Telegraph Hill
Coit Tower
Doyle Dr.
Bay St.
Palace of Fine Arts
Russian Hill
Lombard St.
Scott St.
Van Ness Ave
Hwy 1
Presidio Military Reservation
Pacific Ave
Nob Hill
Chinatown
SAN FRANCISCO

Flowing into the Bay are the San Joaquin and Sacramento Rivers, which are navigable all the way up to Stockton and Sacramento respectively. The tributaries, canals and deltas of these two rivers comprise what is known as the Delta Region. There are many miles of navigable waters which provide protected cruising; if one is planning to visit these waters reference should be made to one of several guides devoted solely to the Delta and Bay areas. As this guide is intended for sailors transiting the coast attention is focused on marina facilities closest to the Golden Gate Bridge, though information on other pleasure boat facilities found elsewhere in the Bay is provided.

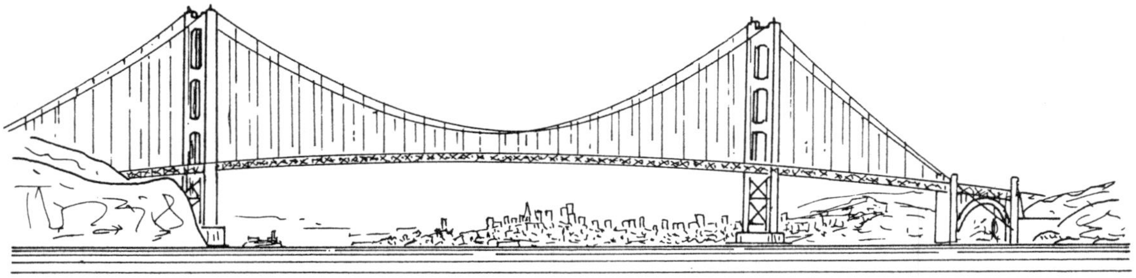

SAN FRANCISCO THROUGH THE GOLDEN GATE

SAN FRANCISCO

San Francisco Marina is almost 2 miles east of the Golden Gate Bridge. A breakwater with a light at the eastern end gives protection to the docks of the Saint Francis Yacht Club, Golden Gate Yacht Club and the west basin of the San Francisco Marina Yacht Harbor. The east basin, also known as Gashouse Cove is just beyond. A few transient berths are available; check with the harbormaster for berth allocation Facilities include laundry, shower, pump-out, fuel and groceries. Propane can be obtained from U-Save Equipment Rental at 3rd and Mariposa. Internet access on 15-minute express terminals is available at the Main Library located at 100 Larkin St. A library card is not needed.

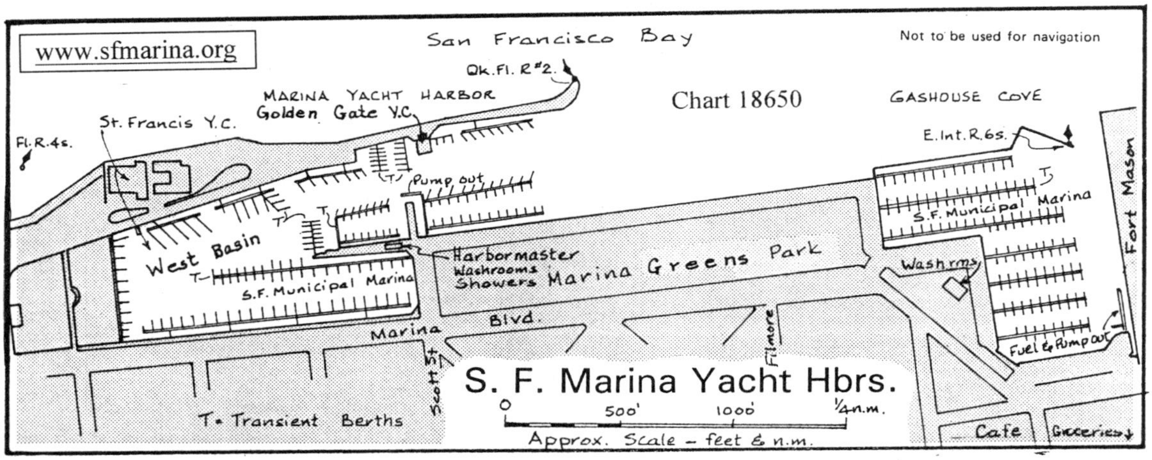

On a rare, clear day the landmarks on the approach to San Francisco are easily identified. (Page 84)

Across from San Francisco's Aquatic Park is Alcatraz Island with Angel Island visible in the left background. (Page 90)

Morro Rock provides a unique landmark at the entrance to Morro Bay, California. (Page 106)

Aquatic Park is located about .5 mile east of Marina Yacht Harbor. A pier curves out from the western side, giving protection from the winds sweeping in from Golden Gate.. Good anchorage for a few days may be taken in 1.5 - 2 fathoms, mud. Fuel, water and marine supplies are available at Fisherman's Wharf just to the east.

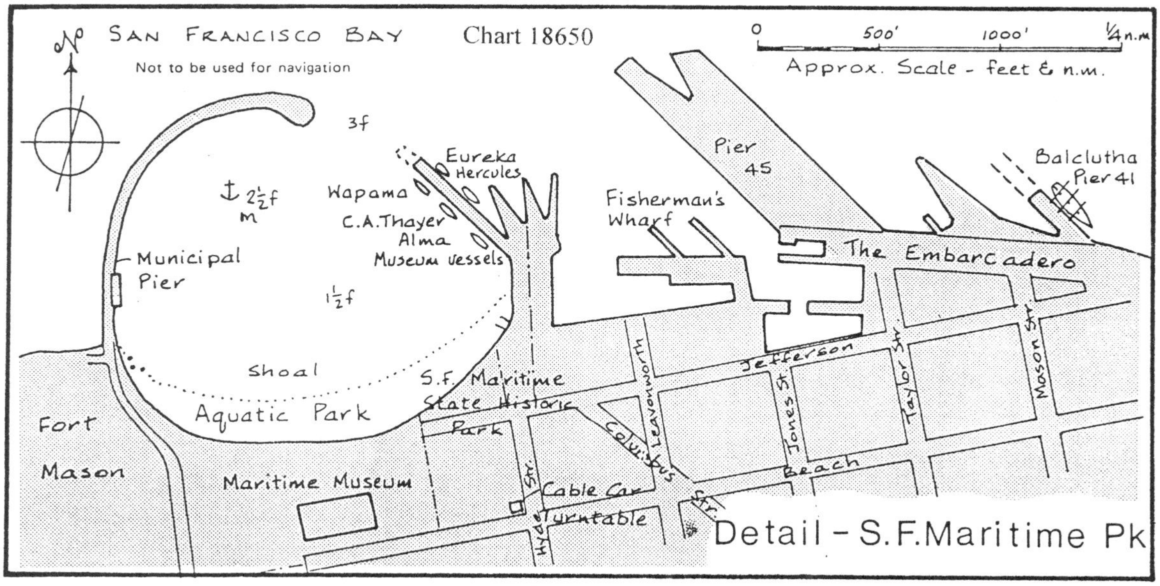

Pier 39 has over 300 slips with moorage for transients with reservations. Office hours are 9 am to 5 pm daily and VHF Ch 16 is monitored. Gashouse Cove is the nearest source of diesel while propane is available at a Rental Equipment facility at 3d and Mariposa. North Point Shopping Center where a Safeway store is located is two blocks away. Internet access is available on the second level of Pier 39. Marine supplies can be obtained from Coast Marine (located on the commercial level) about 6 or 7 blocks distant near Fisherman's Wharf on Jefferson or about 4 miles away at West Marine at Second and Townsend.

www.pier39marina.com

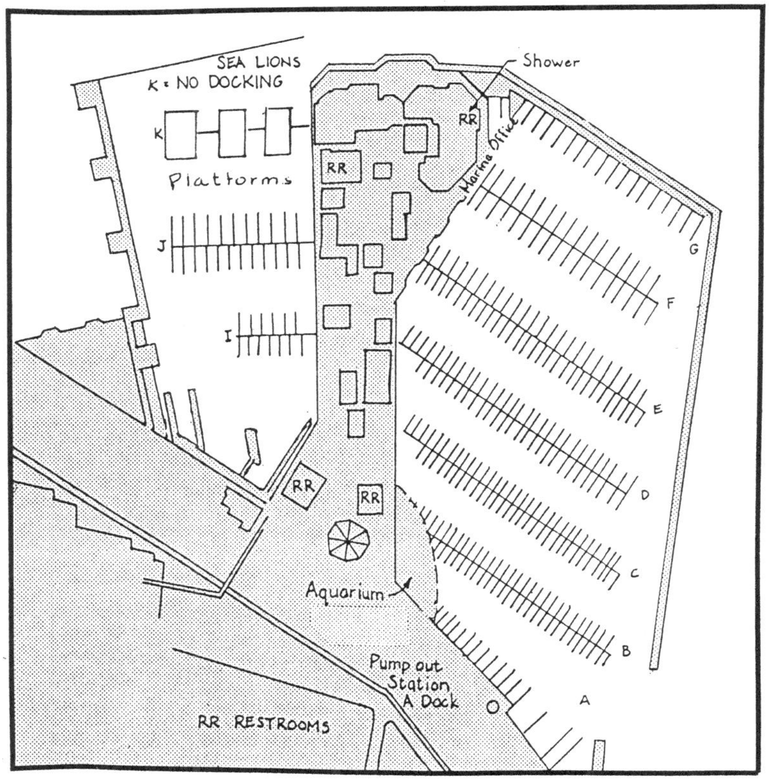

Chart 18649, 18650

South Beach Harbor is tucked around the southern corner of San Francisco at Pier 40 just beyond the Bay Bridge and beside the new Pac Bell Baseball Park. This marina has 700 slips and 20 guest berths with visitor's dock that can accommodate vessels up to 36m (120ft.). Facilities include a pump-out station, showers and laundry with a first-rate restaurant nearby.

The South Beach Yacht Club clubhouse is next door on Pier 40. Street car service provides unique access to the colorful sights of downtown San Francisco.

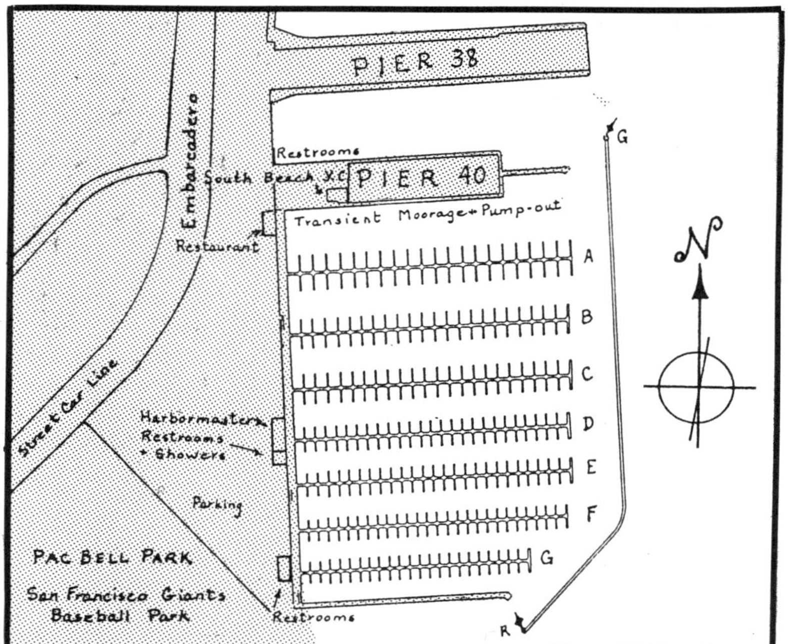

Chart 18649, 18650

USEFUL CONTACTS IN SAN FRANCISCO

Boatyard:

San Francisco Boat Works, 835 China Basin St.	Tel: (415) 626-3275
www.sfboatworks.com	
Chandlers	
Coast Marine and Industrial Supply, 398 Jefferson St.	Tel: (415) 673-1923
West Marine, 101 Townsend Street	Tel: (415) 543-3035
Propane Sources	
U-Save Equipment Rental, 1800 – 3rd St. (3rd and 16th)	Tel: 415) 864-2811
Action Rental, 565 South Van Ness Ave.	Tel: (415) 826-1830
(Between 16th and 17th – Kerosene and Propane)	
U-Haul Co., 100 Valencia	Tel: (415) 864-5020
1575 Bayshore Blvd.	Tel: (415) 467-3830
Hertz Equipment Rental, 435 South Van Ness Ave.	Tel: (415) 865-4444

Internet Access

Kinko's Copies ($12 per hour) at the following locations:

3225 Filmore	1967 Market	1597 Sloat	25 Stanyan
201 Sacramento	1800 Van Ness	367 Pine	50 Fremont

Starbuck's (Broadband Hotspot) at the following locations:

2132 Chestnut	1231 Market	833 Mission (4th and Mission)
201 Spear St	500 Beach	490 Brannon

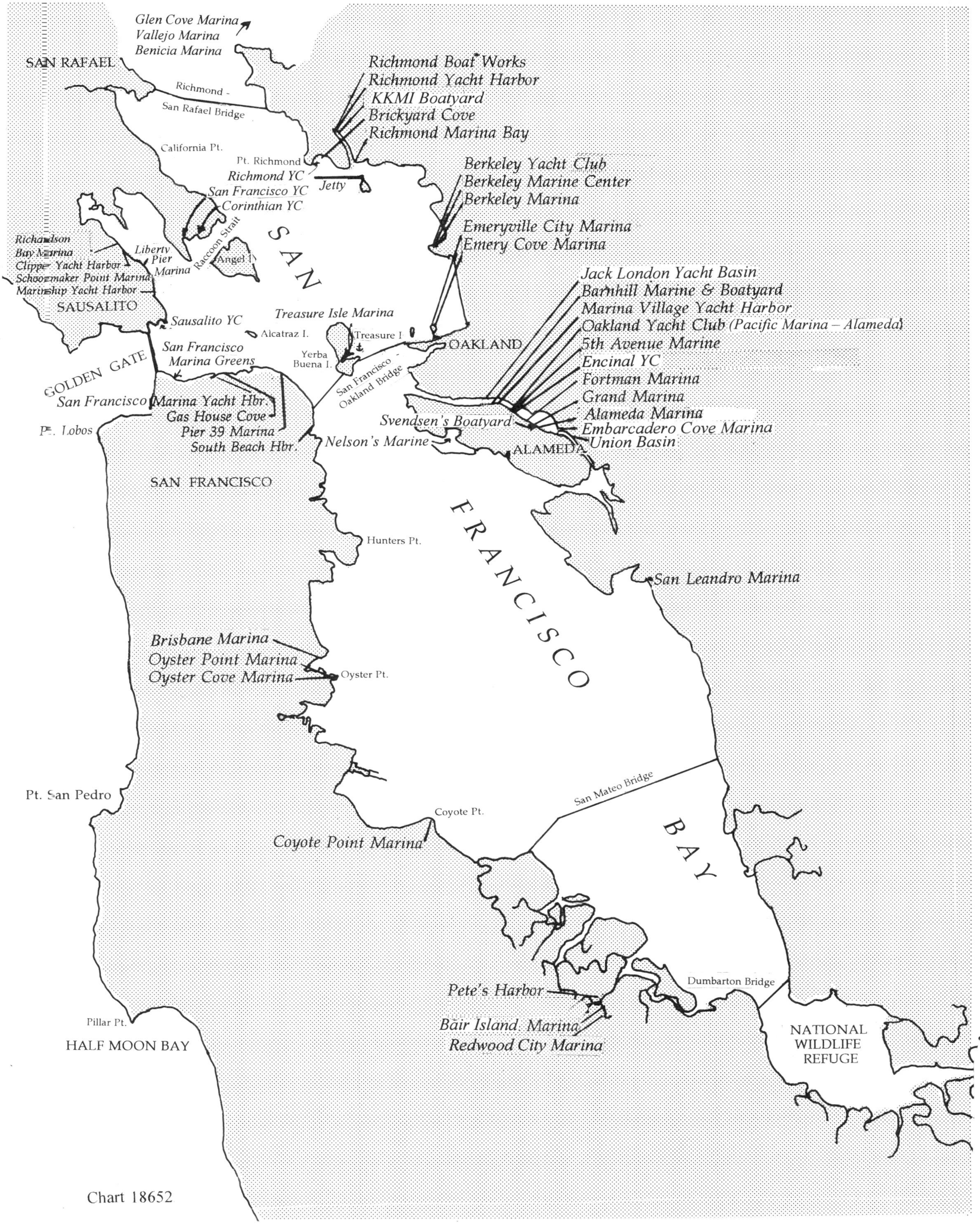
Glen Cove Marina
Vallejo Marina
Benicia Marina
SAN RAFAEL
Richmond - San Rafael Bridge
California Pt.
Pt. Richmond
Richmond Boat Works
Richmond Yacht Harbor
KKMI Boatyard
Brickyard Cove
Richmond Marina Bay
Richmond YC
Jetty
San Francisco YC
Corinthian YC
Berkeley Yacht Club
Berkeley Marine Center
Berkeley Marina
Emeryville City Marina
Emery Cove Marina
Raccoon Strait
Angel I.
Liberty Pier Marina
Richardson Bay Marina
Clipper Yacht Harbor
Schoonmaker Point Marina
Marinship Yacht Harbor
SAUSALITO
Sausalito YC
Alcatraz I.
Treasure Isle Marina
Treasure I.
Yerba Buena I.
San Francisco - Oakland Bridge
OAKLAND
Jack London Yacht Basin
Barnhill Marine & Boatyard
Marina Village Yacht Harbor
Oakland Yacht Club (Pacific Marina – Alameda)
5th Avenue Marine
Encinal YC
Fortman Marina
Grand Marina
Alameda Marina
Embarcadero Cove Marina
Union Basin
Svendsen's Boatyard
Nelson's Marine
ALAMEDA
GOLDEN GATE
San Francisco Marina Greens
San Francisco
Marina Yacht Hbr.
Gas House Cove
Pier 39 Marina
South Beach Hbr.
Pt. Lobos
SAN FRANCISCO
SAN FRANCISCO BAY
Hunters Pt.
San Leandro Marina
Brisbane Marina
Oyster Point Marina
Oyster Cove Marina
Oyster Pt.
Pt. San Pedro
San Mateo Bridge
Coyote Pt.
Coyote Point Marina
Dumbarton Bridge
Pete's Harbor
Bair Island Marina
Redwood City Marina
NATIONAL WILDLIFE REFUGE
Pillar Pt.
HALF MOON BAY
Chart 18652

If you are spending more than a few days in the Bay it is well worth checking out several yachting centers where a large number of marine services are available including boatyards, chart and electronic supplies, marine chandleries, mechanical and electrical repairs, sailmakers and custom canvas goods. The following notes provide brief descriptions of marine facilities as one proceeds around the bay.

SAUSALITO

Sausalito is an area often preferred by coastal cruisers because of its clean air, lack of congestion and quiet pace. Originally an artists' colony, the spectacular homes clinging to steep cliffs are a sharp contrast to the cheek-by-jowl houses of San Francisco. Located 2 miles north of the Bridge, it is reached by way of a channel marked by lights and buoys. Waters outside the channel are shoal, making it necessary for deep-draft vessels to remain within the clearly marked boundaries.

This is a pleasure boat haven where every boating need can be met. Anchorage can be taken in the southern part of Richardson Bay, but watch your depth sounder when looking for a place to drop the hook as it is very shallow outside of the anchoring area.

Richardson Bay Marina (formerly Kappas Marina) is the northernmost facility in Richardson Bay. Club Nautique operates a sailing school and yacht rental agency from this location and there are no slips available for visiting yachts.

Marina Plaza Harbor has just over 100 slips and can accommodate vessels up to 14.5m (48'**)**. Transients may have moorage for a maximum of 10 days. Within walking distance are a large grocery store, restaurants, chandlery and laundry.

Schoonmaker Point Marina is on the opposite side of the channel from the platform with 160 berths where vessels up to 60m (200') can be accommodated. Excellent facilities include dry storage, three-ton hoist and moorage for transients on a hot-berth basis.

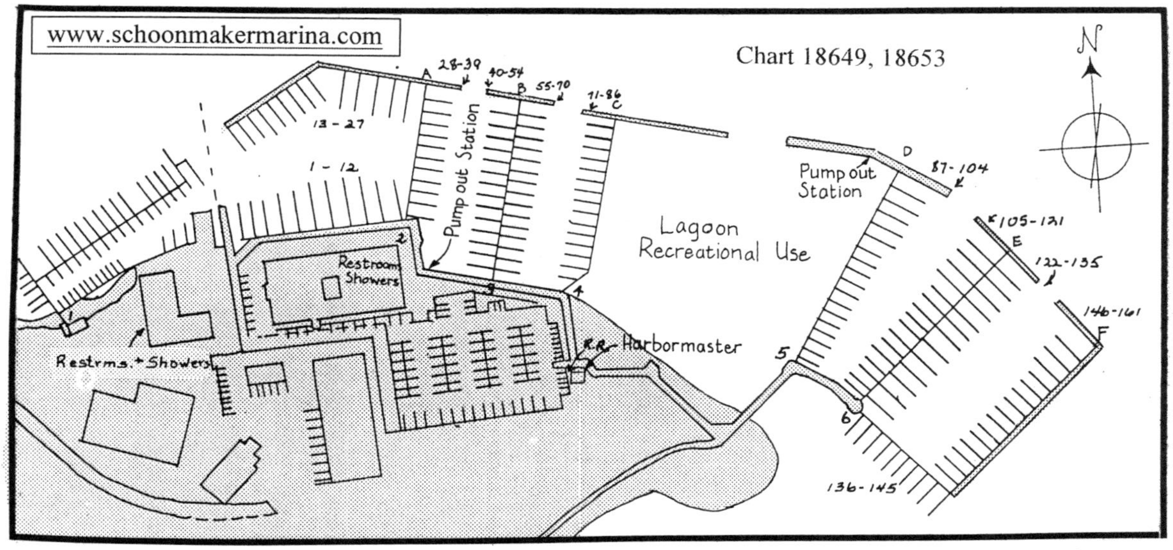

Detail of Schoonmaker Point Marina

Libertyship Pier, between Clipper Yacht Harbor and Schooonmaker Point Marina has some transient slips.

Clipper Yacht Harbor is a large, fully modern marina with over 700 slips. vessels. It offers complete services including a fuel dock where propane is available. Call by phone to check on availability of transient moorage. Grocery stores are three blocks away.

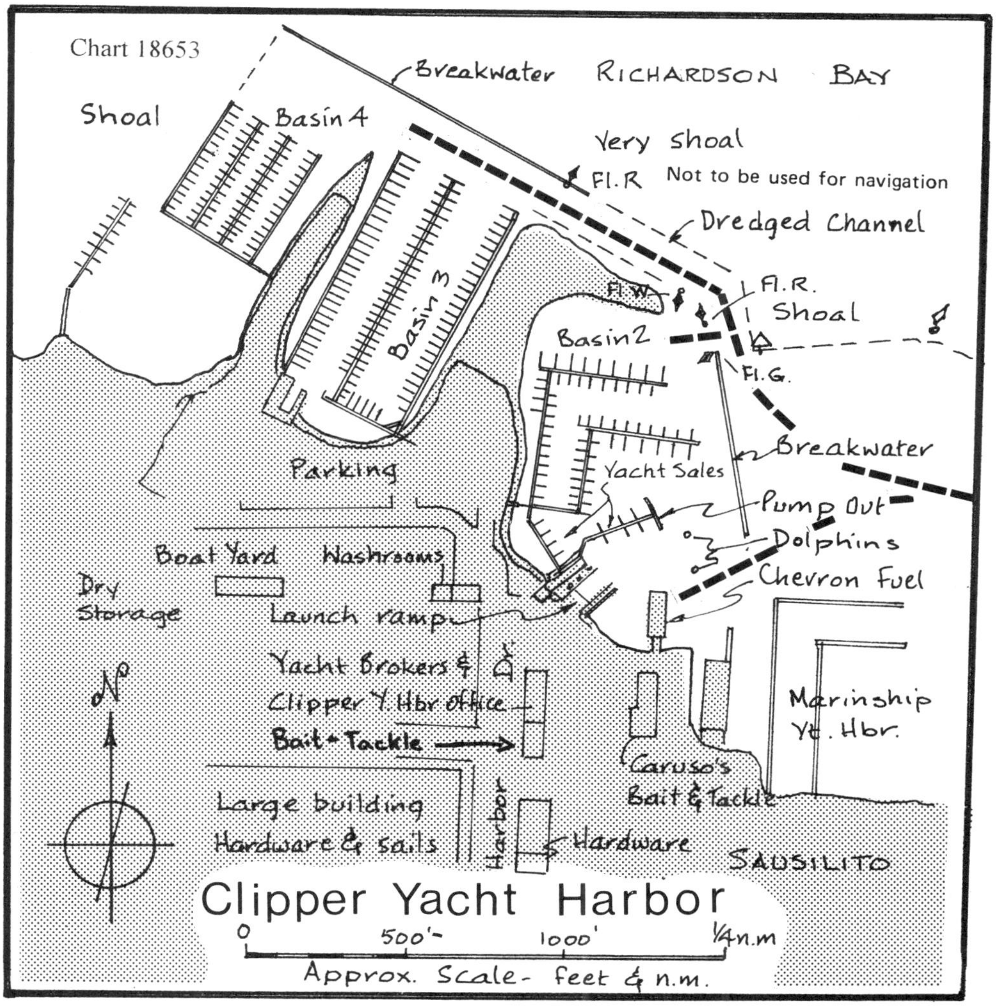

USEFUL CONTACTS IN SAUSALITO

Boatyards

Anderson's Boatyard Inc.	Tel: (415) 332-5432, 800-310-5432
Richardson Bay Boatworks	Tel: (415) 331-0742

Propane Source

Clipper Fuel Dock

Internet Access

Main Library downtown

Kinko's, 2400 Bridgeway (across from Mollie Stones Mkt.) Tel: (415)331-2093

Cruisers who have one or two weeks for travel within San Francisco Bay have a wide choice of marinas and yacht clubs that can be visited for a few days at a time before moving on to another facility. Each place has its own charm and personality.

BRISBANE

Brisbane Marina has 570 slips and (2004) the most reasonable moorage rates in the bay for transients at $15 per day for a vessel 11m – 14m (36' – 45'). For slip availability call on VHF Ch 16. Maximum depth in the channel is 2.7m (9'). It has showers but no laundry. The nearest fuel is just across the way at Oyster Point and the pump-out is free. Internet access is 3.5 miles distant at the coffee house in town. There is a restaurant in Radison. E-mail – Twarburton@ci.brisbane.ca.us

SOUTH SAN FRANCISCO

Oyster Point Marina facilities include fuel, pump-out, showers, bait shop and a nearby restaurant. Maximum depth in the approach is 1.8m (8') and current charges are $.55 per foot per night. The nearest grocery is a taxi trip away. Website – www.smharbor.com

Oyster Cove Marina is a small marina patronized mainly by small power boats.

SAN MATEO

Coyote Point Marina has 564 slips with five berths reserved for guests with charges of $.30 per foot per night. Facilities include a fuel dock, pump-out, showers and a restaurant serving lunch and dinner. Maximum depth in the approach is 2.1m (7'). Website – www.coyotepointmarina.org

REDWOOD CITY MARINAS

Pete's Harbor has moorage for transients with a one-week maximum stay. Rates for 40' or larger vessels are $25 per night, less for smaller boats. Monthly rates are available and liveaboards are accepted.

Bair Island Marina has some transient slips available at $.50 per foot per night plus electricity. Vessels up to 15m (50') are accommodated in this fully modern facility. A playground is on the premises. The nearest grocery store is about a mile away; restaurants and a movie theatre are nearby. Hiking and biking to view the wildlife refuge make this an interesting stop-over. Website – www.Bairislandmarina.com

Port of Redwood City Marina has 185 slips with transient berths sometimes available at $.50 per foot per night for a maximum of 30 days. Depth in the channel entrance and marina is 3m (10') and facilities include a fuel dock, laundromat, pump-out, launch ramp and a restaurant.

SAN LEANDRO

San Leandro Marina has 450 slips with 10 available to guests at $10 per night. Vessels up to 18m (60') can be accommodated; depths vary from 2m (6') to 1.4m (8') at low water. Facilities include fuel, pump-out and showers. Groceries and laundry services are one mile away. Internet access is available. Website – www.ci.san-leandro.ca.us/slmarina.hyml

Caution if you are heading west to Berkeley or Emeryville: parts of this side of the Bay are **extremely shallow** so check charts carefully for safe depths and channels and navigate with care and attention in the area. The Berkeley Pier (extending 3 miles into the Bay) was built for the World's Fair on Treasure Island prior to WWII. It is slowly disintegrating but the cement piers that are **only visible at low water** are still in place. Do not attempt to pass through the area marked for the Berkeley Pier ruins except at the small gap that has been cleared right next to the section of a restored fishing pier near the shore.

BERKELEY (Due east of the Golden Gate Bridge)

Berkeley Marina has over 1,000 slips and 12 transient berths. Facilities include fuel, pump-out, launch ramp, dry storage, restaurants, bait shops and a deli. The marina has small boatyard, fuel docks, a 25-ton hoist and a marine supply store. Public transportation from both facilities gives access to San Francisco. Website – www.ci.berkeley.ca.us/marina

USEFUL CONTACT IN BERKELEY

Boatyard

Berkeley Marine Center, Chandlery and Fuel Dock	www.berkeleymarine.com
(Do-it-yourselfers are welcome.)	Tel: (510) 843-8195

EMERYVILLE

Emeryville Marina has over 400 slips with some guest slips ($16 - $22 per night and liveaboards. Facilities include fuel dock, pump-out, laundry, showers and launch ramp. Watergate Market is a 5-minute walk away. Wireless internet service is available. Bus service and a shuttle (during the week) connects to Bart.

Emery Cove Marina has 430 slips, 15 guest slips ($20 per day) and 26 liveaboards. Maximum depth in the channel is 2.7m (9'). Wireless internet access is available using the I-dock system--bring your own computer. A grocery store and bus service are within two blocks and there is a free shuttle service to Bart. Several restaurants are nearby. There are two pump-out stations in the facility and fuel is available at City Docks. Website – emerycove.com

RICHMOND

If you wish to get out of the direct path of the summer westerly winds and accompanying fog heading north to harbors in Richmond is a good choice. Richmond is on the northeast shore of San Francisco Bay, northeast of Angel Island. The harbor is protected by a jetty that extends from the west end of Brooks Island.

After passing through Raccoon Strait (between Angel Island and Tiburon) the entrance is almost directly ahead on a northwest heading. Approach Buoy "R" 6R4s and the RW beacon Pile at the end of the jetty. The first facility seen will be the Richmond Yacht Club, followed by Brickyard Cove. By continuing through Point Potero Reach channel and around the first turn a channel to starboard leads to Richmond Marina Bay. But by remaining in the main channel and continuing northward, then east. you will arrive at Richmond Yacht Harbor and the KKMI boat yard.

Brickyard Cove has 250 slips and offers some anchorage spots when transient berths ($30 per day) are full. It is wise to call ahead to check on the availability of space. The channel has a maximum depth of 2.4m (8'). Laundry and showers are on site and a mobile pump-out service is available. The complex includes a laundry, dry storage, restaurant, sailmaker, canvas shop and brokerage.

Richmond Yacht Club provides moorage to yacht club members with reciprocal privileges. It is in a protected harbor and has the usual amenities.

Richmond Marina Bay Yacht Harbor can accommodate vessels up to 36m (120') at very reasonable rates. It has a fuel dock, a small convenience store, deli and restaurant and provides i-dock internet connection. It is about 5 miles from chandleries and boatyards. Bus transportation is available with connection to Bart and AMTRAK trains. Website – www.marinabayyachtharbor.com

USEFUL CONTACTS IN RICHMOND

Boatyards

Bay Ship and Yacht — Tel: (510) 237-0140
310 West Cutting Blvd. (Website – www.bay-ship.com)
This yard is for large ships and includes a propeller shop.

KKMI (Keefe Kaplan Maritime Inc.) — Tel: (510) 235-5564
530 West Cutting Blvd. (Website - www.kkmi.com)
Experts in wood, metal and fiberglass; "Do-it-yourselfers" OK.

Richmond Boat Works (Yanmar dealer) — Tel: (510) 232-5800
616 W. Cutting Blvd. (at north end of Richmond Channel)
Rigging, painting, general maintenance, engine repairs

Sanford Wood Marine — Tel: (510) 236-6633

Propane Source

Chevron Pro-Gas, 5000 Seaport Ave. — Tel: (510) 237-7077

ALAMEDA/OAKLAND ESTUARY

Another area that tends to be warm and calm is the Alameda/Oakland Estuary where a full range of services and supplies are available. To reach the Estuary proceed in a southeasterly direction in the Bay and pass under the Bay Bridge which joins Oakland to San Francisco. Beyond the bridge on the eastern shore is the Oakland Outer Harbor with many container docks. A short distance to the south is the estuary entrance.

The entrance to the estuary is clearly marked with lighted buoys that unfortunately blend in to the shore side lights at night forcing one to take extra care to identify them. The entrance has a series of large container docks on the Oakland side. The former Alameda Naval Air Base airstrip is closed and is known as Alameda Point. The first yacht harbor seen is on the north (Oakland) side at Jack London Square where a fuel dock and bait shop mark the entrance. All of the harbors (except 5th Avenue) on the Oakland side of the estuary are under the control of the harbormaster at Jack London Square. Though there are no dedicated guest docks, transient slips are frequently available on a hot-berth basis. Public transportation by bus and an AMTRAK station make Jack London a convenient stop. When proceeding in the estuary you will pass Mariner Square and the houseboats at Barnhill Marine.

ALAMEDA

Marina Village Harbor is a large facility on the Alameda side. It has 750 slips and can handle vessels up to 22m (72 ft.). Guest berths are available a $20 per night and facilities include pump-out, shower and laundry. Grocery and drug stores are two blocks away and a marine supply store is .5 miles distant. The marina monitors VHF Ch. 16.

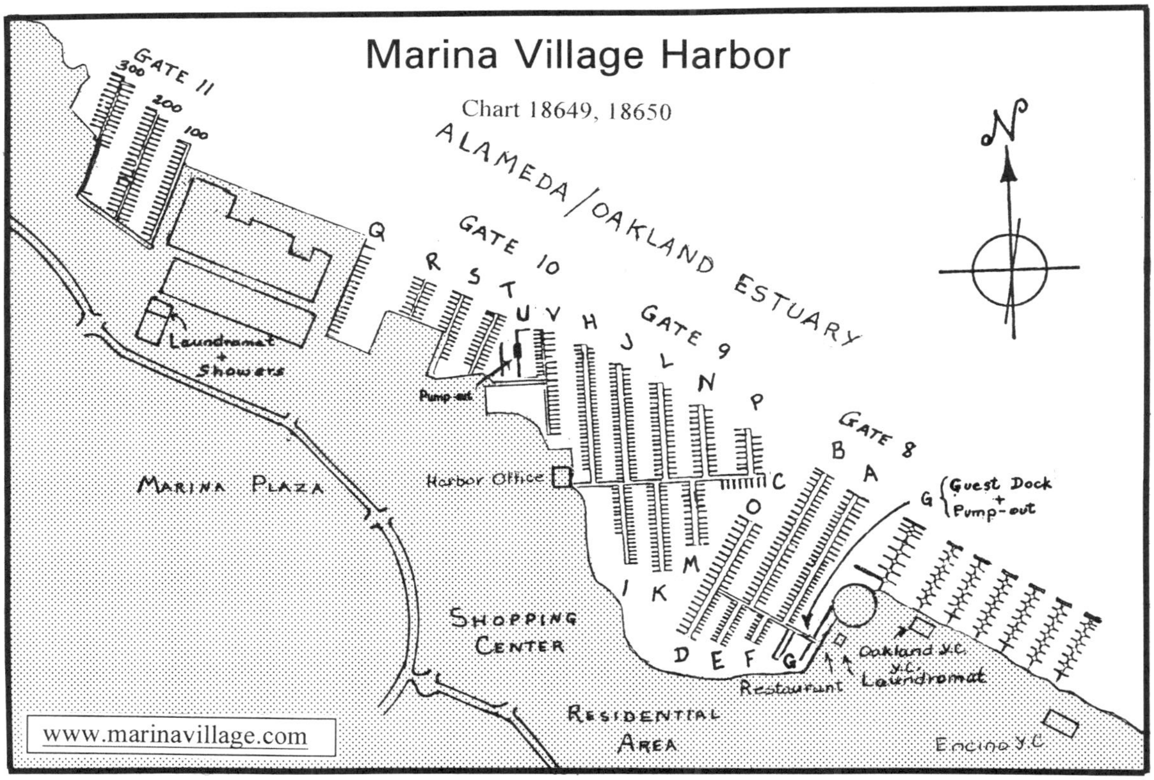

Fortman Marina has 486 slips and can accommodate vessels up to 18m (60 ft.). Transients may find moorage for a limited time when space is available. To arrange moorage call VHF Ch 68, telephone (510) 522-9080 or Fax (510) 522-4803. Facilities include showers, restrooms, electricity and pump-out.

Website – www.fortman.com E-mail – fortman@dsp.net

Grand Marina has over 400 slips with space for guests when space is available. The office monitors VHF Ch 71 and is open daily from 9 a.m. to 5 p.m. For information call (510) 865-1200. Facilities include heated showers, laundry, cable TV, pump-out, fuel, launch ramp, and 60-ton Travelift. Close by are Mariner Boatyard, a chandlery, electronic and mechanical services, a canvas shop, metal fabricator, bait shop and convenience store.

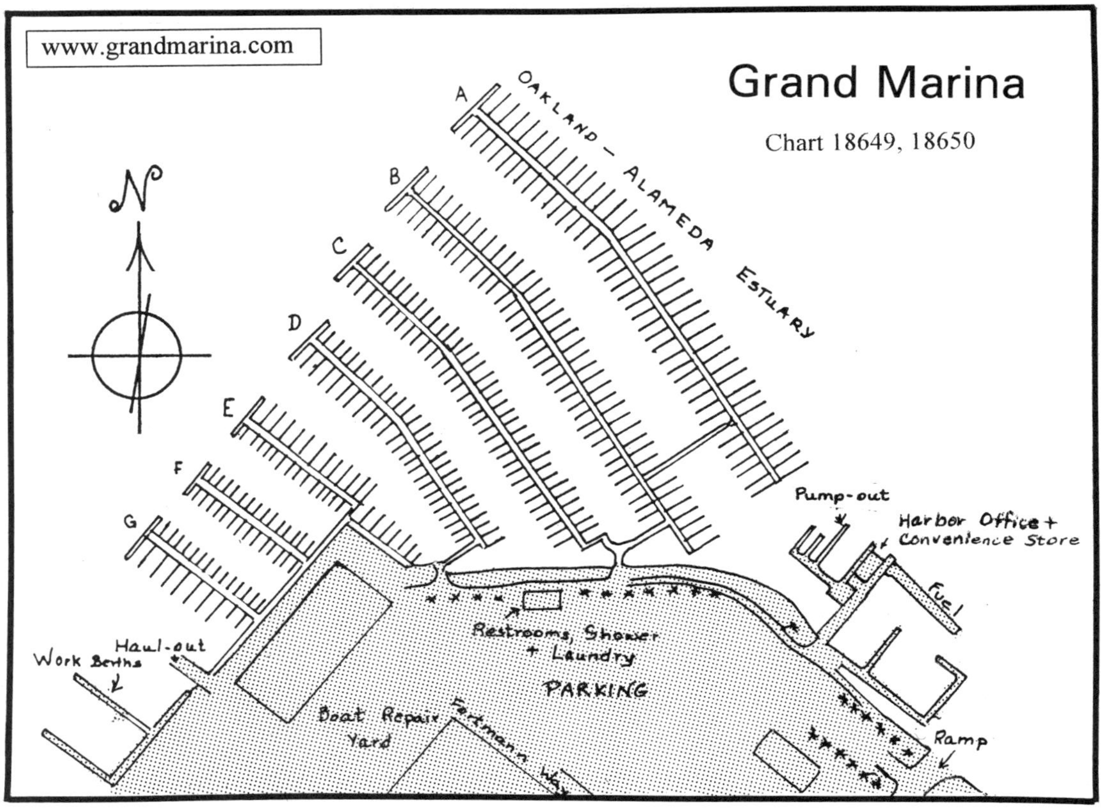

Alameda Marina has 530 slips and is located on the water side of one of the larger marine complexes on Alameda Island along with Svendsen's Boatyard and Chandlery. The marina offers guest berthing on an "as available" basis. Fees charged cover electricity, showers and live aboard fees. A laundry is on site and nearby are two sailmakers, canvas shops, chart and navigation supplies and independent mechanical and electrical services. Minimal groceries are available two blocks away while major grocery shopping is two miles distant. Internet access is found at Java Rama or Starbucks two miles away.

The **Oakland Yacht Club** in Alameda at Pacific Marina and **Encinal Yacht Club** welcome yachts from reciprocating clubs. They are both located at Pacific Marina which is east of Marina Village.

Ballena Isle Marina with over 450 slips is also on Alameda Island but on the South Bay side along with Nelson's Boat Yard at Alameda Point. Facilities include pump-out, showers, laundry, deli, fuel and propane. Grocery stores are less than a mile distant. A restaurant with dinner service is closed on Monday. Transient berths are available at reasonable rates. Clearly marked channels for vessels with a maximum draft of 2.1m (7 ft.) lead to these facilities. Due to shallow water it is important to stay in the channels when approaching the marina. Website – www.ballenaisle.com E-mail – slips@ballenaisle.com

USEFUL CONTACTS IN ALAMEDA

Boatyards

Mariner Boatyard, Inc. at Grand Marina
2021 Alaska Packer Place Tel: (510) 521-6100

Nelson's Marine at Alameda Point Website - www.Nelsonsmarine.com
(Allows limited number of do-it-yourselfers with certain limitations)
1500 Ferry Point Tel: (510) 814-1858

Stone's Boatyard
2517 Blanding Avenue Tel: (510) 523-3030

Svendsen's Boat Works (Alameda Marina) Website – www.svendsens.com
1851 Clement Avenue Tel: Chandlery: (510) 521-8454
Boat yard: (510) 522-8889

Chandlers

Bosun's Locker - Ballena Bay
Nelson's Marine - Alameda Point
Svendsen's Boat Works – Alameda Marina
West Marine - Buena Vista and Constitution Way

Propane Source

Ballena Isle Marina, 1150 Ballena Blvd., #111 www.Ballenaisle.com
Tel. (510) 523-5528 1 – 800 – 675 - slip

Internet Access

Jara Rama – DSL Line, 1333 Park Street
Kinko's, 750 Atlantic Avenue
Main Library, 2200A Central Avenue
Starbuck's (Broadband – hot spot), 700 Atlantic Ave. and 1364 Park Street

OAKLAND

The Port of Oakland operates a number of marinas on the eastern shore opposite Alameda. The office is located at 38 Webster St. and hours of operation are 8 am to noon. For information on the following call 800-675-DOCK or (510) 834-4591.

Website – www.portofoakland.com/realestateforrent/berths

Jack London Marina (East, Center and West Basins) have over 160 slips. Facilities include a fuel dock, showers, pump-out and a farmer's market on Sundays. A laundry is at the Centre Basin.

Embarcadero Cove Marinas (North Basins I and II) has 109 slips and can accommodate vessels up to 13m (52'). Facilities include restrooms, showers and pump-out.

Union Point Basin has 92 slips and can handle vessels up to 18m (60') on end ties. Facilities include restrooms, showers and pump-out services. Several restaurants are nearby.

USEFUL CONTACT IN OAKLAND

Boatyard

British Marine, #11 Embarcadero Cove Tel: (510) 534-2757, 800-400-2757

TREASURE ISLAND

Treasure Isle Marina has 110 slips and a few guest slips available by reservation only. Maximum depth at the guest dock is 1.5m (5'). An anchorage is nearby but there is no access to the island through the marina. Website – www.treasure-isle.com E-mail – slips@treasure-isle.com

You could spend many days visiting the varied attractions in the Bay and surrounding areas. The downtown points of interest include Fisherman's Wharf, Chinatown, the Palace of Fine Arts, the Cannery, Alcatraz, Coit Tower, Ghiradelli Square, the shops and restaurants of Pier 39, Cable Cars and even the unique architecture of the houses of San Francisco. By traveling further afield, interesting day trips can be taken to the wine country in Napa Valley, Sausilito and Muir Woods Park and to the state capital at Sacramento, where many museums and historic buildings are found.

YACHT CLUBS WITH GUEST DOCKS

St. Francis Y.C.	(415) 563-6363
Golden Gate Y.C.	(415) 346-2628
Sausalito Y.C. (Sausalito)	(415) 332-7400
San Francisco Y.C. (Belvedere)	(415) 332-9922
Corinthian Y.C. (Tiberon)	(415) 435-4771
Richmond Y.C. (Richmond)	(510) 237-2821
Berkeley Y.C. (Berkeley)	(510) 540-9167
Oakland Y.C. (Alameda)	(510) 522-6868
Encinal Y.C. (Alameda)	(510) 522-3272
Vallejo Y.C. (Vallejo)	(707) 643-1254
Benicia Y.C. (Benicia)	(707) 746-6600

The Trans America Pyramid is 260 meters (853 feet) tall and has been a San Francisco landmark since 1972.

HALF MOON BAY (PILLAR POINT HARBOR)

A popular stop-over, Half Moon Bay allows southbound boats a rest from the excitement of San Francisco and northbound vessels a short trip prior to entry at the Golden Gate. Half Moon Bay is about 20 miles south of San Francisco and 46 miles north of Santa Cruz. This well protected harbor lies in a bight east of Pillar Point with the coast curving in a southeasterly direction in a long beach backed by sand dunes.

Pillar Point is easy to recognize by the conspicuous radar tower near the tip of the point. By giving the point a clearance of 1.25 miles you can avoid the numerous rocks and kelp that lie in an arc from southwest to southeast of the point. Buoys "1" and "3" mark the seaward edge of this dangerous area. About 1.5 miles SSE of the point is lighted buoy "2" which marks the northern extremity of Southeast Reef. This 1-mile long region of submerged and exposed rocks lies to the southeast and is marked at its southern end by lighted gong buoy "1S" (6 sec.). Do not confuse this gong buoy "1S" with another gong buoy "1" mentioned in a following paragraph which has a 2.5 second light.

The outer part of the harbor is protected by two long, angular breakwaters: the west breakwater extending from Pillar Point, has a white light (visible over the reef) and fog signal at the end, while the east breakwater, built out from the beach, is marked by a red light at the seaward end. The inner harbor is protected by two breakwaters reaching out from the shore and a detached breakwater which lies across the southern side. There is a light at each end of the detached breakwater.

When approaching from the north, pick up lighted gong buoy "1" (2.5 sec.) about .75 mile southeast of Pillar Point. Proceed in an ESE direction for about 1 mile to bell buoy "3" before turning north to the harbor entrance. On making an approach from the south you must identify and clear Southeast Reef before picking up bell buoy "3" and continuing north to the entrance. Don't confuse the white and green light at the airport about one mile to the north with the flashing 5- second light at the end of the breakwater.

Dense fog can appear quickly, making the harbor entrance hazardous, especially if one's navigation is rather approximate. It is essential to update the vessel's position frequently as this is a tricky entrance, especially in foggy conditions.

Pillar Point Harbor is a pleasant place with showers, restrooms and restaurants. To arrange moorage contact the harbormaster on VHF Ch 16 and 74 or telephone (650) 726-4382 or Fax (650) 726-7740. Some end ties, berths and mooring buoys are available to transients at a reasonable cost. Anchorage may be taken as indicated and a dinghy dock is convenient near the ramp leading to the harbormaster's office. Fuel is available from 8 a.m. to 5:30 p.m. Marine supplies are available at The Gear Store, Tel (650) 728-3055. Propane is available in town, about 4 miles away. Computer modem access is available at the harbor office.

Pillar Point Marsh, a wildlife habitat, is a comfortable walk away where herons, egrets and several other species of waders can be seen. Another enjoyable hike is via the 5-mile Coastline Trail along the shores of the bay to the old town of Half Moon Bay, established in 1840. Dinner ashore can be enjoyed at a variety of restaurants.

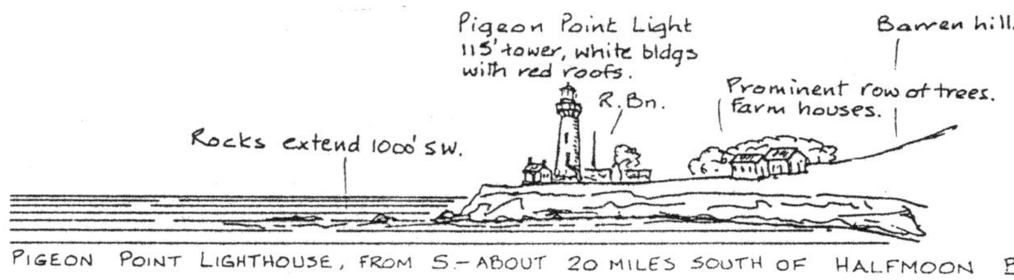

PIGEON POINT LIGHTHOUSE, FROM S – ABOUT 20 MILES SOUTH OF HALFMOON BAY

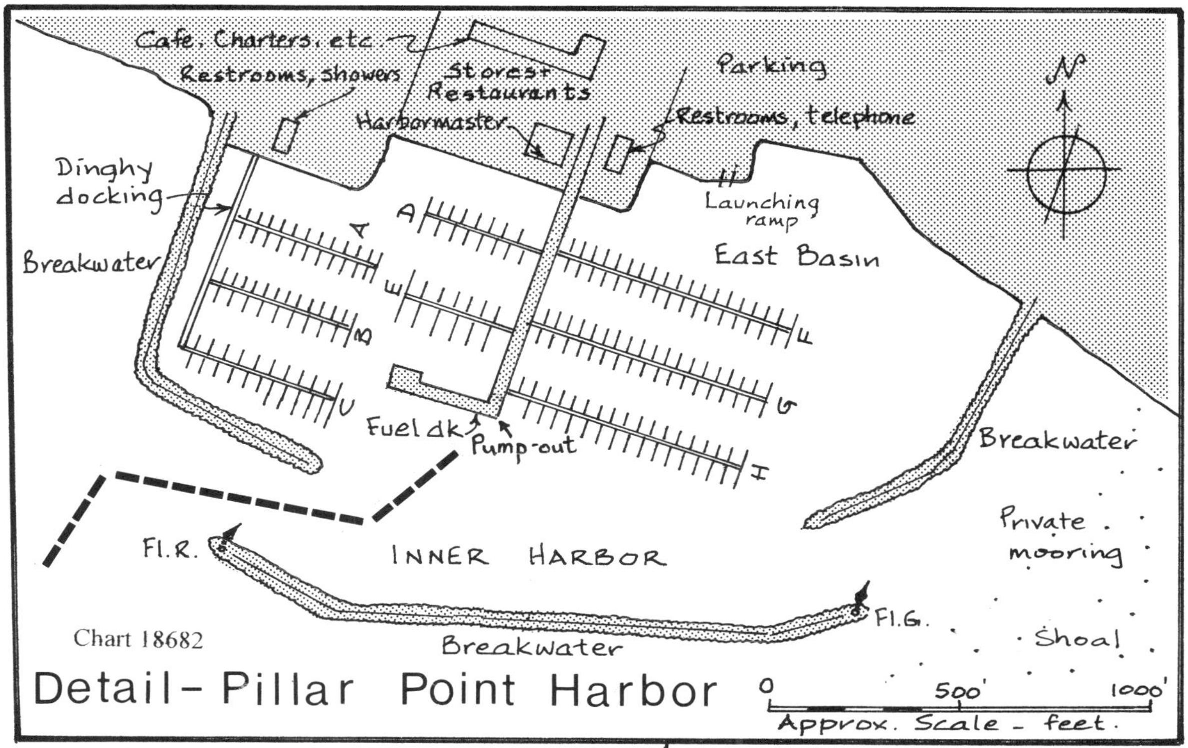

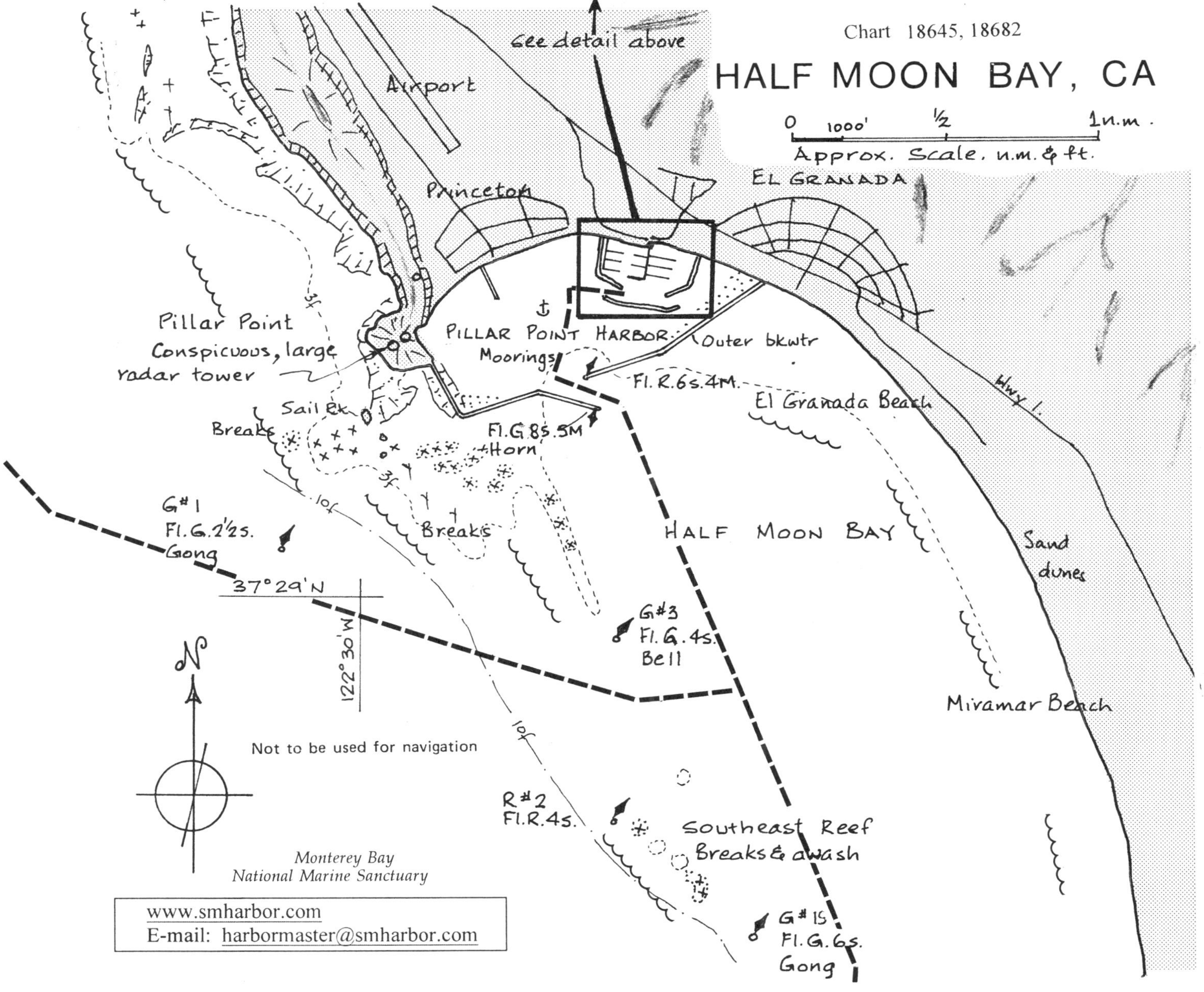

www.smharbor.com
E-mail: harbormaster@smharbor.com

SANTA CRUZ HARBOR and SOQUEL COVE

This year-long vacation area owes its popularity to magnificent beaches and a climate that is made for enjoyment of water sports and beach activities. Santa Cruz lies at the northern end of Monterey Bay between Point Santa Cruz on the west and Black Point on the east. It is about 45 miles south of Half Moon Bay and 18 miles from Moss Landing. Easily identifiable landmarks are the Casino, amusement park and the .4-mile long pier.

Founded in 1791 by the Franciscans, Santa Cruz became one of the series of 21 missions linking San Francisco and San Diego. Another note of historical interest is that in 1881 loganberries originated here, in the garden of Judge J. Logan.

With the exception of the swimming and pier areas indicated on the sketch, anchorage may be taken anywhere within the harbor limits in 6 fathoms, sand, though in northwesterly weather the surge is uncomfortable. A maximum speed of 5 knots within the harbor limits is closely monitored. Boats are prohibited from the outermost 100 yards of the pier and the swimming areas, which are clearly marked by buoys.

About three miles to the east around Soquel Point, is Soquel Cove, which marks the northern part of Monterey Bay. Extending into the bay is a long wharf where Capitola Boat & Bait sells ice, fishing supplies and souvenirs. The Capitola Bay Marina consists of about 50 seasonal (May 1 - September 30) **mooring buoys** set in an arc off the end of the wharf. Rental arrangements may be made by calling the Harbormaster on VHF Ch 11 or by phoning (831) 462-2208. In addition to daily shore boat service from sunrise till 9 p.m,. a dinghy dock is available for easy access to the wharf. An open air shower is behind the store on the wharf and laundry. Restaurants and shops of every kind are found in Capitola.

From July to September, heavy fog is experienced in Monterey Bay and along the coast 10 to 15 days per month. Close to shore the best conditions are normally found in the early afternoon, with clouds increasing and descending in the evening and sunrise bringing the commencement of gradually improving conditions. In the vicinity of Santa Cruz and Monterey Harbor, fog may reduce visibility to less than .5 mile up to 8 days per month during the summer months.

Prevailing winds are west to northwesterly during the summer and east in late fall and winter months. The strongest winter winds are usually from the south. Gusts of up to 70 knots can be experienced during winter and early spring.

The Monterey Bay National Marine Sanctuary is the largest protected coastal area in the U.S., stretching more than 300 miles from the Gulf of the Farallones National Marine Sanctuary near San Francisco to south of San Simeon. It covers the area from high tide to 53 miles offshore. Within the sanctuary, the discharge of any materials is prohibited, as well as injuring or harassing marine mammals, turtles and seabirds.

SANTA CRUZ, CA

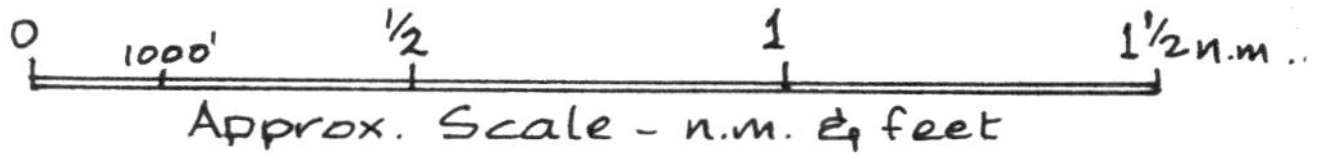

Not to be used for navigation

Chart 18680, 18685

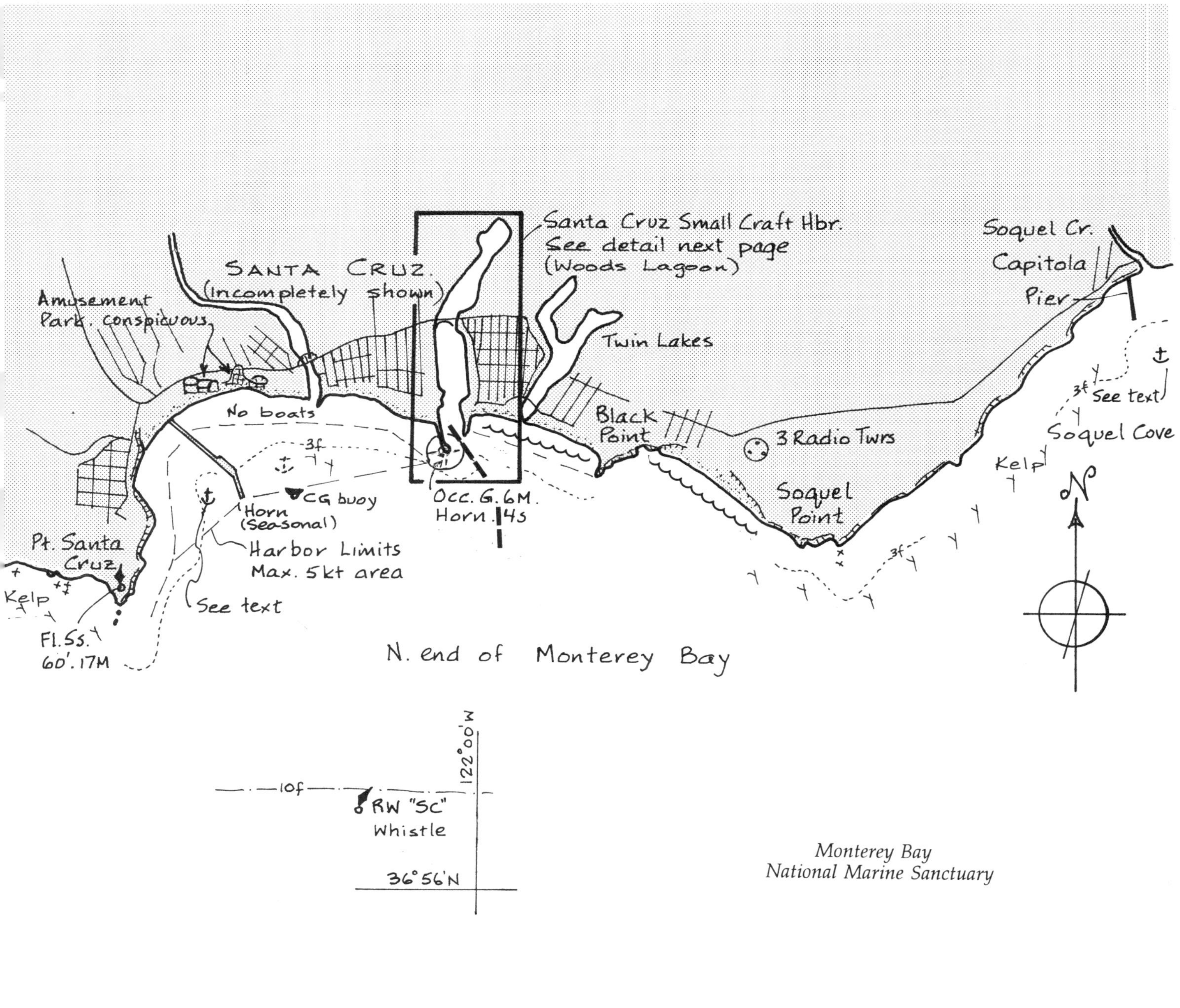

Monterey Bay
National Marine Sanctuary

SANTA CRUZ SMALL CRAFT HARBOR

Entry into the Small Craft Harbor (Woods Lagoon) is made by way of a channel between two jetties. The shorter eastern jetty extends outward from Twin Lakes Beach. The longer western jetty turns southeastward near the tip, where a lighthouse with a green light and fog signal are located. A bridge separating the lower basin from the upper basin has a clearance of 5.5m (18 ft.). Do not enter if a heavy swell is running, as the shoaling in the vicinity of the entrance causes dangerous breakers.

It is recommended that transient vessels communicate with the Port Director on VHF Ch. 9 or 16 to check shoal and sea conditions at the entrance prior to entering the harbor. A patrol boat also monitors the same channels and will guide vessels into the harbor on request.

Shoaling in the vicinity of the entrance is a constant challenge for the harbor, especially during winter months. The dredge *Sea Bright* is usually working near the entrance from November to April and when dredging is being done the entrance buoys are removed. If it is necessary to contact the crew they can be called on VHF Ch. 9 or 16. It is interesting to note that shoaling during the winter storms of 1969-70 completely blocked the entrance and caused boats within the harbor to be temporarily land-locked. Then, in 1977 shoaling extended about three-quarters of the way across the entrance before dredging cleared the channel.

The **Small Craft Harbor** can accommodate 950 wet berths and 275 in dry storage. On average there are about 50 coastal travelers per day passing through this area during the summer months. The end ties are reserved for transients, but as a result of heavy traffic no reservations can be made for these berths, which are filled on a first-come, first-serve basis. Weekends and holidays often find transient boats rafted up. The harbor office is open from 8 a.m. to 5 p.m.daily and VHF Ch. 16 is monitored 24 hours. For information telephone (831-474-6161, Fax (831) 474-9558 or after 5 p.m. call the cell phone at (831) 212-4261.

A fuel dock, pump-out station, showers, restaurants, restrooms and Harbormaster's Office are all conveniently located as shown. Laundry facilities are a long block away on Seabright Avenue. Marine repairs and supplies are available from a number of shops near the harbor. Contact Harbor Marina at (831) 475-3131 to arrange for hauling your vessel. Marine supplies are available at Harbor Marine located in the harbor or at West Marine, a taxi ride away. A few groceries are available at convenience stores in the harbor but for major shopping plan on a good walk. Computer modem access is available at the Harbor Office.

For a change of pace, the crew might enjoy a day at the waterfront amusement park which was established in 1907. It isn't every day that you can take a ride on a landmark, in this case the roller coaster that was so conspicuous from sea. Surfers find Twin Lakes Beach a good place to catch a wave.

Detail of

Santa Cruz

Small Craft Harbor

Not to be used for navigation

www.santacruzharbor.org
E-mail: scpd@santacruzharbor.org

Chart 18685

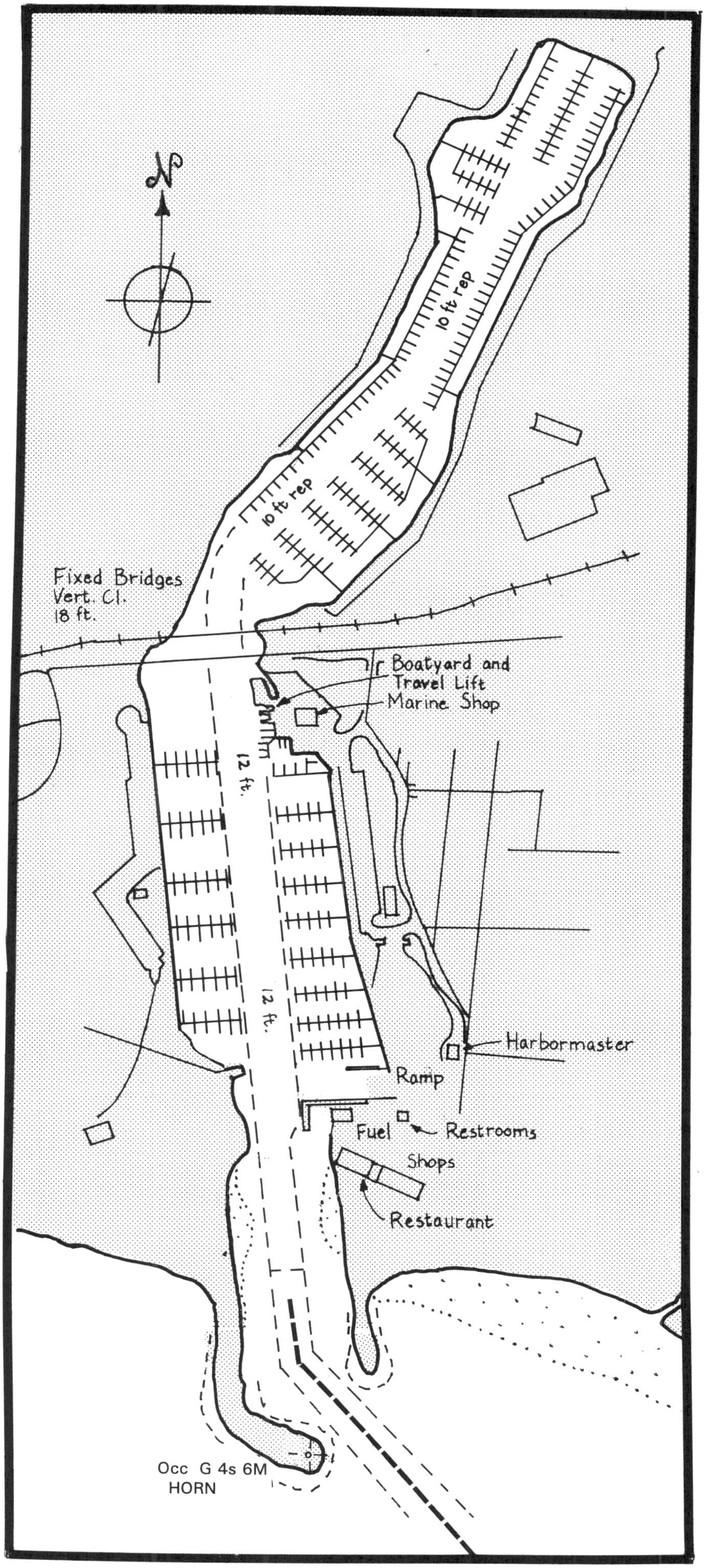

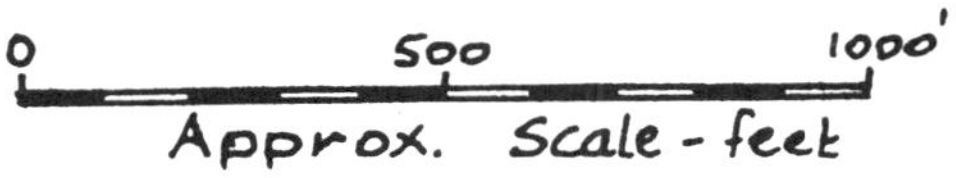

MOSS LANDING

This **harbor of refuge** is roughly mid-way along on the east side of Monterey Bay, about 18 miles distant from Santa Cruz and 15 miles northeast of Monterey.

There are several prominent features which easily identify Moss Landing: two huge power plant stacks (about 160m/528 ft.), and two radio towers located 4 miles north of the town.

When approaching, pick up lighted bell buoy "MLA" which is about .75 mile southwest of the entrance to the channel. A private buoys almost a mile northwest of the harbor entrance marks a submerged pipeline which supplies fuel to the power plant. A short distance northeast of the south jetty is a fog signal. A 5-mile per hour speed limit in the harbor is strictly enforced and anchoring is not permitted.

Two lighted, short (4.5m/15 ft.) jetties protect the entrance to the channel which leads to a turning basin within the harbor. When entering, favor the north side of the channel, as shoaling occurs on the south side. Once inside, there is a buoyed channel to the North Harbor which is maintained by the **Elkhorn Yacht Club**. Moorage for 48 hours is available to yacht club members with reciprocal privileges. The clubhouse and slips are located at the far end. At dockside you are treated to a continuous wildlife show as otters, seals, sea lions, cormorants, brown and white pelicans, and several other species of birds frequent the area.

Turning to the south, a dredged channel (well marked with lights, buoys and a daybeacon) leads to the **Moss Landing Harbor.** The Harbormaster monitors VHF Chs. 16 and 9; the office is open from 8 a.m. to 5:00 p.m., daily. It is advisable to contact them prior to entering the harbor. For information call (831) 633-2461 or Fax (831) 633-4537. If entry is made after office hours check in with the Harbor Patrol. Facilities include showers ($50 key deposit), laundromat, fuel, bilge and sewage pump-out and marine repairs. Propane is available at Moss Landing Boat Works and ice can be obtained at fish buyers' facilities. A few groceries are sold at M & L Liquors but for major shopping it is 2.5 miles to Castroville. Haul-out facilities are at Gravelle Boat Yard, telephone (831) 633-2417. Many cruisers have found this to be a friendly place to visit.

By taking a dinghy trip beyond the highway bridge you can explore a maze of tidal creeks in Elkhorn Slough, an ecological reserve and one of the largest wetlands in California. A walk near any of the local marshes (even along the road bordering North Harbor) is enjoyable whether or not you are a bird-watcher.

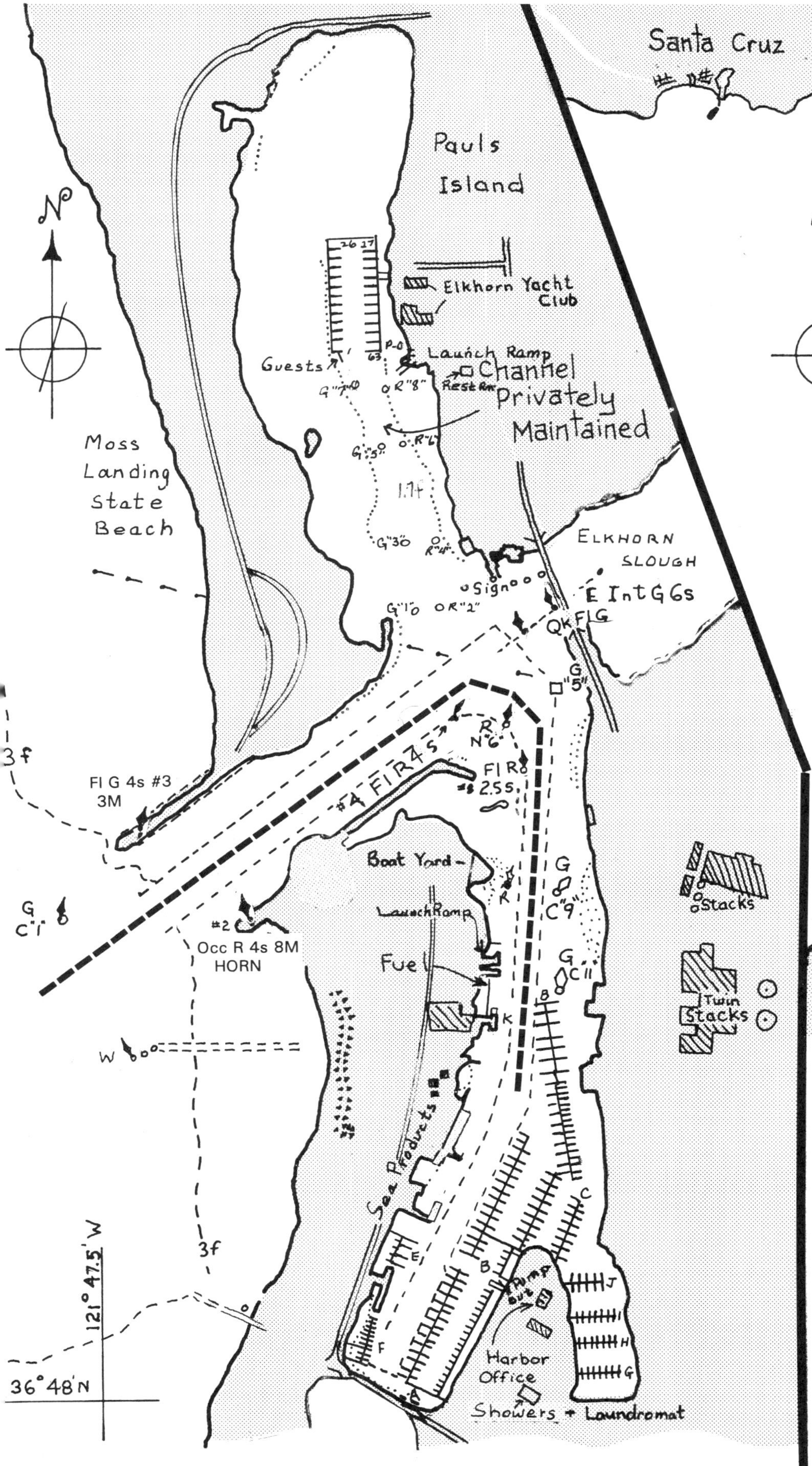

MOSS LANDING HARBOR

0 500 1000'
Approx. Scale - feet

Chart 18685

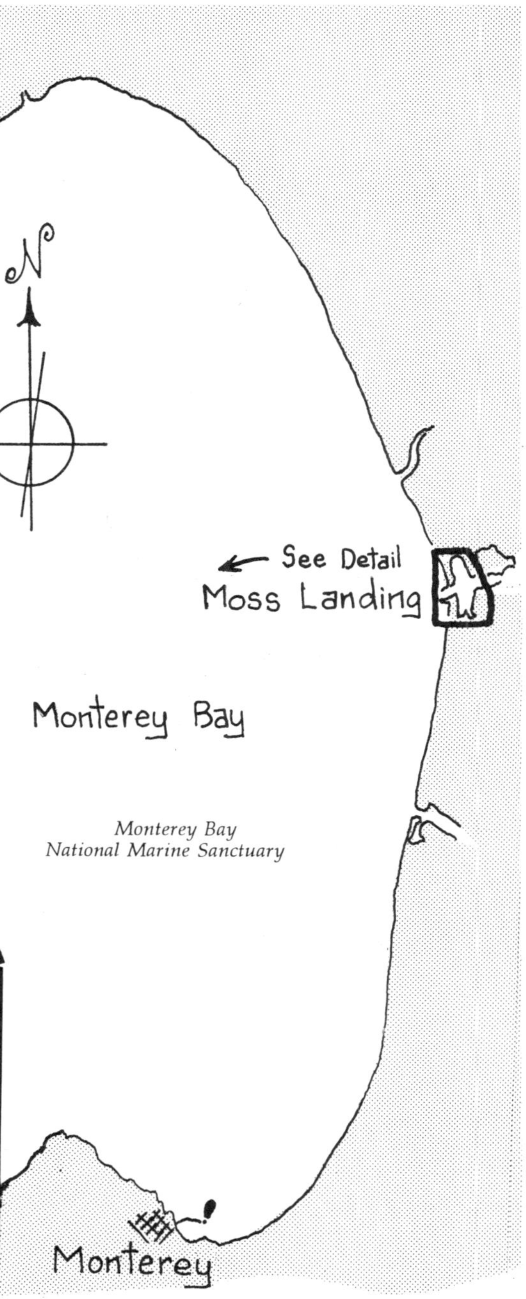

LOCATION SKETCH

0 5
Approx. Scale - n.m.

Not to be used for navigation

Chart 18680, 18685

www.mosslandingharbor.dst.ca.us

MONTEREY

This picturesque city is located within the hook at the southern end of Monterey Bay and is about 20 miles from Santa Cruz and 12 miles from Moss Landing. Other adjacent populations centers are Del Monte and Seaside to the east and Pacific Grove to the NW.

The entrance to Monterey Harbor is almost 4 miles southeast of the red #2 bell buoy located .5 mile off the outermost rocks off Point Pinos. The red #4 bell buoy located about .3 miles off Point Cabrillo is about 1 mile NW of the entrance to the harbor. A series of buoys indicates the route to be followed when entering the harbor. Forecasts are available on W1.

A 518m (1,700 ft.) rock breakwater (Coast Guard Pier) gives the harbor protection from NWly winds. A light and fog signal are at the tip of the breakwater, while at the inner end is the Coast Guard work yard and the office is across the street. The Coast Guard can be contacted on VHF Ch 16 or at (831) 647-7302. Built out from the southern shore is Wharf #2 which forms the eastern boundary of the harbor. Within the harbor is Wharf #1 (Fisherman's Wharf) which houses many small shops, cafes and boat charters.

The **Monterey Municipal Marina** (website: www.monterey.org)has over 400 berths and is located conveniently close to downtown. When you are about 5 miles from Monterey, call the Harbormaster on VHF Ch. 5A or 16 to arrange for moorage on a first-come, first serve basis—no reservations are taken; phone (831) 646-3950, Fax (831) 5674 or email scheibla@ci.monterey.ca.us Office hours are 8 to 5 daily; after hours call (831) 594-7760. A refundable deposit is required for each key needed for showers and gates to the wharves. Pump-out facilities are on the west side of Wharf #2, north of the concrete frontal wall; call the Harbor office to activate. Convenience stores are handy but for grocery shopping Safeways is about .5 mile distant. The library provides internet access, the office has computer modem access and wireless internet service is available in many locations, call (831) 277-6686 for information. Moorage for members of reciprocal yacht clubs is available at the yacht club docks when space permits. To arrange moorage call (831) 372-9686. Good but rolly anchorage may be taken in the area east of Wharf #2, 2 - 5 fathoms, sand and clay. Anchorage may be taken east of the municipal buoys for up to 30 days in a six-month period

If you prefer to be closer to the tourist attractions near Fisherman's Wharf, **Breakwater Cove Marina** has transient berths where reservations can be made by calling VHF Ch 9, phoning (831) 373-7857 or Fax (831) 373-2294. Fuel is available at the Breakwater Cove fuel dock. Propane can obtained at Monterey Beacon (831-649-5105), 2191 N. Fremont, at Hodges Rent-All, 528 Foam, (831-372-7381), The Compass, West Marine Express and at the 76 gas station. Block Ice is available at La Casa Bodega, 500 Del Monte Ave. (6 a.m. - 11 p.m.)

From a historical stand point alone, this is a stop-over of major significance. The area was first visited by Europeans in 1602 by Vizcaino, and Monterey was the capital of California while under Mexican rule until 1846. *The Path of Glory* walking tour is a fascinating visit to various original and reconstructed buildings related to Monterey's colorful past. In addition, numerous art galleries feature local artists, many of whom are known world-wide. The Monterey aquarium is spectacular, featuring an immense underwater viewing gallery.

The well-known 17-Mile Drive commences in Pacific Grove and follows the coast (via Cypress Point and the Pebble Beach Golf Course) to Carmel. When westerlies are blowing good anchorage can be taken in Carmel Bay at **Stillwater Cove** just to the east of Pebble Beach. Watch for scattered rocks near the entrance to the cove. Leave if southwest winds over 10 knots develop making this a lee shore..

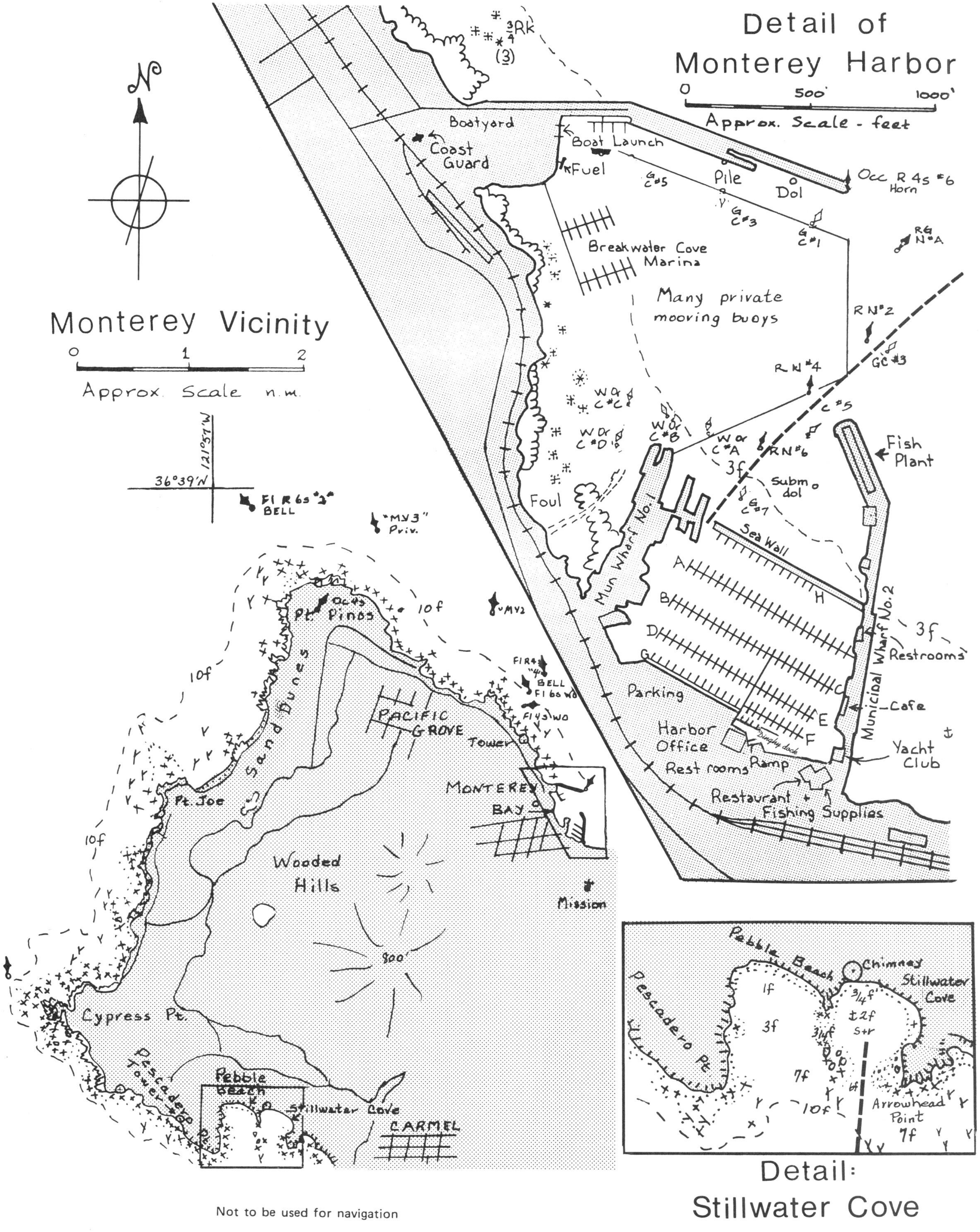

Not to be used for navigation

SAN SIMEON BAY

For cruisers interested in visiting the unique Hearst San Simeon State Historical Monument, the anchorage at San Simeon Bay is ideally located. The bay is a circular bight northeast of San Simeon Point, 85 miles southeast of Monterey and 23 miles northwest of Morro Bay.

When the vessel is within 6 miles of San Simeon Point (on an approach from the north) one sights Point Piedras Blancas, which has a light on a white conical tower at a height of about 43m (142 ft.). San Simeon Bay is about 27 miles northwest of Point Buchon. The bay can be readily identified by the conspicuous buildings of Hearst Castle perched high on the hill.

When approaching the bay, pick up lighted bell buoy "1," which is about .5 mile southeast of Point San Simeon and proceed into the bay. Because rocks and kelp reach out from the point almost to the buoy, it must be passed to port. Anchorage may be taken in the bight south of San Simeon Creek 5 to 8 fathoms, sand. Private mooring buoys are seasonally located off the private beach. There is good protection from northwesterlies, though it is open to the south and there is some surge.

A sportfishing pier extends out from the northern shore of the bay. Surf breaking on the shore sometimes presents a challenge to avoid getting swamped or rolling the dinghy. The village of San Simeon has a general store dating from 1852 where a few groceries and souvenirs can be obtained and an open-air eating establishment is nearby. A better source for groceries is seven miles away at Cambridge. Hearst built the white adobe building with a bell tower for storing some of his artistic collection. Fuel and kerosene may be obtained from the gas station which is operated by the store.

Hearst Castle, a California State Historical Monument built from 1919 to 1947, is prominently located on high ground about 3 miles northeast of the town. The 165-room castle was the home of William Randolf Hearst, a colorful and wealthy newspaper publisher. Thousands of tourists visit the site, making it advisable to make reservations in order to take any of the four different tours of the castle. Within California reservations may be made from 9:00 to 6:00 by calling 1-800-446-PARK, (or from outside California at (619) 452-1950). Tour buses leave throughout the day from the parking lot near the store. Even without a tour, the interesting displays in the Visitor Center can give you an indication of what opulence could mean in the 1920s and 1930s.

California cruising offers the peace of a quiet anchorage such as San Simeon Bay (Page 104) to . . .

the congestion of a large facility such as the Municipal Downtown Marina at Long Beach (Page 138)

The sparkling white sand near the eastern end of Cuyler Harbor on San Miguel Island, contrasts with the deep blue waters of the bay. (Page 116)

The beautiful waters of Emerald Bay on Catalina Island, California make it a favorite spot for many cruisers. (Page 154)

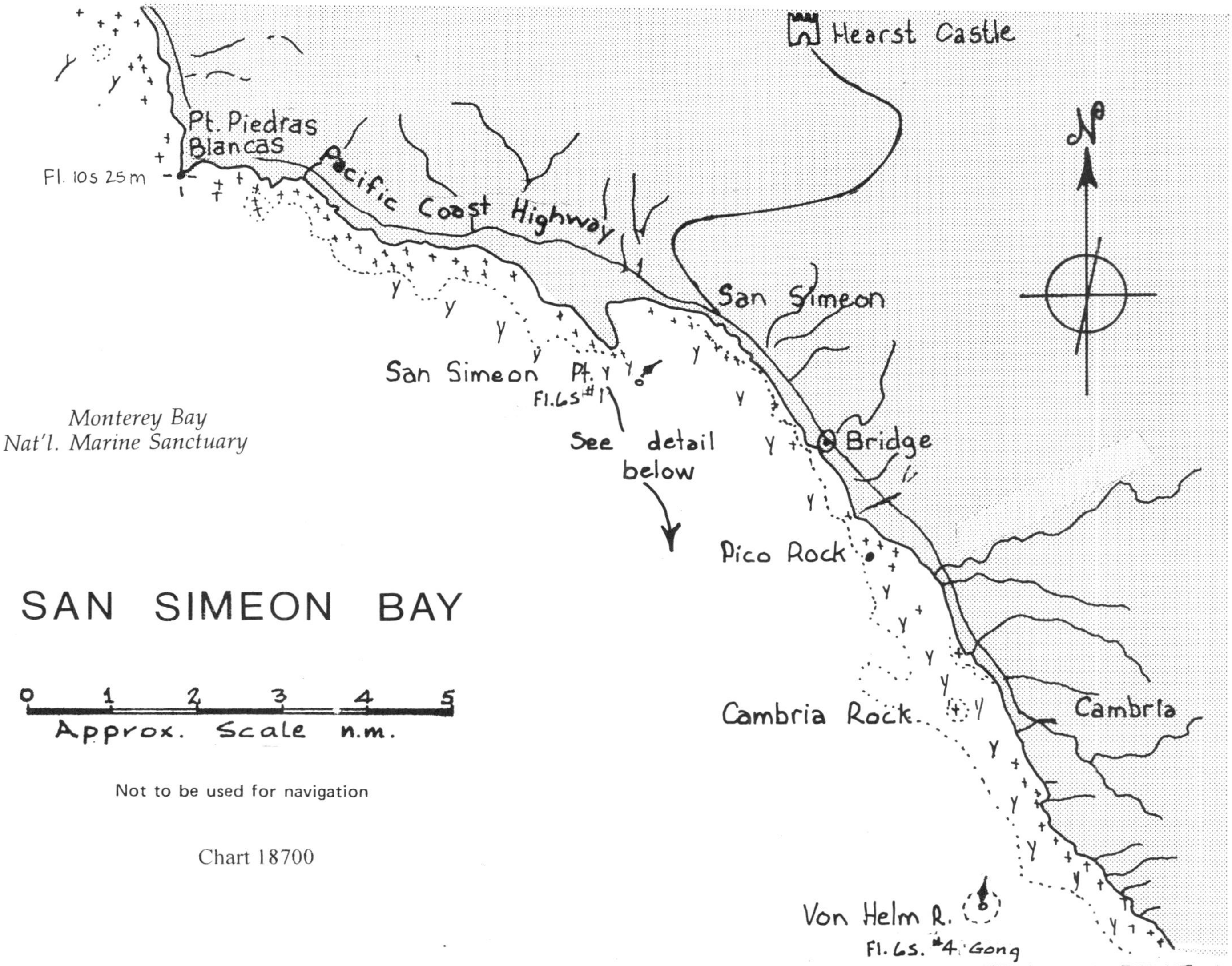
Hearst Castle
Pt. Piedras Blancas
Fl. 10s 25m
Pacific Coast Highway
San Simeon
San Simeon Pt.
Fl. 6s #1
See detail below
Bridge
Pico Rock
Monterey Bay
Nat'l. Marine Sanctuary
SAN SIMEON BAY
0 1 2 3 4 5
Approx. Scale n.m.
Not to be used for navigation
Chart 18700
Cambria Rock
Cambria
Von Helm R.
Fl. 6s. #4 Gong

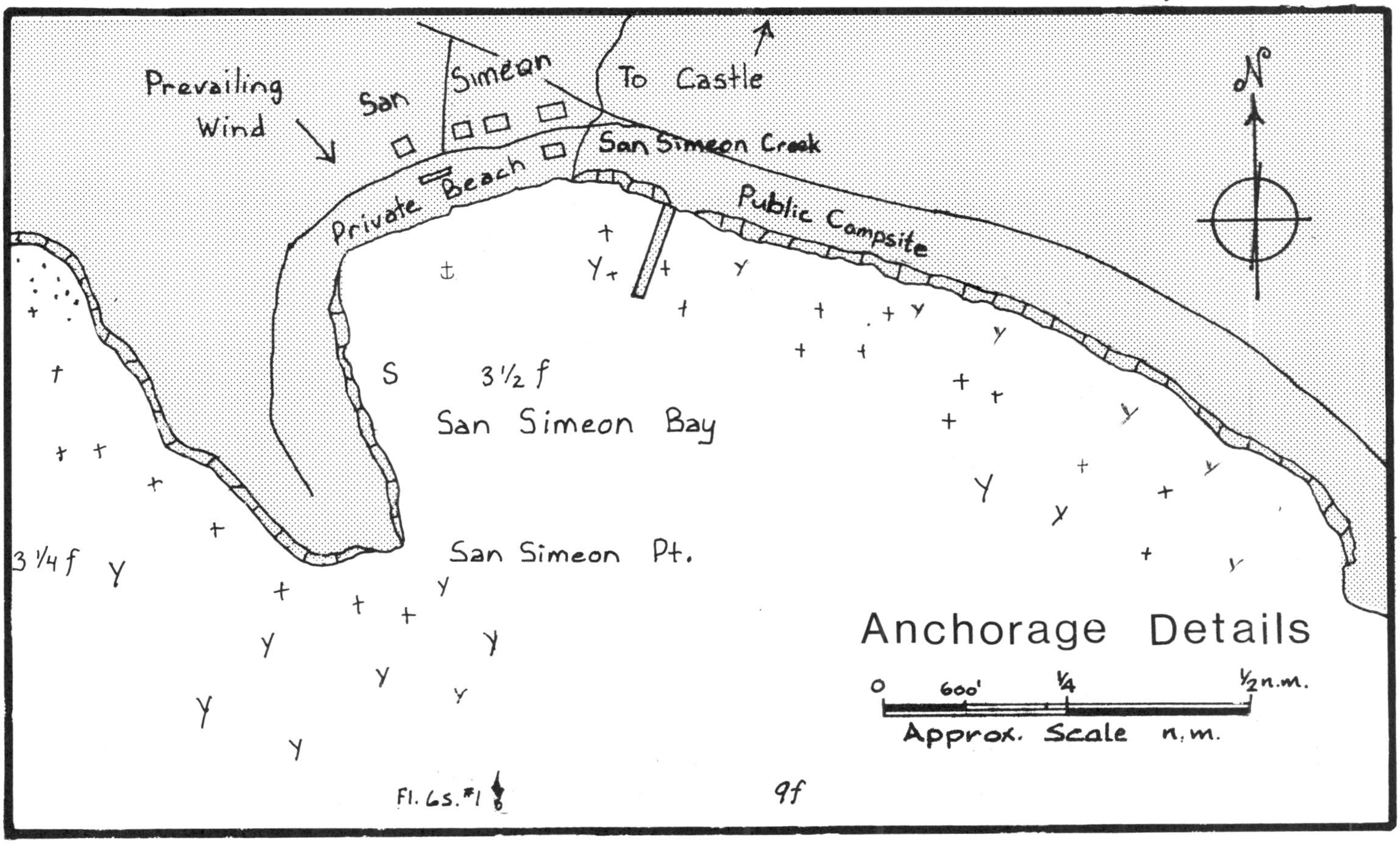
Prevailing Wind
San Simeon
To Castle
San Simeon Creek
Private Beach
Public Campsite
S
3½ f
San Simeon Bay
3¼ f
San Simeon Pt.
Anchorage Details
0 600' ¼ ½ n.m.
Approx. Scale n.m.
Fl. 6s. #1
9f

MORRO BAY

Located 23 miles south of San Simeon Bay and about 20 miles from Port San Luis, Morro Bay is a convenient stop-over. This area has three remarkable features: first, its dubious distinction as being one of the foggiest areas of the Pacific Coast, second, the dangers of its harbor entry and third, the unique Morro Rock, a bulky lava plug 58m (580 ft.) high. It is said that this monolithic landmark reminded a Spanish explorer of a Moor's turban so he named it "Morro Rock."

On a clear day Morro Rock and three nearby 137m (450 ft.) power stacks are excellent landmarks. Two 550m (1,800 ft.) breakwaters protect the entrance channel. The north breakwater extends south of Morro Rock and has a light and foghorn at its tip; the south breakwater is built out from the sand dunes that form the peninsula protecting the bay.

Prior to approaching the entrance it is essential that you contact the Coast Guard or Harbor Patrol on VHF Ch. 16 or telephone (805) 772-2167 to check entrance conditions. Special weather broadcasts are given on Ch. 16 and 22A. Twenty-four hour weather reports are available by calling (805) 772-4141. Breakers across the bar can be as dangerous as any on the coast, so this entrance must not be taken lightly. Steep, breaking seas form off the end of the breakwaters when heavy wind or swell is from the west to northwest especially at low water. Approach the entrance from offshore whistle buoy "MB," easy to identify with its white and red vertical stripes. Follow a course that is about 18m/60 ft. from the outer end of the north breakwater before turning into the channel entrance. The channel buoys are often moved to indicate the best water at the time, consequently the position shown on any chart are subject to change. Favor the north side of the channel until past Buoy #6, then a mid-channel course may be followed. During flood tides, currents in the north end of the bay set towards the city T-pier.

Anchorage may only be taken between Markers # 10 and #12; many permanent private mooring buoys are located beyond. Strong tidal currents necessitate a well-set anchor with plenty of scope. You may get permission to tie to the 15m (50 ft.) floating dock near buoy #12. The **Morro Bay Yacht Club** has been designated as a host facility for transients. It allows authorized club members to side tie at the dock and at the first row of buoys opposite the Yacht Club. The Y.C. harbormaster monitors VHF Ch. 16 or 68 from 4:30 to 6:00 p.m. or can be reached at (805) 772-3981.

All basic cruising repairs and supplies are available. The Coast Guard and Harbor Patrol are located in the gray two-story building across the street from the power stacks, near the first T-pier. Fuel is sold at the Shell dock (NE of #10) and pump-out stations are at Morro Bay Marina (E of #12), Tidelands Park and at the South "T" Pier (E of #8 and #10). Propane is available at Morro Bay RV and Marine. Restrooms and showers are at the harbor office, Yacht Club and Tidelands Park. Haul-out facilities at Coastal Boatworks can handle vessels up to about 12m (40 ft.). The nearest grocery stores are about five blocks away at Vons, Albertsons and Spencers. Internet services are available at Two Dogs, about two blocks distant, and computer modem access is available at the harbor office.

A walk over the sand dunes is often rewarded by finding some shells and in town, a self-guided tour leads to colorful murals on commercial buildings. A longer walk to the south leads to Morro Bay Estuary, temporary home to a large variety of birds that use this as a stop-over on their migratory passage.

The **State Park Marina** east of White Point is filled with local vessels, and transient berths are not available; in any case, it's a long distance out of the way.

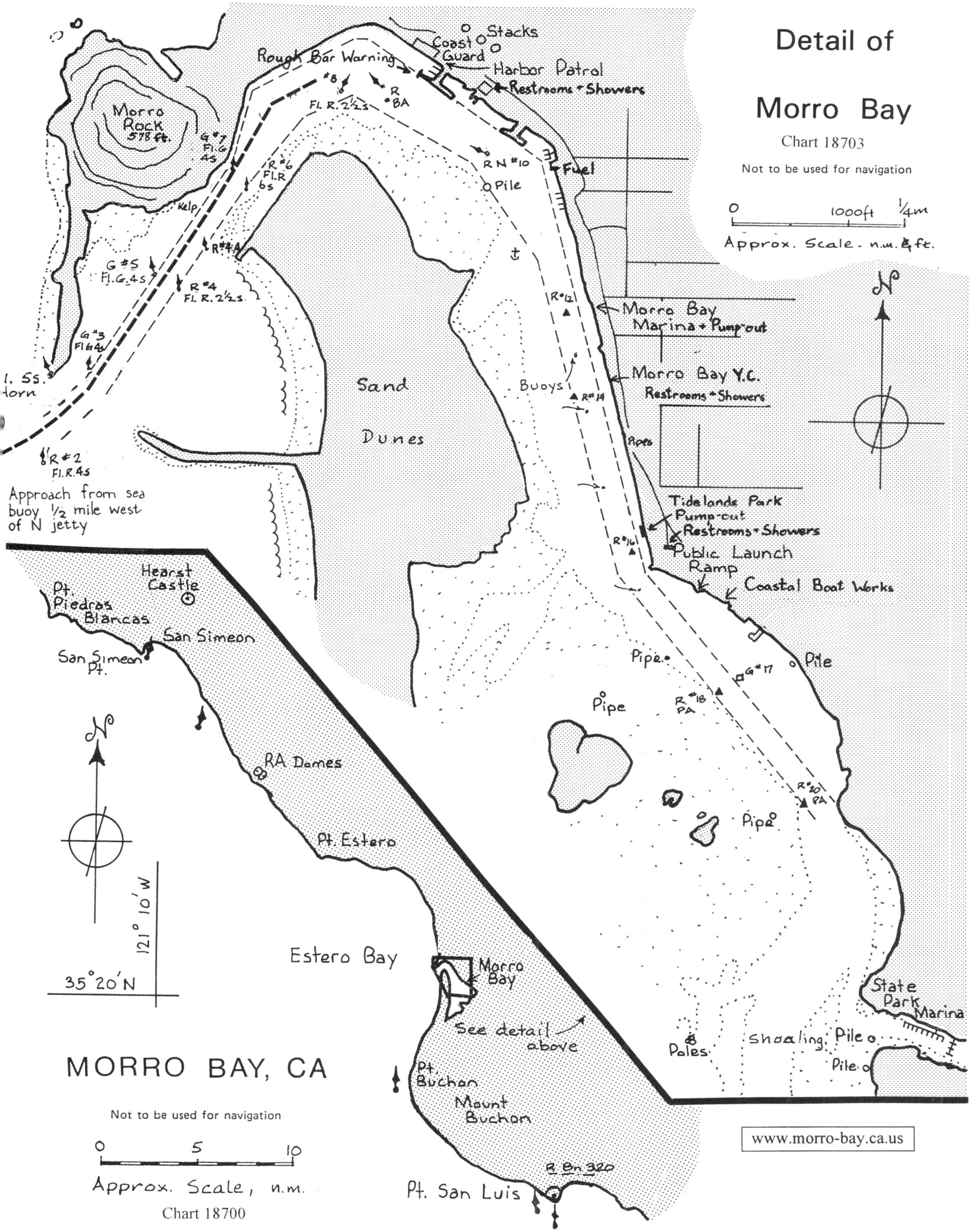

Detail of
Morro Bay
Chart 18703
Not to be used for navigation
0
1000ft
1/4m
Approx. Scale - n.m. & ft.
Stacks
Coast Guard
Harbor Patrol
Restrooms + Showers
Rough Bar Warning
Morro Rock 578 ft.
G #7 Fl.G 4s
#8
R "BA"
Fl. R. 2½s
R #6 Fl.R 6s
R N #10
Fuel
Pile
Kelp
R #4A
G #5 Fl.G.4s
R #4 Fl. R. 2½s
G #3 Fl G 4s
Fl. 5s. Horn
R #2 Fl.R.4s
Approach from sea buoy ½ mile west of N jetty
Sand
Dunes
R #12
Morro Bay Marina + Pump-out
Buoys
R #14
Morro Bay Y.C.
Restrooms + Showers
Pipes
Tidelands Park
Pump-out
Restrooms + Showers
R #16
Public Launch Ramp
Coastal Boat Works
Pipe
Pile
G #17
Pipe
R #18 PA
R #20 PA
Pipe
State Park Marina
Poles
Shoaling
Pile
Pile
Hearst Castle
Pt. Piedras Blancas
San Simeon
San Simeon Pt.
RA Dames
Pt. Estero
Estero Bay
Morro Bay
See detail above
121° 10' W
35° 20' N
Pt. Buchon
Mount Buchon
R Bn 320
Pt. San Luis
MORRO BAY, CA
Not to be used for navigation
0
5
10
Approx. Scale, n.m.
Chart 18700
www.morro-bay.ca.us

PORT SAN LUIS

This popular haven for pleasure and fishing boats is about 20 miles (around Point Buchon) from Morro Bay and over 100 miles (around Points Arguello and Conception) distant from Santa Barbara. San Luis Obispo Bay is a wide bight east of Point San Luis that provides excellent protection from the northwest winds of summer but southeasterly winter winds from November to March sweep the bay. This bold headland has a fog signal and light. The town of Port San Luis is tucked beyond the point on the west shore of the bay.

A number of rocks in the vicinity must be identified. Westdahl Rock, about 1.3 miles southwest of the San Luis Obispo Light is marked by a lighted bell buoy. Souza Rock, about 2 miles southeast of the light is identified by a lighted gong buoy. Howell Rock, about .5 mile east of the light is marked by a buoy. Lansing Rock (covered 5.5m/18 ft.) and Atlas Rock (covered 3.9m/13 ft.) are about .7 and .5 mile east of the light, respectively. Avoid the east part of the bay, where there are many rocks and much kelp.

When approaching from the north, vessels must clear the buoy marking Westdahl Rock and the lighted whistle buoy of the Point San Luis breakwater before proceeding into the anchorage. When approaching from the south, the vessel should pass about 91m (100 yards) west of the buoy marking Souza Rock, proceed about 2 miles until past Lansing Rock and then continue to the mooring area. The Port San Luis breakwater is 732m (2,400 ft.) long. Commercial fishermen use the wharf .5 mile to the north of Point San Luis.

Contact the Harbormaster on VHF Ch. 16 for use of one of the nine white seasonal mooring buoys for transients that can handle vessels from 11m to 26m (36 ft. to 85 ft.) for $10. Reservations are recommended on weekends. Buoys # 1 to #6, southwest of the first pair of fairway markers, and the first three buoys # 7 to #9 northeast of the second set of markers are reserved for visiting boaters. Seasonal yellow buoys for vessels less than 11m (36 ft.) located closer inshore (west of the Calpoly Pier) rent for $6. The buoys are rented on a first-come first-serve basis and vessels may stay for up to two weeks. Register and prepay in the office as soon as your vessel is secure. The orange buoys are for private use.

The marina office is open from 8 a.m. to noon and 1 p.m. to 4:30 p.m. The office monitors VHF Ch. 16, working channel, VHF Ch. 12. For information call (805) 595-5409 or Fax (805) 595-5404.

Anchorage may not be taken in the areas where buoys are located or on the west side of Harford Pier. The recommended anchorage area is between the County Wharf (Avila Pier) and the Calpoly Pier although patches of weeds and grass have been reported that make holding poor in spots. Shore facilities include washrooms, showers and some marine supplies and a launch ramp for small boats off the sand beach. Diesel is available on the west side of Harford Pier and gasoline is sold at Portside Marine (805) 595-7214 for shallow draft boats only. Portside Marine operates a 50-ton mobile hoist and boatyard (805) 595-7895. The nearest source of propane is 10 miles distant at Pismo Beach.

This is a beautiful spot to rest, for the hills form a nice backdrop to the anchorage. Tourist attractions include nearby hot springs and a county park at Avilo Beach. At the end of the old dock, originally built in the 1800s, is a restaurant that can provide a change from eating on the boat. The Coast Guard at Morro Bay provides service for Port San Luis and can be reached at VHF Ch 16 or (805) 772-2167.

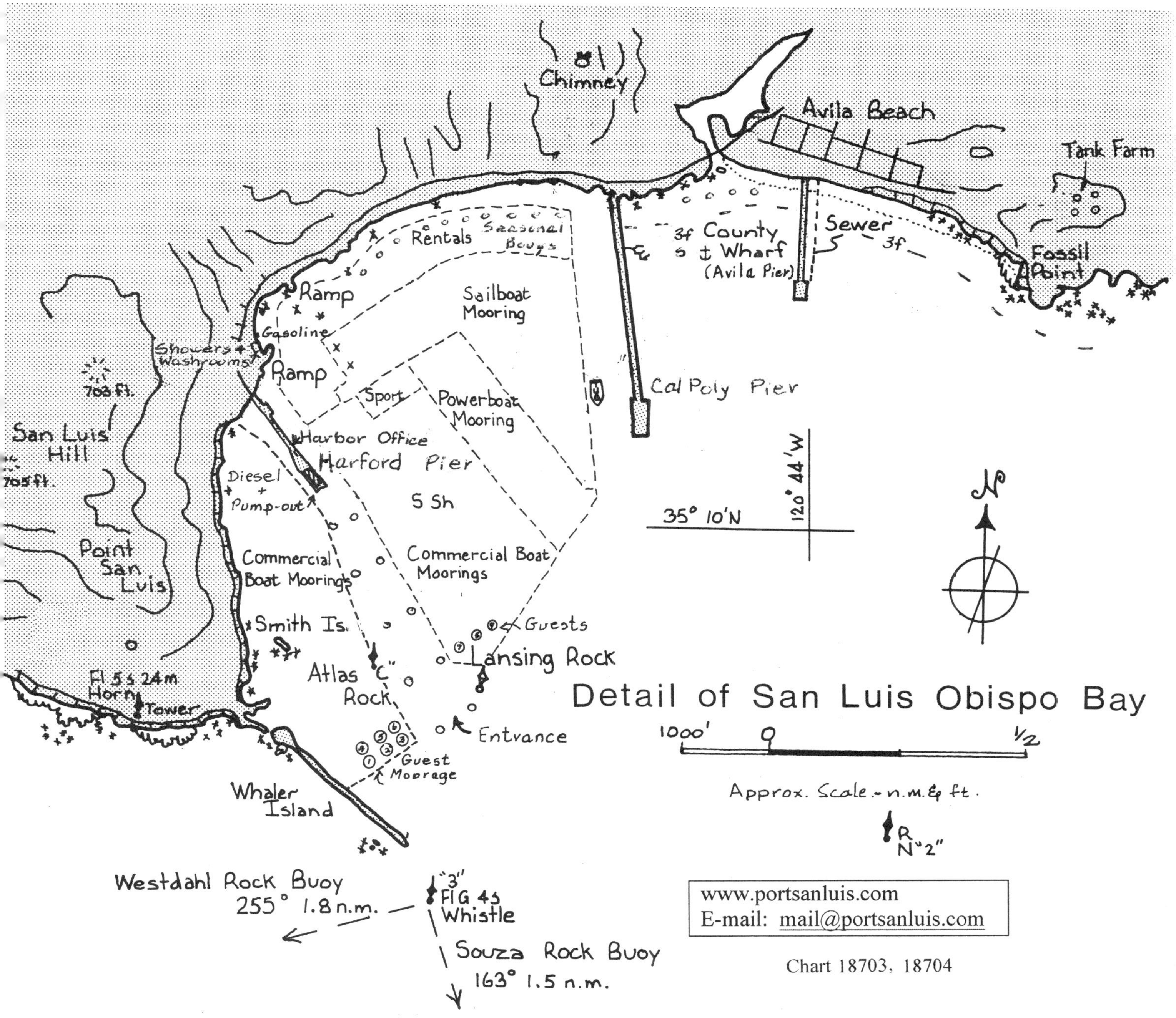
Detail of San Luis Obispo Bay
Chimney
Avila Beach
Tank Farm
Fossil Point
Sewer
3f
County Wharf
(Avila Pier)
Rentals
Seasonal Buoys
Ramp
Gasoline
Showers + Washrooms
Ramp
Sailboat Mooring
Sport
Powerboat Mooring
Cal Poly Pier
703 ft.
San Luis Hill
705 ft.
Harbor Office
Harford Pier
Diesel + Pump-out
5 Sh
Point San Luis
Commercial Boat Moorings
Commercial Boat Moorings
Smith Is.
Guests
Lansing Rock
Atlas Rock
"C"
Fl 5s 24M Horn
Tower
Entrance
Guest Moorage
Whaler Island
35° 10'N
120° 44'W
1000'
0
1/2
Approx. Scale - n.m. & ft.
R N"2"
Westdahl Rock Buoy 255° 1.8 n.m.
"3" Fl G 4s Whistle
Souza Rock Buoy 163° 1.5 n.m.
www.portsanluis.com
E-mail: mail@portsanluis.com
Chart 18703, 18704

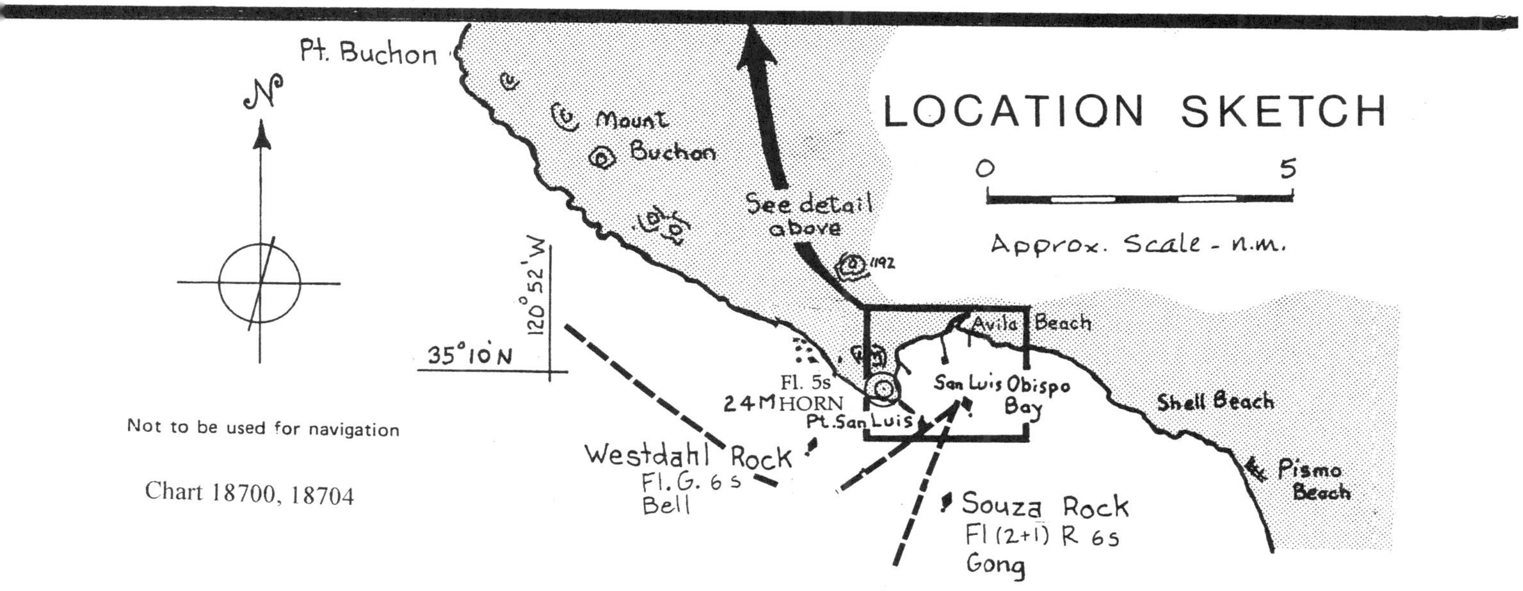
LOCATION SKETCH
Pt. Buchon
Mount Buchon
See detail above
1192
0
5
Approx. Scale - n.m.
120°52'W
35°10'N
Avila Beach
San Luis Obispo Bay
Shell Beach
Pismo Beach
Fl. 5s 24M HORN
Pt. San Luis
Westdahl Rock
Fl. G. 6s Bell
Souza Rock
Fl (2+1) R 6s Gong
Not to be used for navigation
Chart 18700, 18704

POINT ARGUELLO, POINT CONCEPTION and COJO ANCHORAGE

An abrupt change in the coastline occurs in this area. South of Point Conception the NW/SE'ly trend of the coast changes to easterly. Partly as a result of the Venturi effect, such geographic turning points experience accelerated winds and rough seas in the area adjacent to the land prominence, and turbulence can often extend a considerable distance offshore. This effect, together with the funneling effect caused by the offshore Channel Islands, also influences the winds in and near the Santa Barbara Channel.

The rounding of this "turn in the road" is a crucial passage in the transit of the coast. The Pilot refers to Point Conception as the "Cape Horn of the Pacific." It seems to be a point of extremes; either there is little or no wind, or there is a great deal of it. Then again, the winds may be exceptionally strong over a large area and then within a relatively short distance one may pass into calm and, when southbound, markedly warmer air.

Point Arguello has the unenviable record of being the foggiest place on the U.S. Pacific Coast: it has visibility of less than .5 mile on 12 to 20 days per month, with July and August having the worst records. In addition, this area marks a surprisingly clear line between the cool, invigorating air to the north and warm (often hot) air to the south. It's the area where southbound sailors can discard their heavy sweaters and don swimsuits. Passage along this part of the coast is usually more easily done at night when the wind tends to be calmer. Monitoring the weather forecast is essential prior to making this passage.

Cojo Anchorage, just 1.5 miles east of Point Conception, is convenient for recovering from the passage when southbound or waiting for the right time to start when northbound. Work your way through the kelp beds to anchor in clear water; though it can be very windy here, the water remains calm. It is difficult not to be alarmed by the night-time roar of trains, for they sound so close. The best place to anchor is opposite a culvert under the railroad tracks in 5 fathoms, hard sand. The large mooring buoys are for oil platform tenders that may arrive at any hour. Little Old Cojo Cove, just over 3 miles east of Point Conception, is foul and should not be entered.

The first time we made the passage from Morro Bay to Santa Barbara we were treated to one of nature's rare displays. The fog was very dense but for several hours during the night our limited range of visibility was lit by broad bands of underwater "sea-lights." The bands of light were about 15m (50 ft.) wide and were separated by unlit bands of about the same width. Several dolphins that were zig-zagging across our bows ignored the lights and added their own trail of phosphorescence to the show. Though rare, this light is produced under the right conditions by tiny organisms known as dinoflagellates. This magical experience stands out as one of the highlights of the passage along the coast.

COJO ANCHORAGE

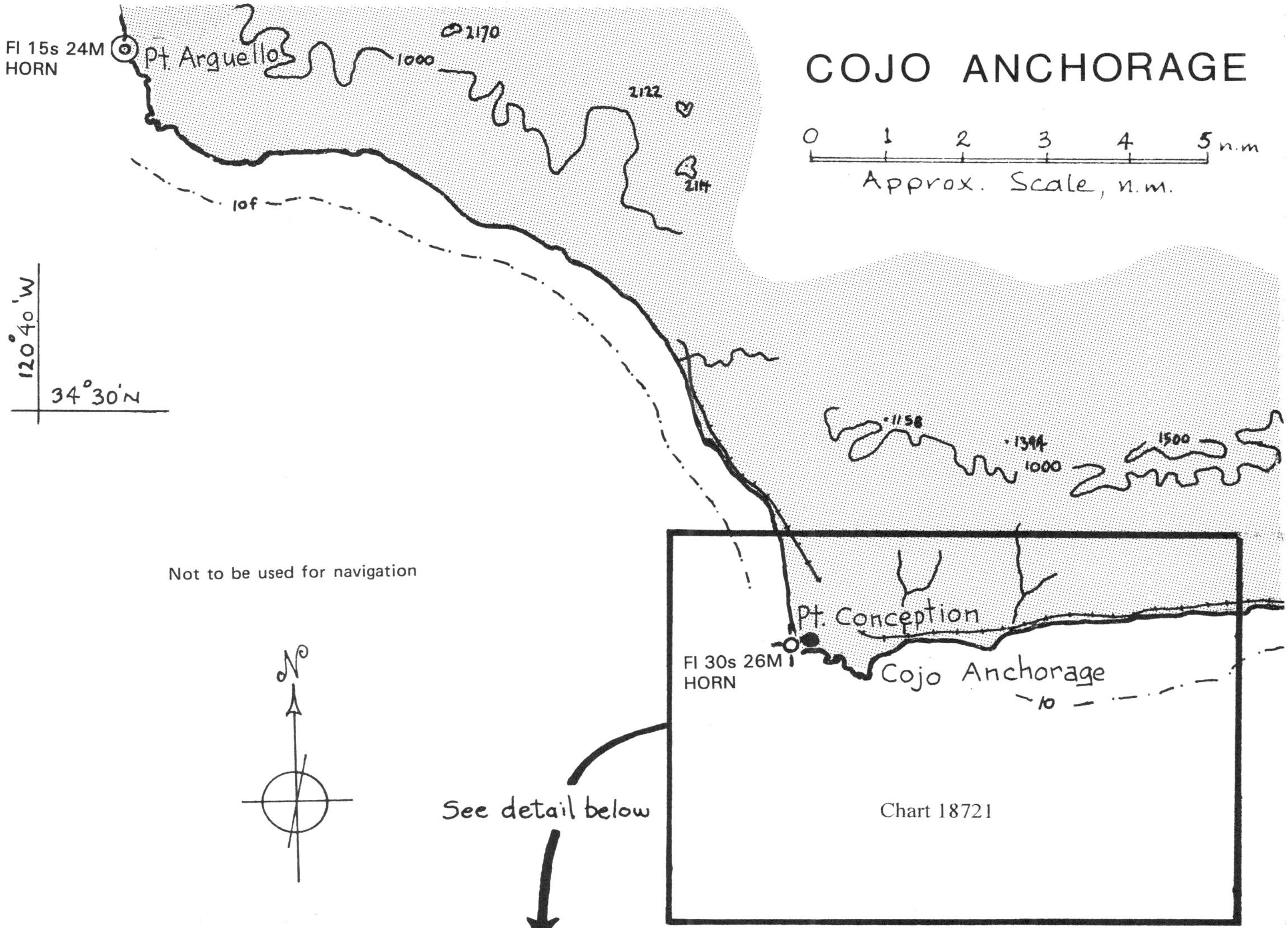

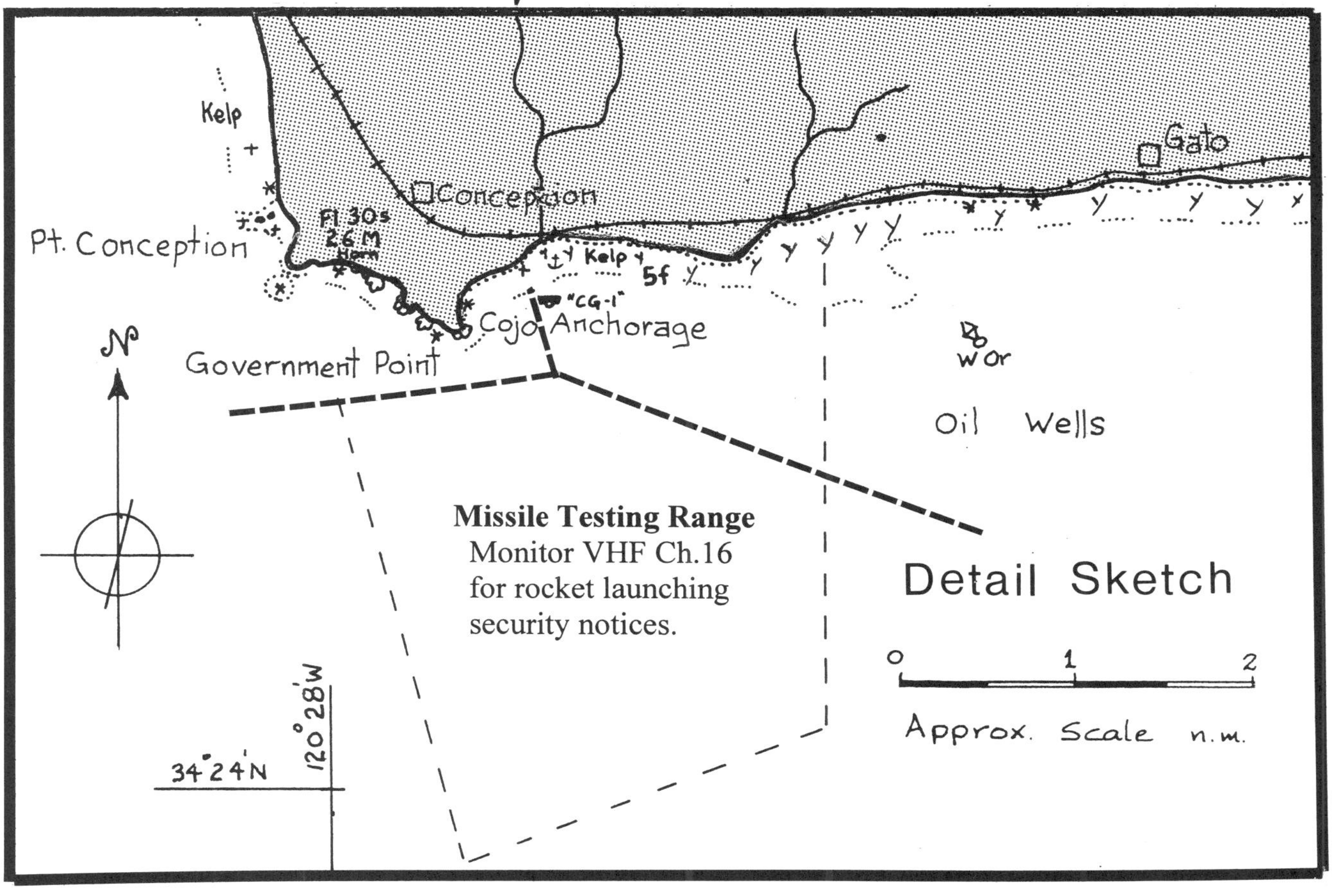

THE CHANNEL ISLANDS

Cruising the Channel Islands provides a view of California that is a refreshing contrast to harbor-hopping from one crowded marina to the next. Though the Santa Barbara Channel that separates the northern islands from the mainland is less than 25 miles in width, its boisterous winds form a buffer between the fabricated high-tech world of urban California and the stark, pristine beauty of the islands. Similarly, the scant 25-mile stretch between Santa Catalina Island and the mainland draws an abrupt line between two different worlds. A good feature of this alternative route is that the cruiser is not committed to either one course or the other, since distances to the mainland are so short that one may travel back and forth to refuel or pick up provisions as desired.

Many possible anchorages may be used in the islands during favorable weather conditions. Only a few anchorages are described here and the names of other spots are shown on the accompanying sketches. For detailed information on all coves one should refer to: *Cruising Guide to Southern California's Offshore Islands* by Brian M. Fagan.

The eight offshore islands form two groups. The Northern Channel Islands, consisting of San Miguel, Santa Rosa, Santa Cruz and Anacapa are 10 to 25 miles off the coast and form an east/west chain almost 60 miles long. The Southern Channel Islands of Santa Catalina, San Clemente, San Nicolas and Santa Barbara are scattered over a large area from 25 to 60 miles offshore.

The Northern Channel Islands are described following Cojo Anchorage as this departure provides the best angle for taking advantage of the prevailing northwesterly winds. Most local sailors depart from their more southerly home ports early in the day, making maximum mileage before the afternoon winds pick up.

A cruise of the islands has much to offer: unique scenery, good fishing, excellent diving and snorkeling and a large variety of marine life to observe. But these waters are no place for triflers—especially the Northern Islands.

The most consistent feature of the weather in the Channel Islands is its variety. Though the prevailing northwesterly winds buffet the northern islands for much of the time, local storms, fog and clear, calm conditions can also be experienced. The infamous Santa Ana winds are more prevalent during winter months. Heavy fog often envelopes the islands at night, posing multiple hazards, for not only is visual navigation impeded but avoidance of the heavy deep-sea traffic and the 24 oil-drilling platforms found in the Santa Barbara Channel is made more difficult. Radar capabilities become essential.

The best time to visit the islands is during September and October when the weather is relatively calm and the crowds from the mainland are greatly reduced by the resumption of another school year.

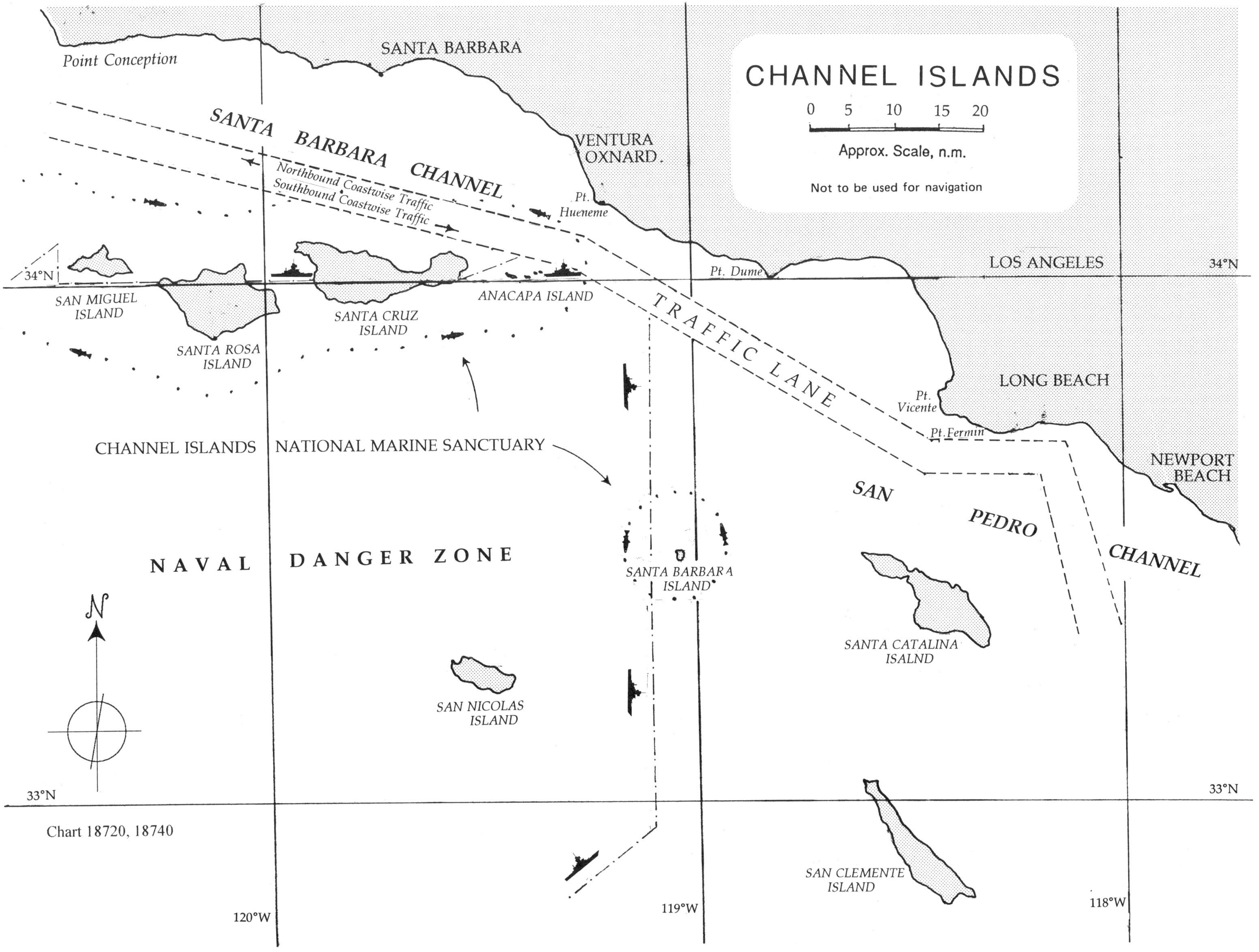
CHANNEL ISLANDS
0 5 10 15 20
Approx. Scale, n.m.
Not to be used for navigation
Point Conception
SANTA BARBARA
VENTURA
OXNARD
SANTA BARBARA CHANNEL
Northbound Coastwise Traffic
Southbound Coastwise Traffic
Pt. Hueneme
Pt. Dume
LOS ANGELES
34°N
SAN MIGUEL ISLAND
SANTA ROSA ISLAND
SANTA CRUZ ISLAND
ANACAPA ISLAND
TRAFFIC LANE
LONG BEACH
Pt. Vicente
Pt. Fermin
NEWPORT BEACH
CHANNEL ISLANDS NATIONAL MARINE SANCTUARY
SAN PEDRO CHANNEL
NAVAL DANGER ZONE
SANTA BARBARA ISLAND
SANTA CATALINA ISALND
SAN NICOLAS ISLAND
SAN CLEMENTE ISLAND
N
33°N
120°W
119°W
118°W
Chart 18720, 18740

THE NORTHERN CHANNEL ISLANDS

The Northern Channel Islands are composed of San Miguel, Santa Rosa, Santa Cruz and Anacapa. These mainly treeless islands, often fringed by kelp and edged with steep cliffs and eroded sea caves, have an almost intimidating air of detachment. However, this rugged demeanor undergoes a miraculous metamorphosis following the spring rainy season, when some areas of the islands are blanketed by an explosion of vibrant blossoms and short-lived greenery.

As a result of the introduction of various animals and consequent overgrazing, damage has been done to the fragile balance of nature in this area. Fortunately, the National Park Service is making a concerted effort to recreate the natural environment so that indigenous plants and animals can once again prosper. The Channel Islands National Park consists of the Northern Channel Islands and most of Santa Barbara Island, with the exception of some privately owned land. Landing is restricted, and permission to go ashore must be obtained from: The Channel Islands National Park, 1901 Spinnaker Drive, Ventura, CA 93001, telephone (805) 644-8262. Park Rangers are resident on each island, though permission to visit the island cannot be obtained from them; they oversee conservation efforts. To arrange for an orientation session and free guided nature walks call the National Park Service at (805) 658-5711 prior to visiting the islands. The waters surrounding each of the above islands compose the Channel Islands National Marine Sanctuary, and strict rules are enforced regarding harassment of wildlife, water pollution and harm to the environment.

Weather patterns are dominated by strong, almost daily, northwesterly winds, that often reach gale force as they accelerate at Point Conception and impinge on the islands. The weather is calmest in the morning with wind and seas gradually building as the day wears on. This daily pattern makes an early morning departure for the islands highly recommended. Avoid the busy shipping lanes where there is heavy commercial traffic. The windy conditions, usually more pronounced in the adjacent waters and 6 to 7 miles offshore, have gained the area such names as "Windy Alley," or "Windy Pass."

In addition to local storms, fog and Santa Ana winds are weather factors that merit close monitoring of weather forecasts. Santa Ana winds blow with such ferocity that you should remain in port until they have subsided. If you are already in the islands then head for the safety of the south shore. A lack of dew on deck and a seaward blanket of smog are clear indications that Santa Ana winds are coming. Continuous broadcasts are given over the following stations:

WX 1 162.55 Mhz - Los Angeles

WX2 162.40 Mhz - Santa Barbara

The Coast Guard San Diego Radio (NMQ) 2670 kHz broadcasts weather information at 0633, 0800,1503, 2000 and 2103 PST. National Weather Service Marine Weather Forecasts may be obtained from VHF W1, W2 or W3.

The frequently cool, windy weather here demands good, wind-proof clothing to prevent hypothermia. In addition, water temperatures in the 60s make the use of wet suits desirable when snorkeling and diving in this divers' paradise. Sturdy hiking boots are needed when walking ashore. Keep to the trails to keep from damaging the fragile plant life and to avoid unexploded ordnance found on San Miguel Island.

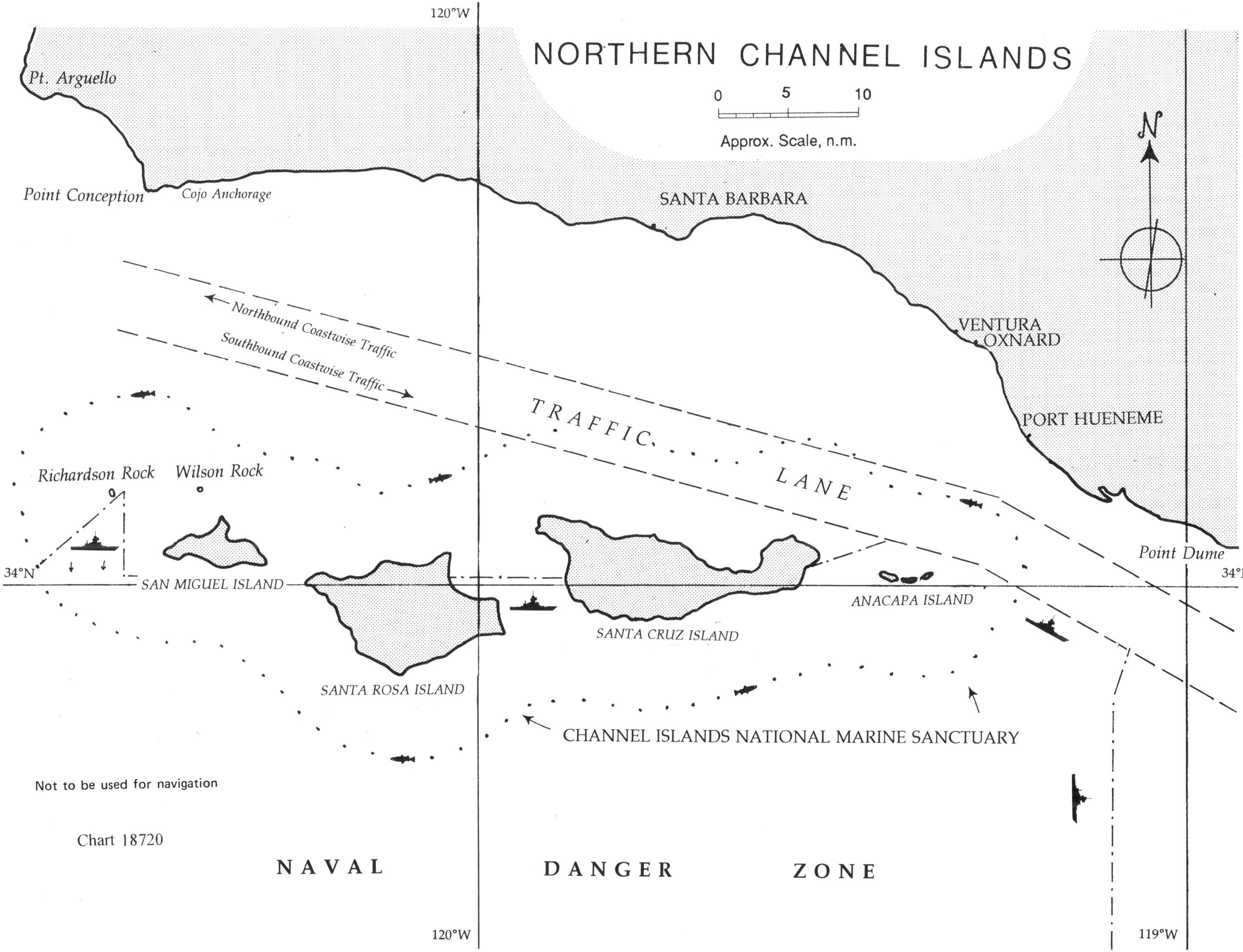
NORTHERN CHANNEL ISLANDS
0
5
10
Approx. Scale, n.m.
N
120°W
Pt. Arguello
Point Conception
Cojo Anchorage
SANTA BARBARA
VENTURA
OXNARD
PORT HUENEME
Point Dume
Northbound Coastwise Traffic
Southbound Coastwise Traffic
TRAFFIC
LANE
Richardson Rock
Wilson Rock
34°N
SAN MIGUEL ISLAND
SANTA ROSA ISLAND
SANTA CRUZ ISLAND
ANACAPA ISLAND
34°N
CHANNEL ISLANDS NATIONAL MARINE SANCTUARY
Not to be used for navigation
Chart 18720
NAVAL
DANGER
ZONE
120°W
119°W

SAN MIGUEL ISLAND

This is the westernmost island of the Northern Channel islands and it is the third largest island in the park. Measuring about 7.5 miles long by 4 miles wide, the island rises to 244m (800 ft.) and lies about 25 miles south of Cojo anchorage. A stop-over here should only be undertaken by a crew that has proven experience in anchoring in windy conditions.

Offshore dangers demand careful navigation and use of Chart #18727. The western end of the island should be given a particularly wide berth for rocks and reefs are spread over a wide area. Richardson Rock lies 5.5 miles northwest of Point Bennett while shoals in the vicinity of Castle Rock lie 1.6 miles NNE of Point Bennett. Another danger off the north shore is Wilson Rock, which lies about 2.2 miles NW of Harris Point; give this rock a berth of at least 1.25 miles to avoid shoals in the vicinity. Anchorage is allowed in only two harbors; landing is permitted, but access to the interior of the island is by permit only.

Cuyler Harbor provides shelter from northwest to southwest winds, but is exposed to easterly winds. The dark landmarks of Harris Point and Prince Island contrast with the white beaches and sand dunes beyond. A clear berth must be given to the sand bar in northern curve of the bay and to rocks and shoals near Middle Rock. Kelp patches can easily be avoided, as ample clear space is available for anchorage in 3 - 4 fathoms, sand. Landing on the beach is easy. The periodic sight of a fishing vessel blown ashore gives a vivid reminder of the need to set the anchor well and to monitor wind changes. Surge curving around the point sometimes makes the anchorage uncomfortable.

Tyler Bight Anchorage, located on the south side of the island near the west end, is shown on old charts as a possible small craft anchorage. However, recent government regulations prohibit anchoring within 275m (300 yds.) of the shore to prevent disturbance of critical rookeries of endangered pinnipeds. Since the best protection is close to the shore, this spot has been eliminated as an anchorage.

Military activity in the **Naval Danger Zone** near San Miguel Island is announced on VHF Ch 16 at 0800 and 1200 local time and/or 1 hour prior to bomb drops. Inquiries regarding military exercises may be made by calling "Plead Control" or the Pacific Missile Test Center on VHF Ch 16.

San Miguel Island, part of the Channel Islands National Park, is an Ecological Reserve. Surrounding waters are in the Channel Islands National Marine Sanctuary and form the most significant pinniped rookery in the islands. Here thousands of seals, sea lions and sea elephants "haul out" and breed. Another fascinating natural feature of the island is the large number of sand castings of plants (caliche), which continue to form. In addition San Miguel Island boasts over 500 undisturbed archeological sites.

It is believed that Juan Cabrillo, the Spanish discoverer of California, died here in 1543. A monument commemorating his voyage of discovery overlooks Cuyler Harbor. In the 1840's cattle, sheep and horses were brought to the island; overgrazing denuded the trees and ground cover, and the ever-present wind has made recovery very difficult.

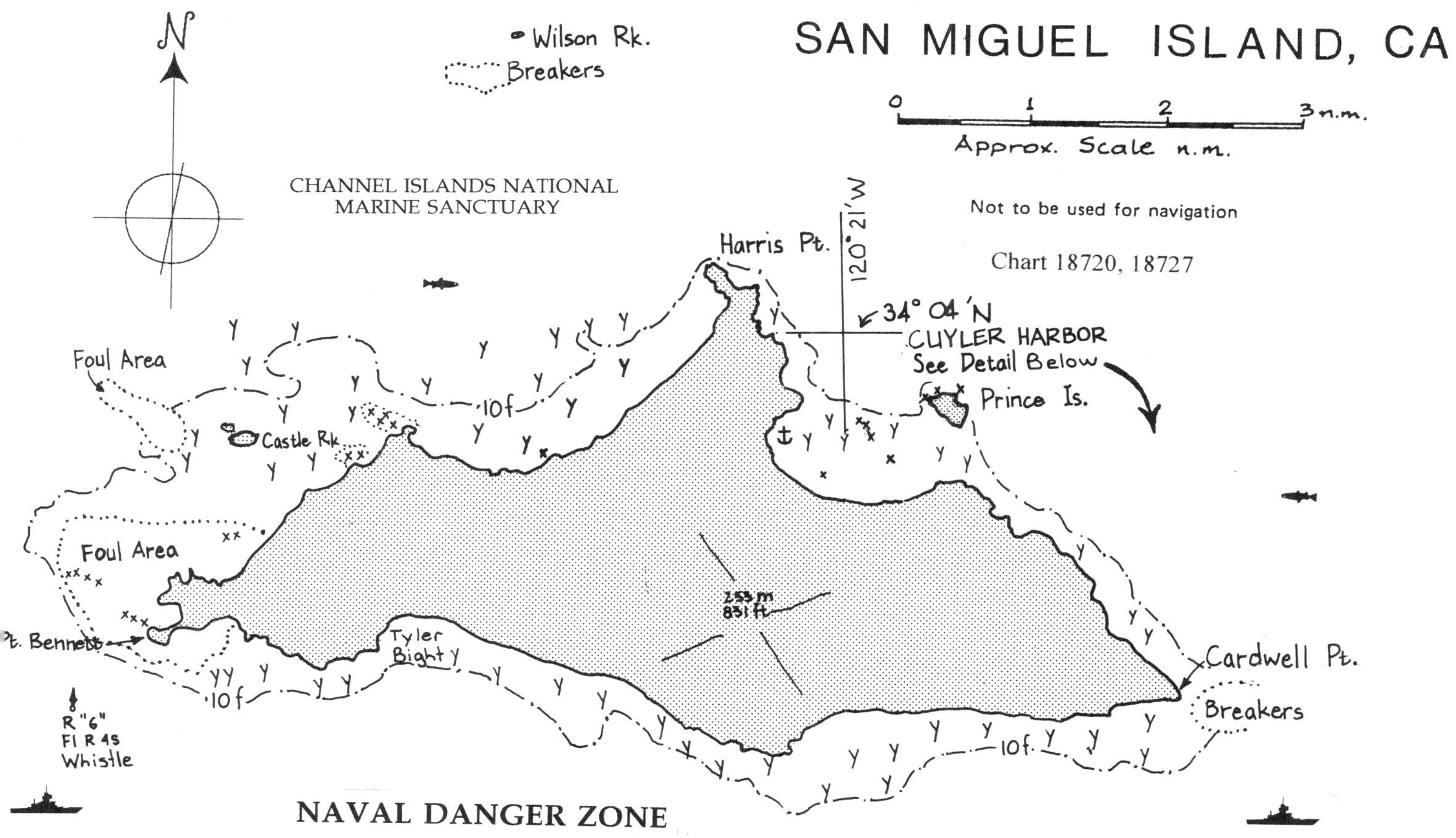
SAN MIGUEL ISLAND, CA
N
Wilson Rk.
Breakers
0 1 2 3 n.m.
Approx. Scale n.m.
CHANNEL ISLANDS NATIONAL MARINE SANCTUARY
Not to be used for navigation
Chart 18720, 18727
Harris Pt.
120° 21′ W
34° 04′ N
CUYLER HARBOR
See Detail Below
Prince Is.
Foul Area
Castle Rk
10f
Foul Area
Pt. Bennett
Tyler Bight
253 m
831 ft
Cardwell Pt.
Breakers
R "6"
Fl R 4s
Whistle
10f
10f
NAVAL DANGER ZONE

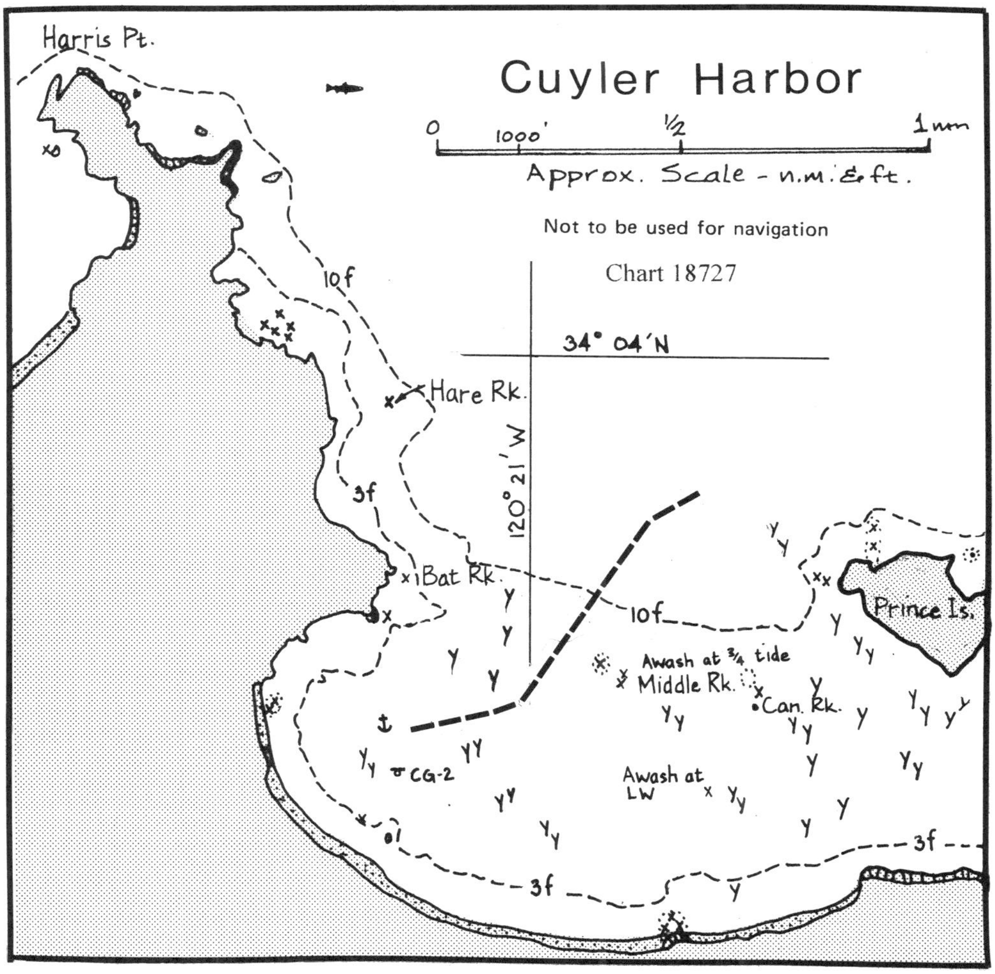
Harris Pt.
Cuyler Harbor
0 1000′ ½ 1 nm
Approx. Scale - n.m. & ft.
Not to be used for navigation
Chart 18727
10f
34° 04′N
Hare Rk.
120° 21′ W
3f
Bat Rk.
10f
Prince Is.
Awash at ¾ tide
Middle Rk.
Can. Rk.
CG-2
Awash at LW
3f
3f

SANTA ROSA ISLAND

This gently sloping, featureless island lies about 30 miles from both Point Conception and Santa Barbara. Just 15 miles long, it reaches its highest point near the center of the island, which rises to 485m (1,589 ft.). Like San Miguel, the island affords limited protection from the winds spawned off Point Conception. The site of one of the largest cattle ranches in the U.S., much of the island is blanketed by short vegetation and sand dunes. Thick kelp surrounds the island.

San Miguel Passage separates the island from Santa Rosa Island. It is a **naval danger zone,** used for bombing practice; though passage is permitted it is best avoided. Announcements of bomb drops are made on VHF Ch 16 at 0800 and 1200 local time.

Except when approaching an anchorage, you should stand off the island by at least 3 miles to avoid numerous rocks and reefs scattered about. Onshore currents on the north side of the island, augmented by the prevailing wind, can push the unwary skipper into dangerous waters. A hazardous area is Skunk Point, is 2.5 miles north of the island's eastern tip, where strong currents and breaking seas occur almost 1 mile off.

Northwest Anchorage in **Bechers Bay** is the only anchorage spot on the northern coast of the island. A broad bight exposed to easterly weather, it does provide protection from northwest winds, though the swell produced can curl around the point, and make the anchorage uncomfortable. A long pier (used for offloading supplies to the ranch and loading cattle for sale on the mainland) is an unmistakable landmark. Anchorage may be taken .5 mile north of the pier between the kelp beds, or off the pier, well clear of the mooring buoy. Though temporary anchorage may be taken across the bay in Southeast Anchorage, it is not advisable as an overnight spot.

On the south side of the island is **Johnson's Lee,** a roadstead anchorage just to the northeast of South Point. A large storage shed and pier are all that remains of a World War II radar station and Air Force base which was manned until 1963. A Ranger from the National Park Service periodically visits the site. Approach to the anchorage is made by threading a route through the kelp, where anchorage may be taken off the pier in 4 - 6 fathoms, sand. It has been reported that poor holding, rocky bottom is located in the northern part of the bight. Gusty winds off the nearby hills are common.

To the east of Johnson's Lee is another spot where protected anchorage may be taken at **Ford's Point**. Camping is permitted from June 1 to January 1.

The island is the site of important fossil beds of such creatures as mammoths and giant mice. In addition, archeological finds have dated habitation on the island by Chumash Indians to some 10,000 years ago. A cattle ranch operated by Vail and Vickers since 1902 has permission to continue operations here until 2011, at which time the entire island will be under the control of the Channel Islands National Park.

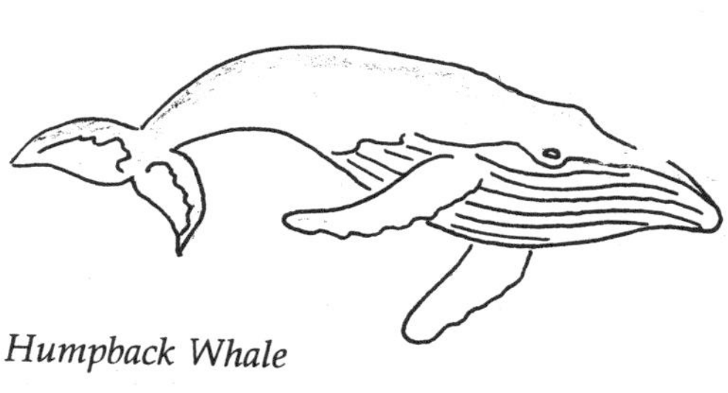

Humpback Whale

SANTA ROSA ISLAND, CA

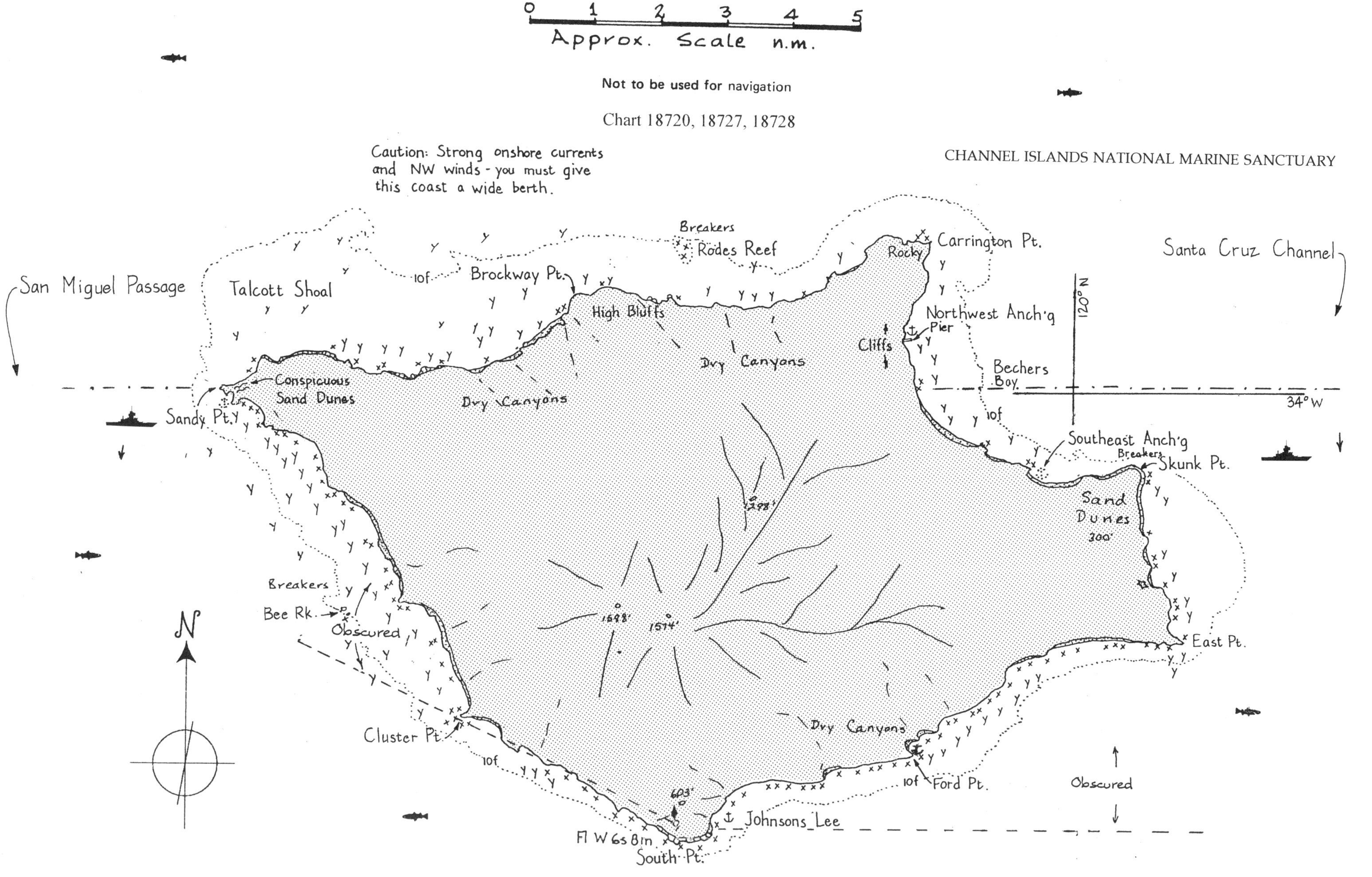

SANTA CRUZ ISLAND

This rugged and almost treeless island lies less than 30 miles from Santa Barbara, Ventura and Oxnard. The largest of the Channel Islands, it rises to 749m (2,434 ft.) in the western part, 549m (1,800 ft.) in the eastern section. As the shores are steep-to, adjacent waters are deep, resulting in less kelp than is found near the other islands. The shoreline is diverse and features steep cliffs, huge sea caves, coves and sandy beaches. The Gherini family owns the eastern tip of the island. Permission to land there may be obtained by writing to Mr. F. Gherini, 162 South 'A' St., Oxnard, CA 93030. The Nature Conservancy controls the remainder of the island, and landing permits may be acquired from them. The island has many spots that can be used as anchorages during settled weather, only a few of which are briefly described here.

The westernmost tip of the island is Fraser Point, a low, fairly flat projection of land having kelp and rocks extending about .5 mile to the east and southeast. An anchorage spot is located a short distance to the southeast gives good protection from northwesterly seas, though strong winds may be blowing. Entrance to **Forney's Cove** requires caution. A gap in the kelp provides the channel leading to anchorage at the head of the cove, where off lying rocks to the west break the effect of northwesterly swell. During Santa Ana conditions, extremely strong wind gusts sweep the anchorage, at which time it is best avoided. About 3 miles east of West Point on the N shore is **Painted Cave**, the largest marine cave in the world-—stretching 365m (1,200 ft.)-—a truly awesome sight which should not be missed.

About 4 miles to the east is the beautiful anchorage of **Lady's Harbor,** suitable only during calm conditions. It is best identified by first locating Arch Rock, 3.5 miles east of the Painted Cave, then proceeding another .5 mile to the double-lobed indentation forming Lady's Harbor on the north and **Little Lady's (Baby's Harbor)** close by. Enter midway between the cliffs. Anchorage may be taken in the middle of the cove in 3 - 4 fathoms, sand. Two anchors are needed to prevent swinging in this very snug spot which is affected by surge. Two miles to the east is **Fry's Harbor**, a popular and more spacious spot where anchorage may be taken off the beach on the west side. Gusty winds blowing down the canyons make the use of bow and stern anchors highly recommended.

Several anchorages on the north shore of the island are located in the large bight. **Pelican Bay** is picturesque and popular in spite of limitations such as exposure to Santa Ana winds and inconsistent holding. Yellow cliffs on the west side of the anchorage pinpoint its location. Anchorage may be taken in 4 - 5 fathoms, sand. A tricky landing may be accomplished in the southeast corner of the anchorage where some rickety steps are found. It is worth a dinghy trip to Hole-in-the-Wall, a tiny anchorage sometimes used by 1 or 2 local vessels, which is just to the west.

About a mile to the southeast is **Prisoner's Harbor**, a picturesque anchorage though it has more easterly exposure. It can be identified by a pier and some metal roofed buildings nearby. Anchorage (fore and aft) with indifferent holding may be taken west of the pier 3 - 6 fathoms sand, rock and eelgrass.

On the southeast side of the island **Smuggler's Cove** has good holding in 4 to 6 fathoms, sand. A few miles to the west on the southern shore is **Albert** anchorage with good protection. Further to the west **Coches Prietos** provides a secure anchorage as well as easy access to the shore where the sandy beaches are popular with family groups and shore parties.

SANTA CRUZ ISLAND, CA

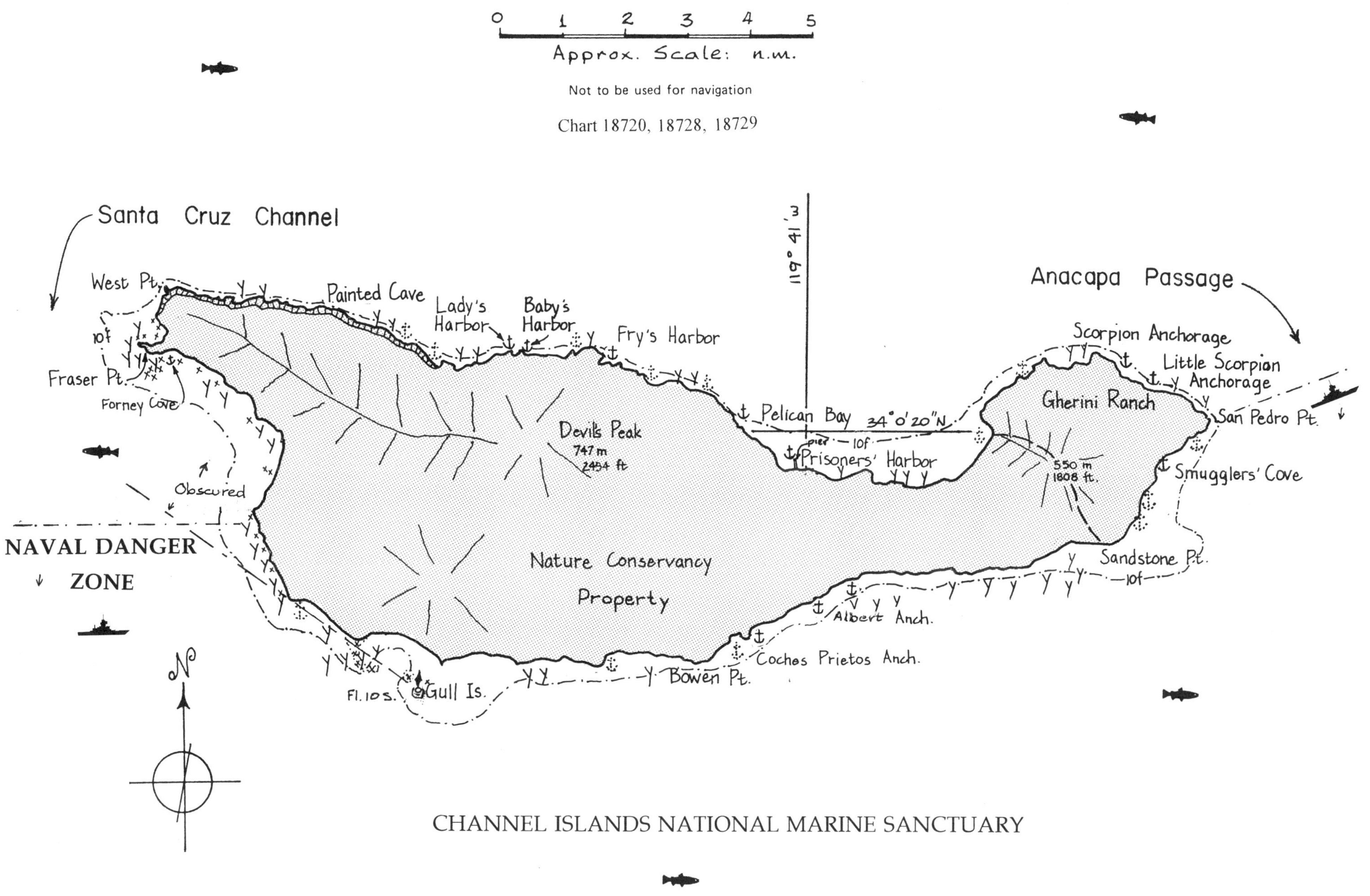

ANACAPA ISLAND

Anacapa is the closest of the Channel Island to the mainland, just 11 miles southwest of Oxnard. It is actually a group of three islands (East, Middle and West Anacapa Island) separated by narrow passages foul with rocks and sand. The precipitous, steep-to shores are laced with caves and steep cliffs, and most of the island is fringed with kelp. The eastern end is marked with the much-photographed Arch Rock (24m/80 ft.) with its spectacular 15m (49 ft.) arch, and a pyramid-like rock just to the south. A cylindrical, unmanned lighthouse over 84m (275 ft.) above sea level is located at the eastern end of the island. A landing platform at the northeastern end of the island makes a convenient dinghy landing. East Anacapa is the site of campgrounds used by many visitors arriving by tourist boats, which operate daily from April to November.

The western part of the island rises to 283m (930') while the middle and eastern islands rise to 99m (325') and 76m (250') respectively. Separating Anacapa from Santa Cruz Island is Anacapa Passage, a 4-mile wide passage which experiences strong tide rips during southeasterly storms and northeasterly winds. The island is famous for its tide pools and excellent diving and snorkeling.

The only anchorage on the north shore, and the most used spot on the island is **Frenchy's Cove**. Though this is a popular destination for week-end cruisers from the mainland, it should only be used during southerly conditions for during northerly winds it becomes a dangerous lee shore. It has the dubious distinction of having the largest number of vessels blown ashore of any place in the Channel Islands. The beautiful, clear nights are sometimes accompanied by strong northeasterly winds, making this anchorage a lee shore. Cautious sailors anchor for a day visit and return to the mainland at night.

Frenchy's Cove is situated just to the west of West Passage, a narrow, foul gap between the Western and Middle Island. Avoid offshore kelp beds when approaching Frenchy's Cove. Setting fore and aft hooks, anchorage may be taken in 3 to 5 fathoms, sand, gravel and some eelgrass off the beach under the cliffs. Landing is possible on the beach in calm conditions; no facilities are ashore. This is the only place on the island where landing is permitted without a permit.

The best anchorage on the southeast side of the Middle Island is **East Fish Camp,** a bight less than .5 mile southwest of the eastern opening. Anchorage may be taken in 5 - 6 fathoms, sand and rock, keeping clear of kelp beds which can sometimes be a problem.

Some protection from northeasterly weather is found in the bight west of **Cat Rock** on the south side of the West Island. The water is clear of dangers in the area about 193m (200 yds.) west of Cat Rock. The western end of the island is a Research Natural Area closed to the public for the protection of the important brown pelican nesting area located here.

Historical Note: Raymond "Frenchy" La Dreau (Ledreau) lived on the island from 1925 to 1956. He was a colorful character who so enjoyed the comings and goings of rum-runners in the early years and yachtsmen in later years that he became the unofficial watchman for the island.

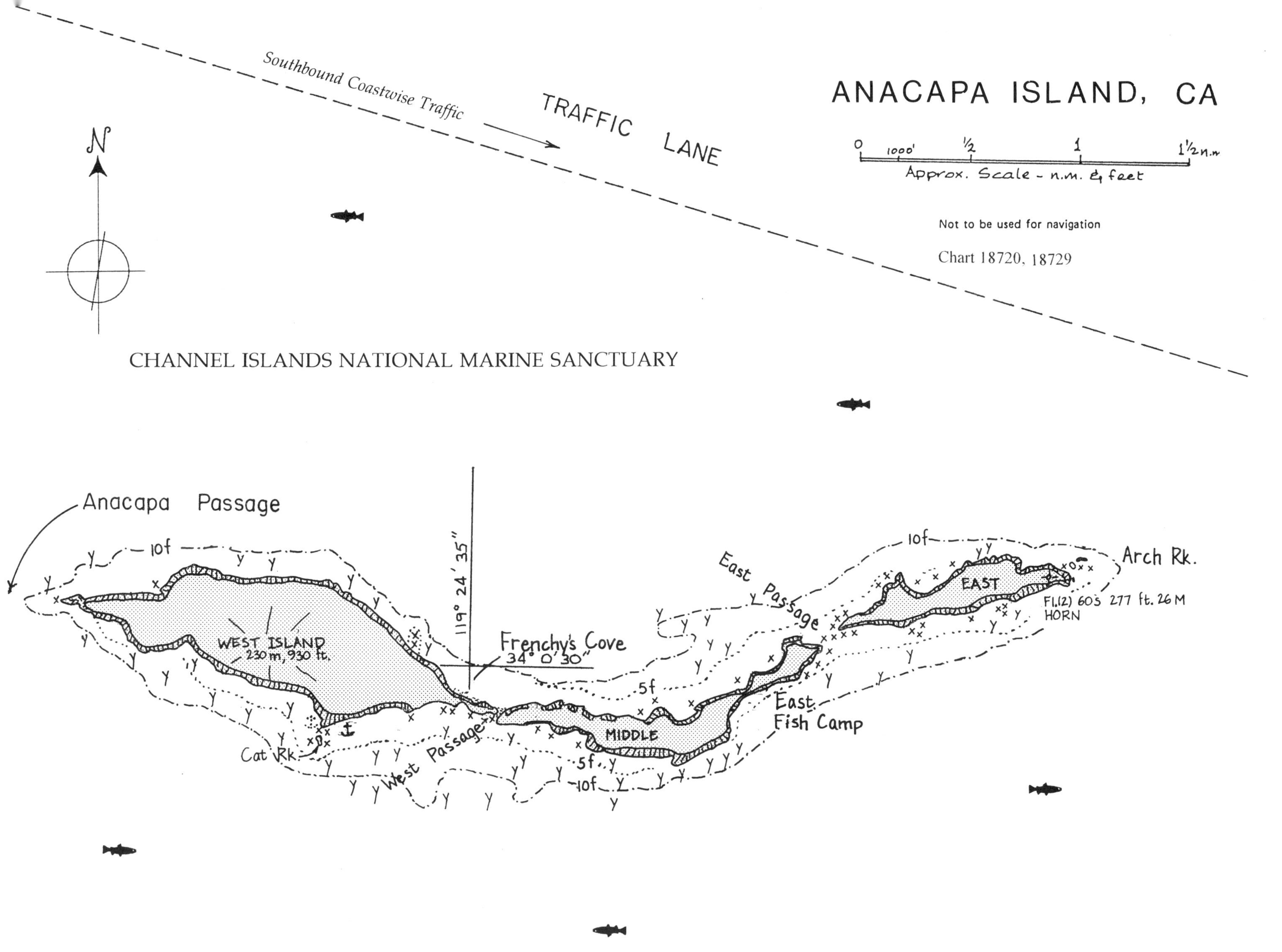

ANACAPA ISLAND, CA
0
1000'
1/2
1
1½ n.m.
Approx. Scale - n.m. & feet
Not to be used for navigation
Chart 18720, 18729
Southbound Coastwise Traffic
TRAFFIC LANE
N
CHANNEL ISLANDS NATIONAL MARINE SANCTUARY
Anacapa Passage
10f
WEST ISLAND
230 m, 930 ft.
119° 24' 35"
Frenchy's Cove
34° 0' 30"
Cat Rk.
West Passage
5f
MIDDLE
East Passage
East Fish Camp
EAST
Arch Rk.
Fl.(2) 60s 277 ft. 26 M
HORN

SANTA BARBARA

This attractive and affluent city marks the northern boundary of the Southern California Coastal region. Within the region, over 200,000 pleasure boats are berthed in harbors as busy as Los Angeles/Long Beach and as intimate as Oceanside and Dana Point. Santa Barbara itself is home to over one thousand recreational and commercial boats, and is a popular destination for cruisers from other southern California locations.

Santa Barbara is 40 miles east of Cojo Anchorage and about 22 miles WNW of Ventura. Santa Barbara Light is on a bluff about 2 miles east of the harbor entrance. Prominent from all directions is a hill about .5 miles northeast of the light. At night a major landmark is the neon light atop a hotel tower on the beach, 1 mile east of town. When approaching from the channel, the city lights are visible behind the harbor. However, if approaching from the west, the lights are obscured by the hill.

A 457m (500 yd.) rock breakwater built out from the shore joins a jetty across a sand bar to protect the harbor from southerly weather. A light marks the junction of the breakwater and jetty. Along the breakwater are over 100 flagpoles making a colorful and distinctive honor guard for the harbor. Stearns Wharf forms the northeast side of the harbor. A number of shops are located on the wharf and a light and fog signal are at the seaward end. Within the harbor a 45m (150 ft.) groin, lit at its tip, extends south from the shore line to form an eastern boundary to the marinas.

Santa Barbara harbor is a man-made basin that fights a continual battle against shoaling at its entrance. There is a long sandbar north of the breakwater light which is difficult to see at night, particularly during a high tide. To enter, follow the buoyed channel which passes between the sand bar and Stearns Wharf before turning southwest into the harbor. The channel buoys are not shown on the sketch as they are frequently altered to mark the changing channel. Fog is a hazard, particularly from August to November. It is inadvisable to enter at night as the many city lights and masts within the harbor can be confusing for a first-time visitor. Be sure to identify all lights and follow the buoyed channel carefully. For current channel conditions contact the harbormaster on VHF Ch. 16; assistance on entering the harbor is given by a Patrol Boat on request.

The Harbor Patrol controls transient moorage and their office is upstairs in the Chandlery building. They monitor VHF Chs. 16 and 12 24 hours, daily and can also be reached at (805) 564-5530 or Fax (805) 897-2588. Guests may first dock at the Harbor Patrol Accommodation Dock at the west end of the harbor, in front of the Chandlery building. Slip fees are payable in advance, on a first-come first-serve basis. Boats leaving during the day may not have a slip when they return if they have not already paid the moorage fee. There are showers and restrooms adjacent to the docks. Restaurants, chandlers, boat services and brokers are on shore adjacent to the marina. Fuel is available at the Harbor Patrol dock. Propane is available at Suburban Propane or at Coast Village Chevron, both of which are a five-minute ride from the harbor on an electric trolley. Internet access is available at Hot Spot on State Street.

A touch screen monitor at the fuel dock provides a current report on coastal weather conditions, an invaluable source of information for northbound vessels planning on transiting Point Conception or for vessels heading for the Channel Islands.

Anchorage may be taken east of Stearns Wharf (no closer than 91m/300 ft.). There is some surge, but the area affords good protection from northwesterly winds but if Santa Ana winds are predicted, move to the marina.

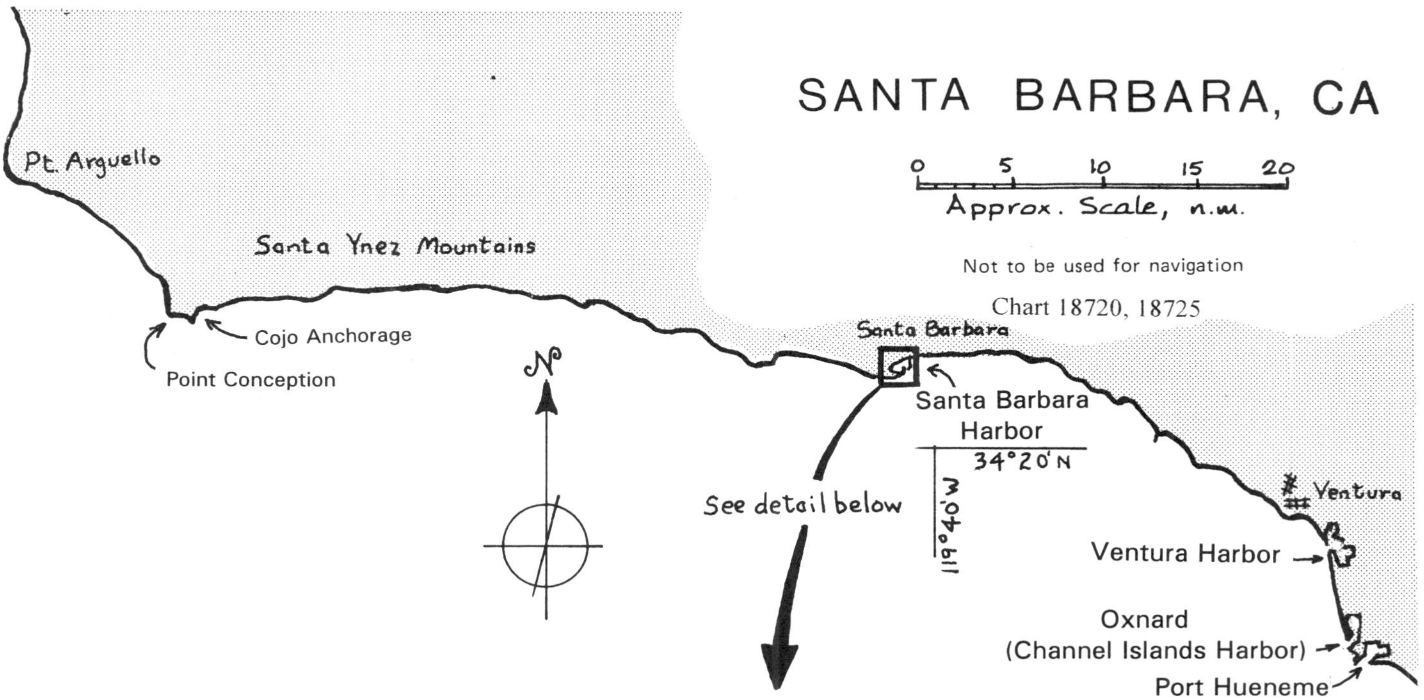
SANTA BARBARA, CA
0 5 10 15 20
Approx. Scale, n.m.
Not to be used for navigation
Chart 18720, 18725
Pt. Arguello
Santa Ynez Mountains
Cojo Anchorage
Point Conception
Santa Barbara
Santa Barbara Harbor
34°20'N
119°40'W
See detail below
Ventura
Ventura Harbor
Oxnard
(Channel Islands Harbor)
Port Hueneme

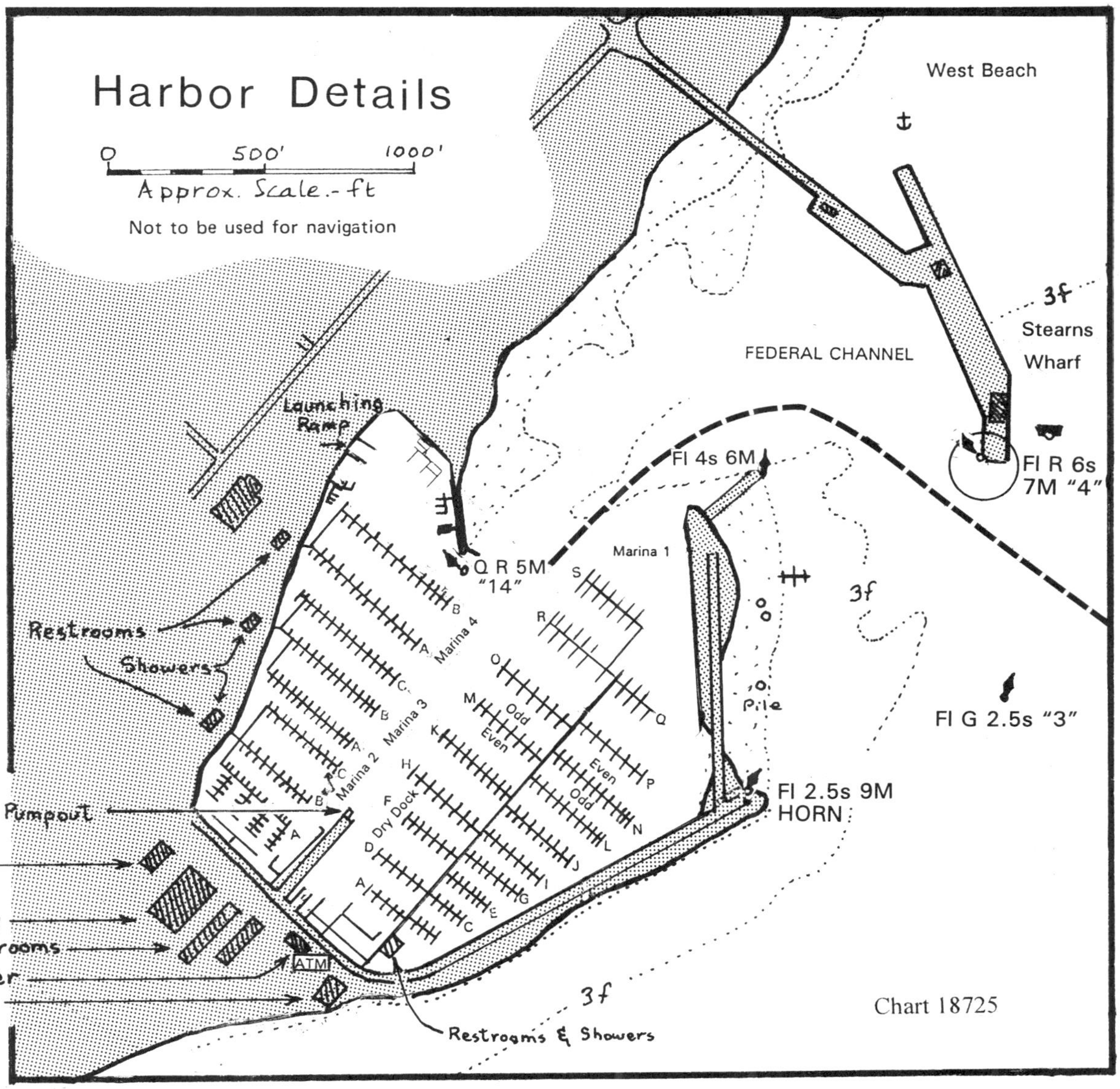
Harbor Details
0 500' 1000'
Approx. Scale - ft
Not to be used for navigation
West Beach
Stearns Wharf
FEDERAL CHANNEL
3f
Launching Ramp
Fl 4s 6M
Fl R 6s 7M "4"
Q R 5M "14"
Marina 1
Marina 2
Marina 3
Marina 4
Restrooms
Showers
Pile
Fl G 2.5s "3"
Fl 2.5s 9M HORN
Fuel Dock & Pumpout
Cafe
Naval Armory
Cafe & Restrooms
Harbormaster
Yacht Club
ATM
Dry Dock
Restrooms & Showers
Chart 18725

VENTURA

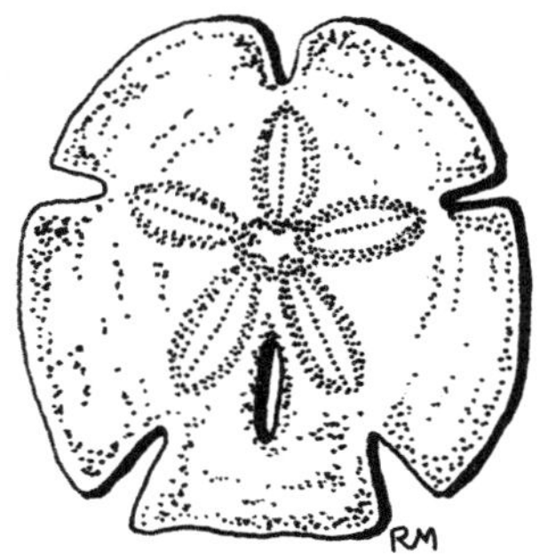

Ventura, 22 miles east of Santa Barbara and 7 miles from Oxnard's Channel Islands Harbor, is a fully developed man-made harbor. This is a good place to stop for vessels transiting the coast. It caters to pleasure craft, with space for over 2,000 vessels as well as commercial fish-handling facilities. Beacon Marine, a marine supply store is within walking distance of the marinas as are restaurants, boutiques and gift shops. A nearby airport and AMTRACK station provide convenient connections for travel.

Transient boaters are advised to visit the city during the week to avoid the heavy weekend marine traffic of local boaters cruising to the nearby Channel Islands of Santa Cruz and Anacapa.

Avoid this harbor if strong westerly winds are blowing, as they cause dangerous breakers at the entrance. Constant shoaling is a problem at the entrance that is controlled by regular dredging. Seven miles to the SSE is the deep-sea commercial terminal of Port Hueneme (Y-nee-mee), that has no small craft facilities.

Landmarks of note are the large white cross on a hill northwest of the town and a large Holiday Inn Motel sign west of the pier. The entrance to the harbor lies between two lit jetties, the southern one also having a fog signal. A 460m (1,500 ft.) detached breakwater (lit only at the southern end) lies about 230m (250 yds.) off the ends of the jetties. The detached breakwater is sand trap and the area around it is very shallow. Both the detached breakwater and the southern jetty experience breaking waves and should be given a wide berth.

When approaching the harbor, pick up Ventura Harbor Entrance lighted whistle buoy "2," which is about .75 mile southwest of the entrance. Proceed directly to the entrance, passing to the south of the detached breakwater before following the buoyed channel. Do NOT approach the entrance from the north, as shoals prohibit passage in this area. Buoys marking the channel are frequently moved to mark the best part of the channel and so are not shown on the sketch. Boaters are advised to contact the harbormaster on Ch. 12, 16, or 73 for current entrance and harbor conditions. A patrol boat is on call 24 hours per day.

Marine supplies, repairs, fishing gear and haul-out facilities are available from boating businesses within the harbor. There are two fuel docks as shown on the sketch. Transient berths are not available from the harbormaster but you may temporarily tie up at his dock while making arrangements for moorage at one of the three large marinas in the harbor. These first-class marinas monitor VHF Ch. 16 and usually have guest slips available. Internet connections are available at most marina offices. It is wise to call ahead if you are arriving on a week-end or holiday.

Ventura Isle Marina (805) 644-5858 or Fax (805) 644-0684

Ventura West Marina (805) 644-8266 or Fax (805) 644-1402

Ventura Harbor Village (805) 644-0169 or Fax (805) 644-1684

Right: The beauty of the days of sail has been recaptured in this realistic replica based at Ventura, California. (Page 126)

Below: Any bar can be dangerous as is shown by the entrance to Oceanside, California. (Page 162)

Even the relatively well protected roadstead anchorage at Shelter Island, San Diego does not forgive poor anchoring techniques or inadequate ground tackle. (Page 168)

Los Coronados (Coronados Islands) in Mexico seen from Point Loma invite one to travel south. (But first, have you got a copy of CHARLIE'S CHARTS of MEXICO?)

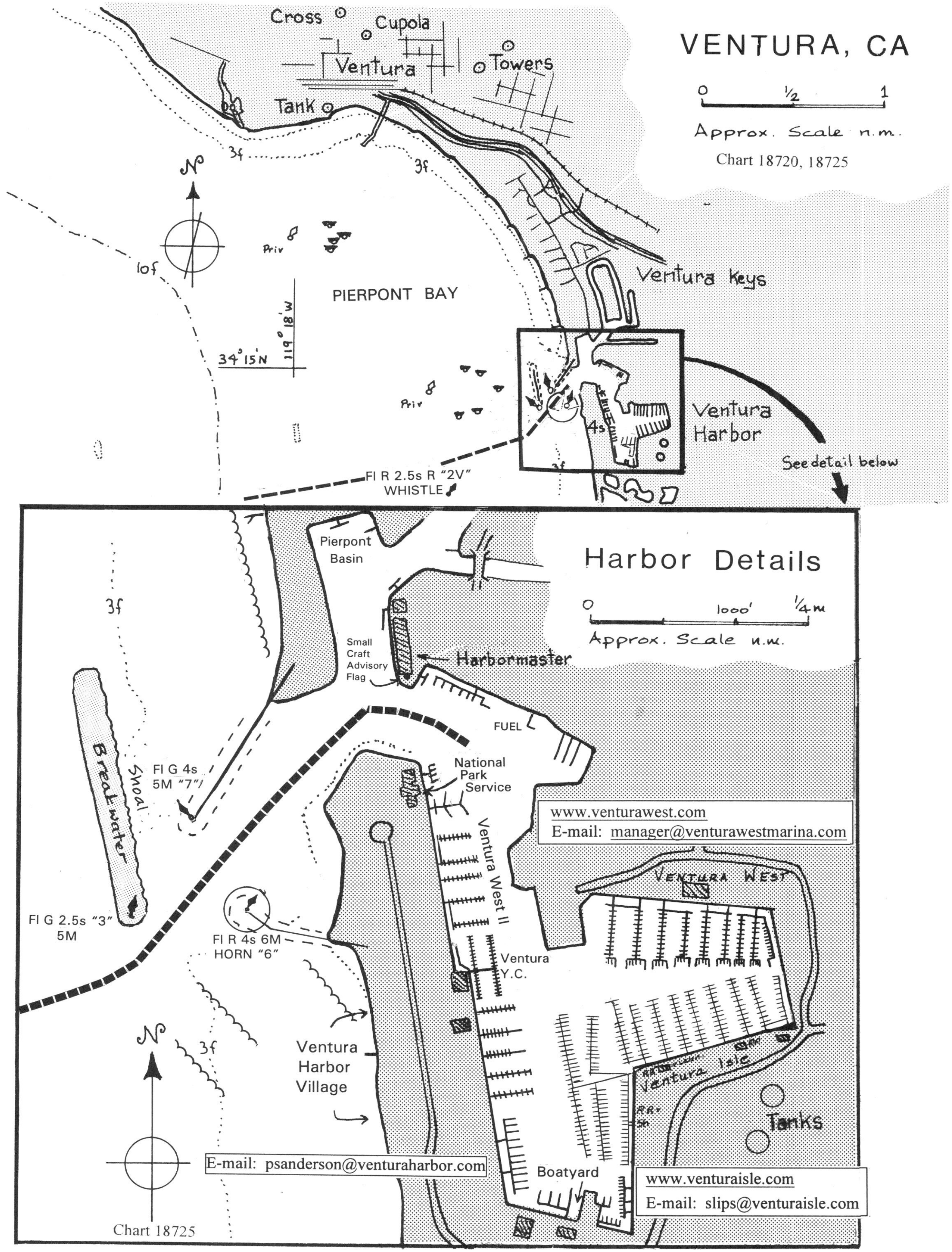
VENTURA, CA
0 1/2 1
Approx. Scale n.m.
Chart 18720, 18725
Cross
Cupola
Ventura
Towers
Tank
3f
10f
N
Priv
PIERPONT BAY
34°15'N
119°18'W
Ventura Keys
Ventura Harbor
4s
See detail below
Fl R 2.5s R "2V"
WHISTLE
Harbor Details
0 1000' 1/4m
Approx. Scale n.m.
Pierpont Basin
Small Craft Advisory Flag
Harbormaster
FUEL
Breakwater
Shoal
Fl G 4s 5M "7"
Fl G 2.5s "3" 5M
National Park Service
www.venturawest.com
E-mail: manager@venturawestmarina.com
Ventura West II
Ventura West
Fl R 4s 6M HORN "6"
Ventura Y.C.
Ventura Harbor Village
Ventura Isle
Tanks
E-mail: psanderson@venturaharbor.com
Boatyard
www.venturaisle.com
E-mail: slips@venturaisle.com
Chart 18725

CHANNEL ISLANDS HARBOR, OXNARD

With close proximity to the beautiful Channel Islands just across Santa Barbara Channel, Oxnard's Channel Islands Harbor is home to some 2,800 pleasure craft. It is devoted to the yachting and sports fishing communities, and four yacht clubs are based here: Channel Islands, Anacapa, Pacific Corinthian and Islander. If possible, transient boaters should time their visit to this busy harbor to avoid heavy local marine traffic transiting the entrance during week-ends and special holidays. The harbor is about 7 miles southeast of Ventura and about 42 miles northwest of Marina del Rey.

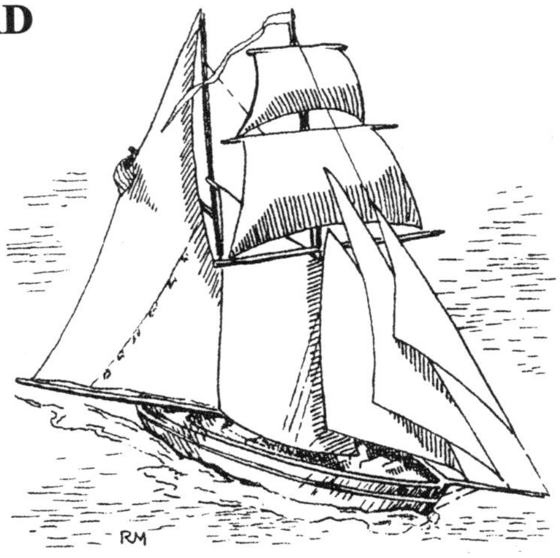

The state training tall ship, *Californian*
(A replica of the 1849 Revenue Cutter, *Lawrence)*

One mile to the SSE is the deep-sea commercial terminal of Port Hueneme (y-nee-mee) which has no small craft facilities. Only in the event of an extreme emergency should a cruising vessel enter this port.

Two jetties extending southwesterly from the beach protect the entrance channel of Channel Islands Harbor. A detached breakwater is seaward of the ends of the jetties. Lights mark the ends of both jetties and each end of the breakwater. The seaward end of the south jetty also has a fog signal. The entrance is around the southern end of the breakwater before following the buoys marking the channel. Shoaling occurs at the northern end of the breakwater (which is known as a sand trap) and southeast of the southern entrance along the shore. Enter only at the southern end of the breakwater and expect heavy pleasure craft traffic on holidays and weekends. Visibility is sometimes hampered by fog and haze occurring from July to October.

With 50 County-owned guest slips, as well as vacant slips in the various marinas, this is a popular stop-over for cruisers. Contact the Coast Guard on VHF Ch. 16 (working Ch. 12) or call (805) 382-3007 (7/24) for advice on channel entrance information if visibility or sea conditions are a concern or if this is your first entry.

For reservations contact the marinas:

Anacapa Isle Marina - (805) 985-6035, E-mail: slips@anacapaisle.com

Channel Islands Marina - (805) 985-7558, E-mail: cimarina@aol.com

Pacific Corinthian Marina - (805) 984-2847, E-mail: paccormar@aol.com

Vintage Marina - (805) 984-3366, E-mail: vintagemarina@idock.usa.com.

Peninsula Yacht Anchorage - (805) 985-6400,
E-mail: peninsulayacht@tcresidential.com

For information regarding the Channel Islands Yacht Club visit their Website at www.ciyc.com

A weekly Farmer's Market is held in the parking lot between Vintage Marina and Channel Islands Harbor Marina. Local events include an annual boat show held about the end of July and a seafood festival that takes place near the middle of September.

OXNARD, CA

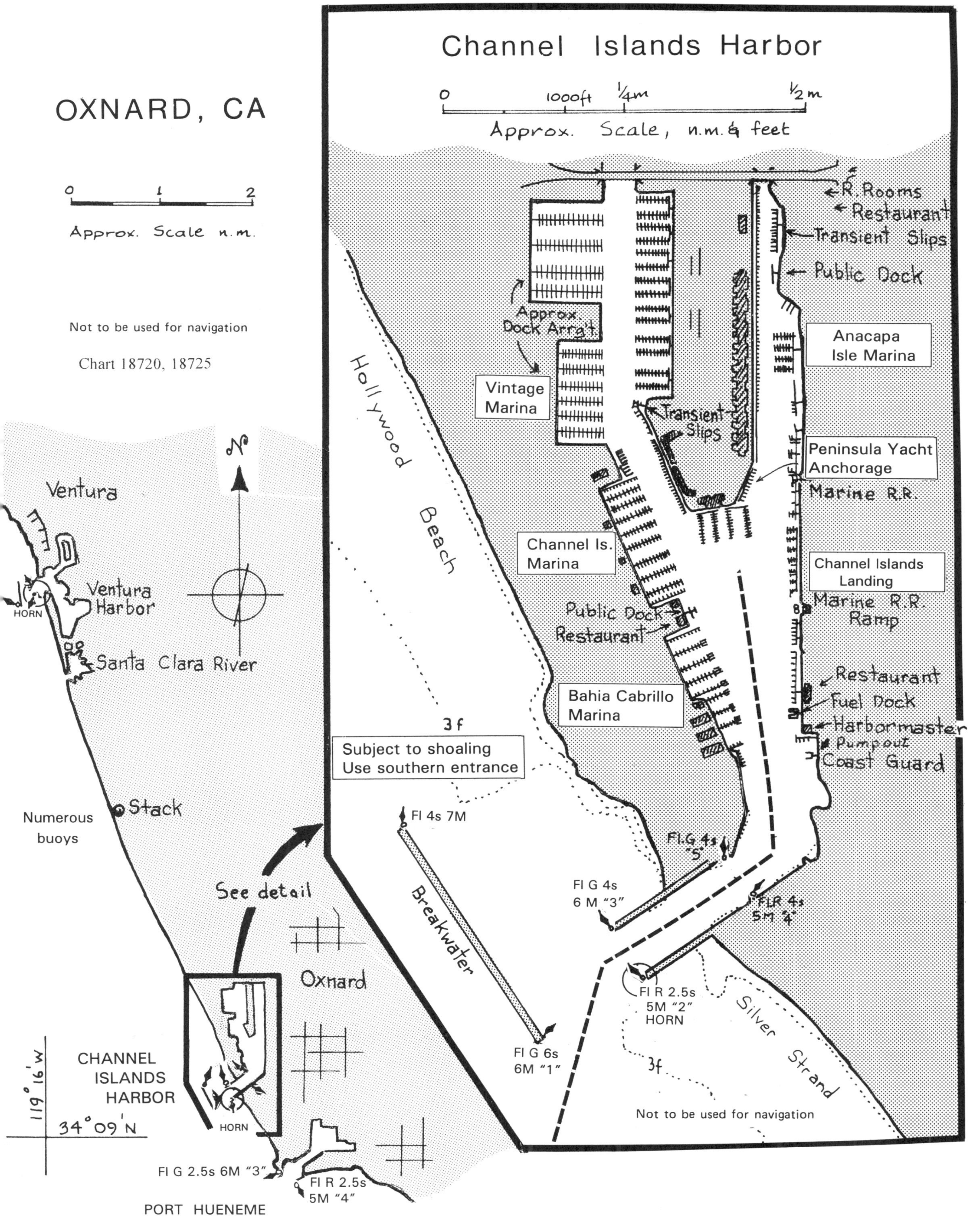

0 1 2
Approx. Scale n.m.
Not to be used for navigation
Chart 18720, 18725
Ventura
Ventura Harbor
HORN
Santa Clara River
Stack
Numerous buoys
See detail
Oxnard
CHANNEL ISLANDS HARBOR
HORN
119° 16'W
34° 09'N
Fl G 2.5s 6M "3"
Fl R 2.5s 5M "4"
PORT HUENEME
Channel Islands Harbor
0 1000ft 1/4m 1/2m
Approx. Scale, n.m.& feet
R. Rooms
Restaurant
Transient Slips
Public Dock
Approx. Dock Arrg't
Anacapa Isle Marina
Vintage Marina
Transient Slips
Peninsula Yacht Anchorage
Marine R.R.
Hollywood Beach
Channel Is. Marina
Channel Islands Landing
Marine R.R. Ramp
Public Dock
Restaurant
Restaurant
Fuel Dock
Harbormaster
Pumpout
Coast Guard
Bahia Cabrillo Marina
3f
Subject to shoaling
Use southern entrance
Fl 4s 7M
Fl.G 4s "5"
Breakwater
Fl G 4s 6 M "3"
Fl.R 4s 5M "4"
Fl R 2.5s 5M "2" HORN
Silver Strand
Fl G 6s 6M "1"
3f
Not to be used for navigation

LOS ANGELES and LONG BEACH AREA

Los Angeles and Long Beach combine to form the largest and most densely populated commercial concentration on the Pacific Coast. Within this 35 miles of coastline are found two of the largest deep-sea ports on the Pacific Coast: Los Angeles has over 100 cargo-sized piers and wharves and Long Beach has more than 70.

Because of the heavy international and inter-coastal traffic, extreme caution must be exercised when transiting this area. Take note of the Traffic Separation Scheme in effect and don't linger in the busy lanes and Precautionary Zones. When crossing traffic lanes, do so quickly and at a right angle. Always give way to large ocean-going vessels as they have difficulty maneuvering in the confined traffic lanes and need a great distance to stop. For them the many pleasure craft hereabouts are an annoying blip on the radar screen and may not even be noticed.

The Traffic Separation Scheme routes were recently moved slightly, and some charts currently available may not show the newest boundaries. Check the local Notice to Mariners for the new boundaries if your charts are pre-2003 editions.

In addition to heavy commercial traffic, a large number of recreational vessels are moored in this area. Weekend traffic of small craft is aptly referred to as the LA Marine Freeway. To satisfy the needs of such a large number of boaters, small craft facilities have been developed in every possible space. Accommodation in over 20,000 wet berths is to be found at Marina del Rey, Redondo Beach (King Harbor), San Pedro, Los Angeles Harbor, Long Beach Harbor and Alamitos Bay.

Demands of a large and steadily increasing boating population result in expensive moorage, often $10 or more per foot per month, with a waiting period as long as 4 or 5 years for certain sizes of boats. Given such pressure from boaters wanting permanent moorage it is easy to understand the very limited number of slips set aside for transients. Municipal docks vary in their allotment of space for transients, though temporary moorage may be obtained from most private marinas on a hot-berthing basis.

This guide lists marina facilities but will not give rates and details related to the many private marinas, since they frequently change. One marina in Marina del Rey increased its rates three times in one year, so it is impossible for a guide such as this to keep up to date. A list of marinas, along with their current telephone number, is given in Appendix III. This list is probably already out of date at the time of printing, but it does give a possible starting point for a cruiser wanting temporary moorage.

In addition to congested commercial and pleasure boat traffic the weather deserves some comment. Though fog is more common during the fall and winter months, visibility is often reduced during summer months by a gray-brown haze (smog). Another concern is the short warning period that weather forecasts can give of the onset of Santa Ana winds. These easterly desert winds may reach 50 knots or more in a very short time, and may extend as much as 50 miles offshore. Monitor local weather forecasts for a warning of this violent winds before departure.

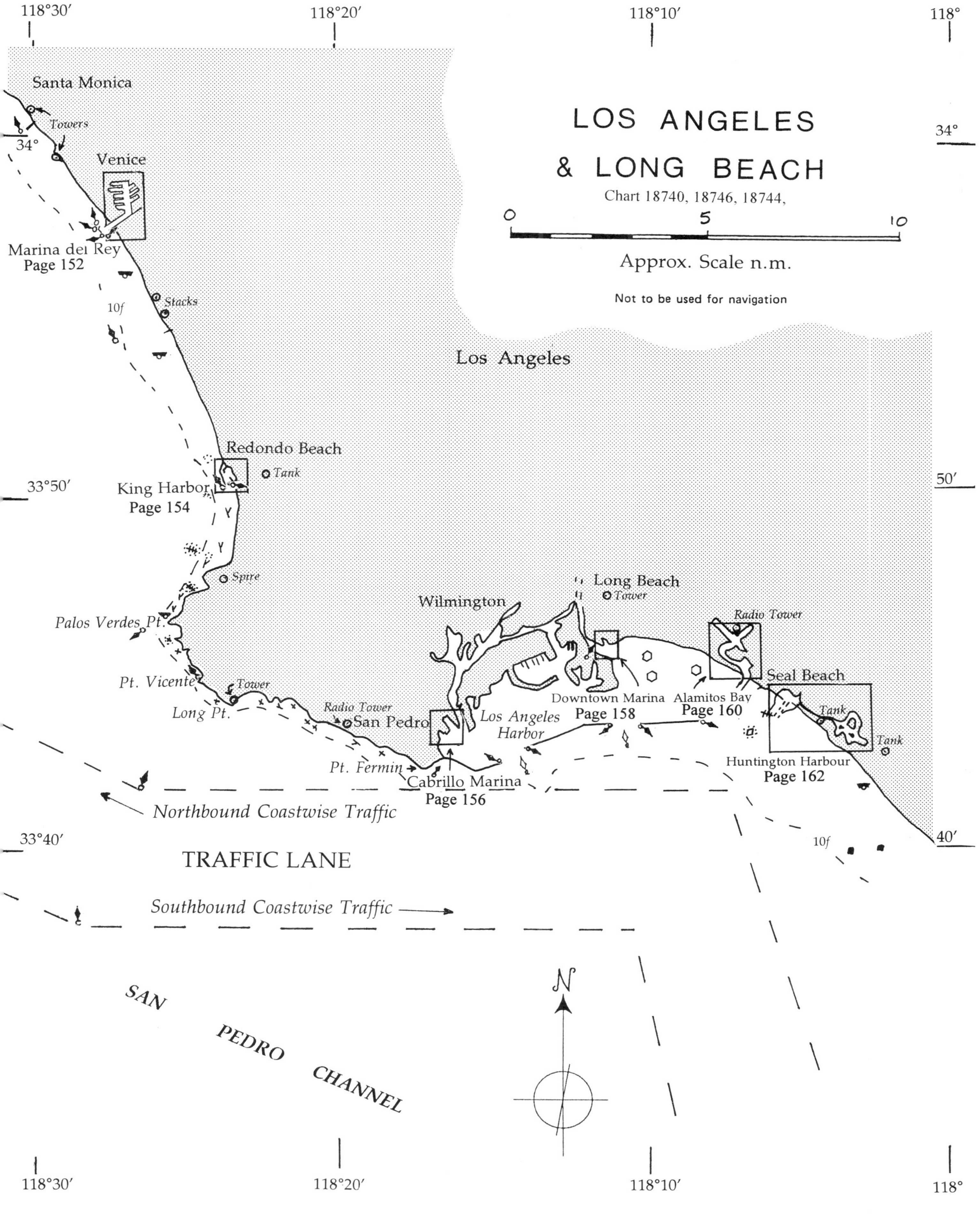
LOS ANGELES
& LONG BEACH
Chart 18740, 18746, 18744,
0
5
10
Approx. Scale n.m.
Not to be used for navigation
118°30′
118°20′
118°10′
118°
34°
33°50′
50′
33°40′
40′
Santa Monica
Towers
Venice
Marina del Rey
Page 152
Stacks
10f
Los Angeles
Redondo Beach
Tank
King Harbor
Page 154
Spire
Palos Verdes Pt.
Pt. Vicente
Tower
Long Pt.
Radio Tower
San Pedro
Pt. Fermin
Cabrillo Marina
Page 156
Wilmington
Los Angeles
Harbor
Long Beach
Tower
Downtown Marina
Page 158
Radio Tower
Alamitos Bay
Page 160
Seal Beach
Tank
Tank
Huntington Harbour
Page 162
Northbound Coastwise Traffic
TRAFFIC LANE
Southbound Coastwise Traffic
SAN PEDRO CHANNEL
N
118°30′
118°20′
118°10′
118°

MARINA DEL REY

Home port to over 6,000 pleasure boats, this is the ultimate marine basin. Administered by the County of Los Angeles, it is an immense center devoted to yachting with 9 yacht clubs and 16 marinas based here. It is favorably located some distance from the activity of commercial traffic in the Los Angeles/Long Beach Harbor area while being conveniently close to the cruising offered in the offshore islands. It is about 42 miles ESE of Channel Islands Harbor and 8 miles NNW of Redondo Beach (King Harbor).

Landmarks in the area include the long Santa Monica pier 3.5 miles to the northwest and the shorter Venice pier 1 mile to the north. Three miles south of the entrance are two prominent stacks at El Segundo and lighted bell buoy "2ES" which is about 2 miles offshore. Two jetties with lights at their seaward ends protect the channel entrance; a fog signal is at the end of the northern jetty. A detached breakwater with lights at both ends is seaward of the jetties.

Marina del Rey is 1.2 miles northwest of Los Angeles International Airport. During restricted visibility, vessels approaching the harbor can use the flight path of departing planes to establish a preliminary line of approach. When you reach the 10-fathom curve, proceed carefully to the 5-fathom curve, then alter course to the northwest and follow the depth contour to the south entrance.

First-time visitors are advised to approach using the northern entrance as shoaling occurs between the south jetty and the end of the breakwater. If approaching from the south, watch depths carefully and stay toward the outer breakwater to avoid shoal areas. Anchoring is not allowed in the restricted area shoreward of the detached breakwater and continuing for 305m (1,000 ft.) into the entrance channel between the jetties. Within the channel, shoaling occurs along the northern jetty. The main channel is dredged and shoal areas are marked with buoys. Buoys within the channel define areas for sail and power vessels. Power vessels must keep the buoys to port when entering or leaving the harbor; sailing vessels are restricted to the center channel. During settled conditions, northbound vessels generally use the north opening and southbound vessels exit via the south opening. But during heavy weather all vessels should use the north opening as it is less affected by shoaling. During foggy conditions do not be misled by the Venice Pier foghorn that is one mile northwest of the Marina del Rey jetty fog horn.

A pergola and an impressive statue of a helmsman are seen at Chace Park on the north side of Basin H where the transient dock is located. During business hours 6:00 to 10:00 p.m. go to the Community Building in Chace Park, adjacent to the transient dock. When arriving after hours register in the morning. In an emergency, anchorage (fore and aft) can be taken in the northern part of the main channel above 33°58.6'N after obtaining permission from the harbormaster by calling (310) 305-9503.

There are two fuel docks in the basin, and all boating needs can be met by the many boat-oriented businesses hereabouts. A launch ramp (open 24 hours daily) is located at the head of Basin D. Propane is available at Marina Boat and RV Storage (310-823-7979). A 24-hour grocery store is in the Marina Shopping Center at Mindanao and Admiralty. Many restaurants and specialty shops line the harbor making interesting shore trips. An internet café is two blocks away at "The Cows End," on Washington Blvd. Marine supplies are available at Ships Store located between Basins C and D on Panay Way and at two West Marine stores southeast of Basin H. Marina identification using the keyed indicators are found on page 201.

MARINA DEL REY, CA

Not to be used for navigation

Chart 18744

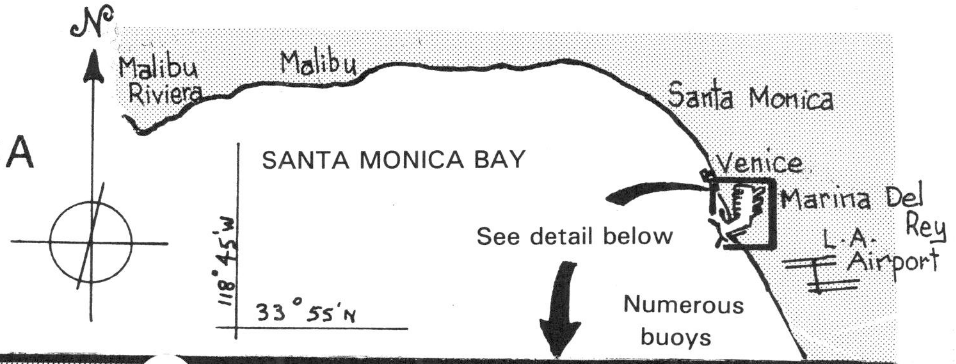

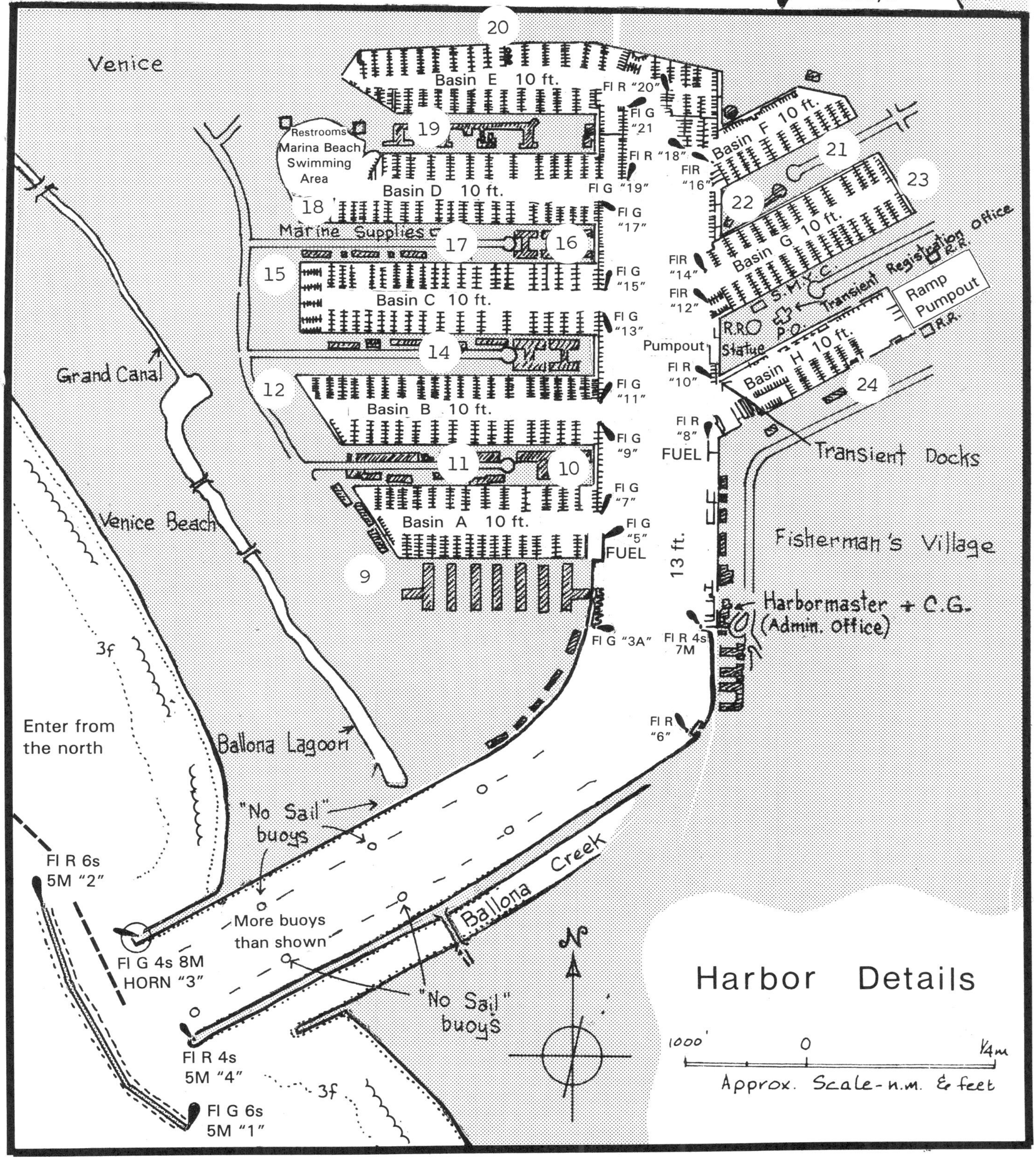

REDONDO BEACH (KING HARBOR)

This man-made harbor was widely televised in January, 1988 as it suffered millions of dollars worth of damage in the violent storm which struck the Californian coast. The breakwater protecting the harbor was breached in several places, boats were sunk and 60m (200 ft.) of the pier were destroyed. Repairs have returned the harbor to its normal status, though a few uncharted wrecks and debris are still to be found.

Just 8 miles south of Marina del Rey and about 20 miles around Palos Verdes Point from Los Angeles Harbor, **King Harbor** is a moorage devoted to pleasure craft. Eight large power plant smokestacks at the north end of the harbor make this an easy spot to identify. A private light is shown from the top of the power plant.

Two breakwaters with lights at their tips protect the entrance to the harbor; the eastern breakwater also has a fog signal at the light. The western breakwater is about 365m (1,200 ft.) long and curves from the north end of the harbor in a roughly SSE direction to give protection to the three basins.

Some areas to avoid: When approaching from the north, do not pass within the 9-fathom line as there are underwater hazards in this area. The areas to the south of the eastern breakwater should also be avoided as shoaling occurs in that location. The sports fishing barges anchored up to 2 miles offshore during summer months should also be given a wide berth. However, the "restricted area" shown offshore and to the south of Marina del Rey does not apply to recreational craft.

When approaching from any direction, pick up the safe-water bell buoy (RW "RB" MoA) which is about 210m (230 yds.) SSW of the tip of the western breakwater, then proceed to the entrance. Fog and haze may reduce visibility, especially during summer months. Natural oil seepage results in gas bubbles, blobs of oil, or a thin film of oil on the water in an area 1.5 to 4 miles offshore. Under normal weather conditions there are no problems in entering or using this harbor.

Though no transient slips are available from the harbormaster, permission to anchor can be obtained from the Harbor Patrol. A large area between the breakwater and western side of the channel (indicated with green channel markers) affords good anchorage. The best area is at the north end of the harbor, but you must watch for shoaling. There is a 72-hour limit, and bow and stern anchors are required. When available, transient berths can be obtained from the private marinas: **King Harbor Marina,** Tel. (310) 376-6926 or Fax (310) 376-9927, **Portofino,** Tel. (310) 798-5869 or Fax (310) 406-2501, **Port Royal Marina** Tel. (310) 376-0431 and **Redondo Beach Marina** Tel. (310) 374-3481, Fax (310) 372-7999. Two yacht clubs are based in the harbor. Fuel, marine supplies, hoists and most cruising needs can be met in the vicinity.

The nearest Internet cafe is a short walk away at Catalina Coffee (126 North Catalina between Diamond and Emerald Streets), a good place for coffee and baked goods open from 6:30 a.m. to 10 p.m.

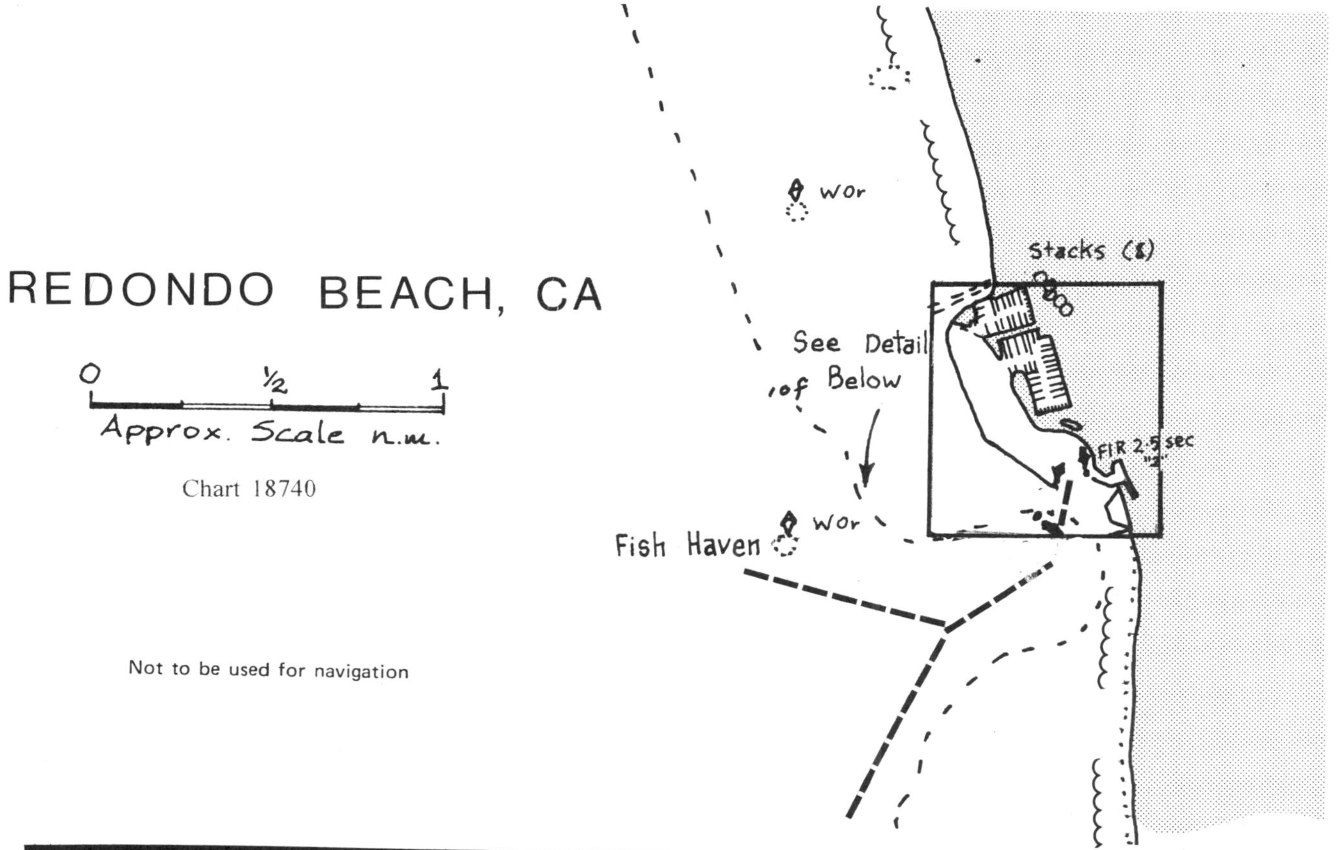
REDONDO BEACH, CA
0 ½ 1
Approx. Scale n.m.
Chart 18740
Not to be used for navigation
Wor
Stacks (8)
See Detail Below
10f
Wor
Fish Haven
Fl R 2.5 sec "2"

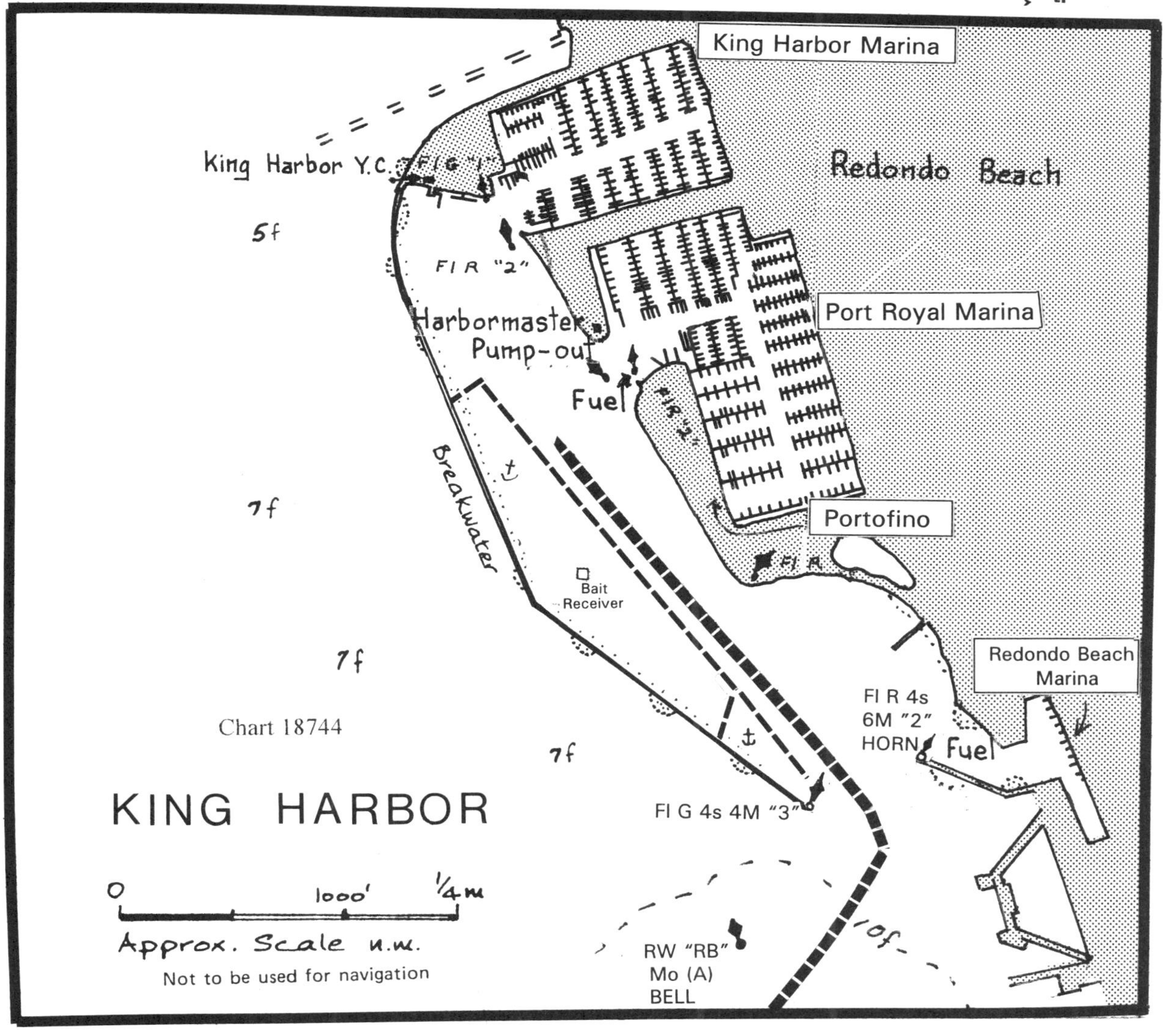
King Harbor Marina
Redondo Beach
King Harbor Y.C.
Fl G "1"
5f
Fl R "2"
Port Royal Marina
Harbormaster
Pump-out
Fuel
Breakwater
7f
Portofino
Fl R
Bait Receiver
7f
Redondo Beach Marina
Fl R 4s 6M "2" HORN
Fuel
Chart 18744
7f
KING HARBOR
Fl G 4s 4M "3"
0 1000' ¼m
Approx. Scale n.m.
Not to be used for navigation
RW "RB" Mo (A) BELL
10f

LOS ANGELES

Los Angeles Harbor has extensive port facilities developed for commercial purposes. Yacht slips, now numbering over 4,000, compete for space in a long-established port. Transient cruising vessels are not a high priority to this busy harbor and as a result there is a grand total of 1 public slip allocated for visiting yachts. Many private marinas have waiting lists for as much as 3 years for permanent moorage. As berths become temporarily available transients may obtain moorage at these marinas on an overnight basis. Unless a cruising vessel has made prior arrangements this is not a harbor to visit expecting to find moorage without a search.

Located just to the east of Point Fermin, Los Angeles Harbor is adjacent to Long Beach Harbor. When fog or haze are not present San Pedro Hill, about 3.5 miles northwest of Point Fermin, is the most noticeable feature in the San Pedro Bay area. Atop the hill are two white radar domes. Point Fermin, the southern end of San Pedro Hill, has a light and the eye-catching pavilion housing the Korean Bell of Friendship is just to the north of the light. Other prominent landmarks are shown on the sketch.

The harbor area is protected by two large breakwaters: San Pedro Breakwater curves out from Point Fermin for almost 2 miles and has a fog signal and Los Angeles Light at its tip. Middle Breakwater (detached) protects the outer Los Angeles and Long Beach Harbors for about 3 miles; it has a light at each end. Before proceeding into the harbor check the local Notices to Mariners for information on construction activities and changes to aids to navigation as well as updates to area charts. Great care must be taken not to impede the flow of traffic when transiting the harbor entrance. Facilities in San Pedro are under the jurisdiction of the Los Angeles Harbor Commission.

A short breakwater gives protection to the moorage area at the western end of **California Yacht Marina** (formerly known as Cabrillo Marina); a launch ramp and short fishing pier are built out from Cabrillo Beach. Moorage is sometimes available for visitors on a hot-berth basis as there is just one slip designated for transient vessels. The office monitors VHF Ch 68 from 8 to 5 or contact them at (310) 732-2252, fax (310) 732-2260 or email at slips@cymcabrillo.com While the approach to the marina is well marked, shore lights at night make it difficult to distinguish the buoy and jetty lights. Proceed with caution when approaching the inner breakwater protecting the marina. A pump-out station is located at the fuel dock.

The only anchorage area in the harbor is off Cabrillo Beach. Permission to anchor for up to 10 days must be obtained from the Port Police who monitor VHF Ch 13 and 16 or phone them at (310) 732-3500. This has been dubbed "Hurricane Gulch" because of the strong, gusty winds often affecting the area from noon to late afternoon. Consequently, you must set the anchor securely with plenty of scope.

Moorage for recreational and commercial vessels is available in Fish Harbor which is bordered by canneries and fish processing plants. Call Al Larsen at (310) 832-0526 to arrange use of a mooring ball.

Some of the varied attractions in Los Angeles include: the Museum of Natural History, Museum of Art, George C. Page Museum of La Brea Discoveries, Los Angeles State Arboretum, Virginia Robinson Gardens, South Coast Botanic Garden and Descanso Gardens.

LOS ANGELES, CA
(San Pedro)

0 1 2 3
Approx. Scale, n.m.

Not to be used for navigation

Chart 18746, 18749, 18751

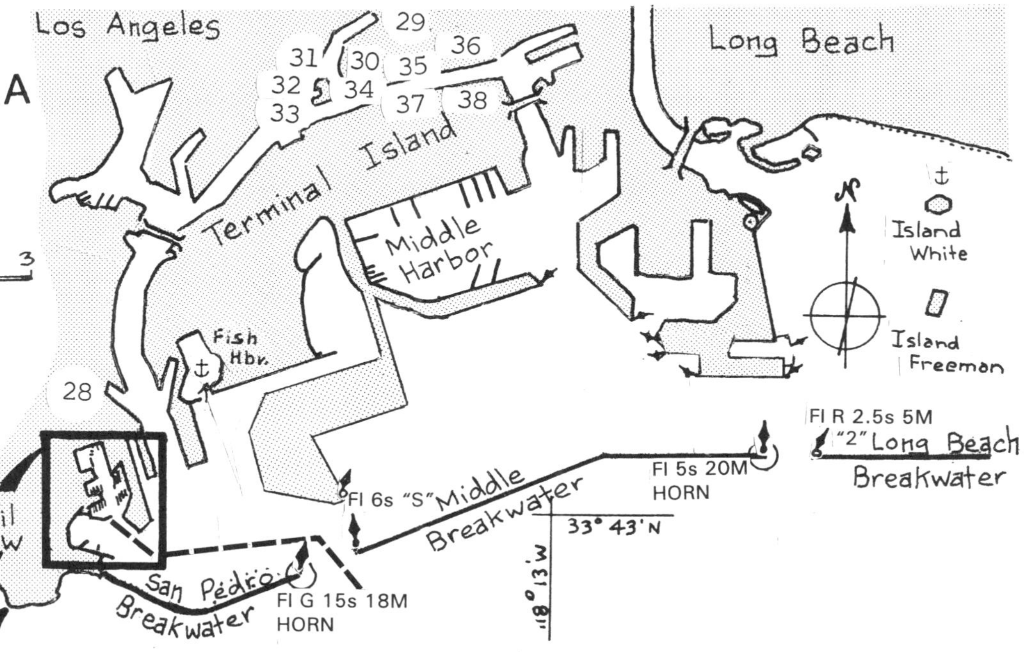

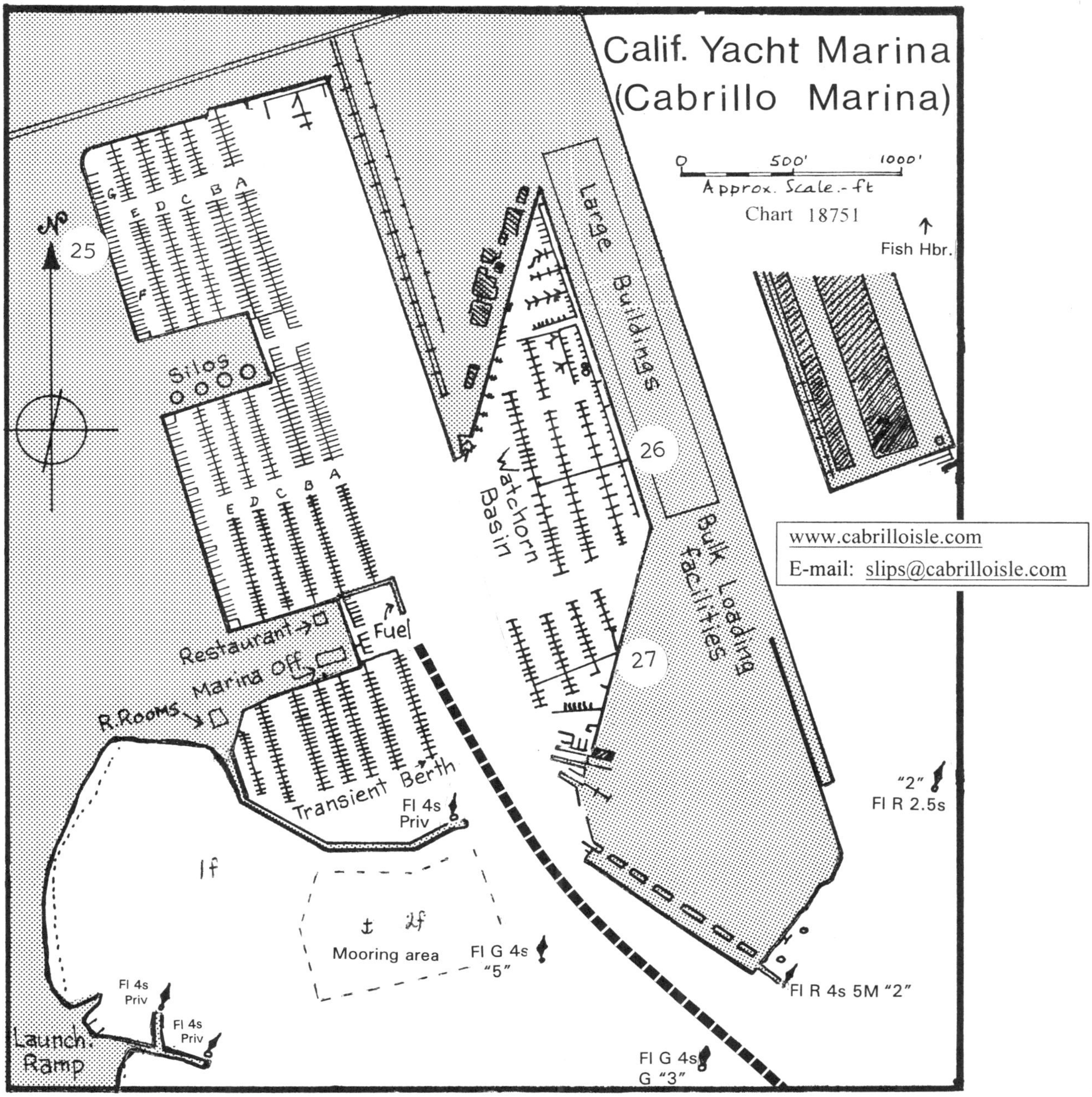

LONG BEACH

The Long Beach Harbor entrance is just 4 miles east of the Los Angeles Harbor entrance. Protecting the harbor is the Long Beach Breakwater which is about 2 miles long and is lit at both ends. The 4-mile long Middle Breakwater gives protection to both Long Beach and Los Angeles Harbors. Terminal Island, a cluster of huge port facilities shared by both harbors is beyond this breakwater. The two most eye-catching landmarks are a large white dome (which formerly housed Howard Hughes' Spruce Goose) and the massive Queen Mary, which sit side by side west of the Downtown Marina.

The entrance to the harbor lies between the eastern end of Middle Breakwater and Long Beach Breakwater. As this is a very busy container cargo port a cruising vessel must take care not to impede the large deep-sea vessels moving about the harbor. In addition, particularly on weekends, a multitude of pleasure craft of every shape and size are found crisscrossing in the confined space making an offshore cruiser long for the open sea.

Daylight approaches to Long Beach are relatively easy but at night the myriad lights in the harbor and on shore make identification of navigation aids difficult. Do not confuse the flashing light at the top of the Port Authority Building which is almost 46m (150 ft) high with the Long Beach Light which is about 15m (50 ft.) high. The breakwater is low and difficult to see from a distance and vessels without radar should approach with caution to make a positive identification of the lights before entering the harbor. If you are uncertain, approach from the east end of the breakwater, between it and two lighted oil platforms which are more easily identified. A bait barge is moored just inside the east end of the breakwater during the fishing season. Although its position is not charted, the barge is usually well marked and lighted. Passenger ferries running from Long Beach to Catalina Island follow a course close aboard the mooring buoys adjacent to Pier J. These vessels usually move quickly without regard for recreational traffic and should be given a wide berth.

There are two municipal small craft facilities in Long Beach Harbor: the **Long Beach Shoreline Marina** (formerly known as the Downtown Marina)on Queensway Bay and **Alamitos Bay Marina** (formerly known as Long Beach Marina) in Alamitos Bay. In addition there are many private marinas, yacht clubs and small craft facilities scattered around the various channels. Over 6,000 pleasure craft moor in Long Beach waters. Extensive renovations to the Long Beach Shoreline Marina commenced in the fall of 2004 and are expected to take two years to complete during which time there will be no moorage available to transient vessels. For information call (562) 570-4950, fax (562) 570-1799 or visit their website at www.longbeach.gov

For future reference after renovations have been completed the entrance to the Long Beach Shoreline Marina is between Island Grissom and the curved breakwater which forms the southwestern edge of the basin. Do not attempt to enter between the northeast side of Island Grissom and the breakwater forming the eastern edge of the basin. A dangerous wreck is .5 mile SSW of the entrance and a sunken wreck is reported about .2 mile to the west.

Anchorage may be taken north of Island White, to the east of Long Beach Shoreline Marina.

LONG BEACH, CA

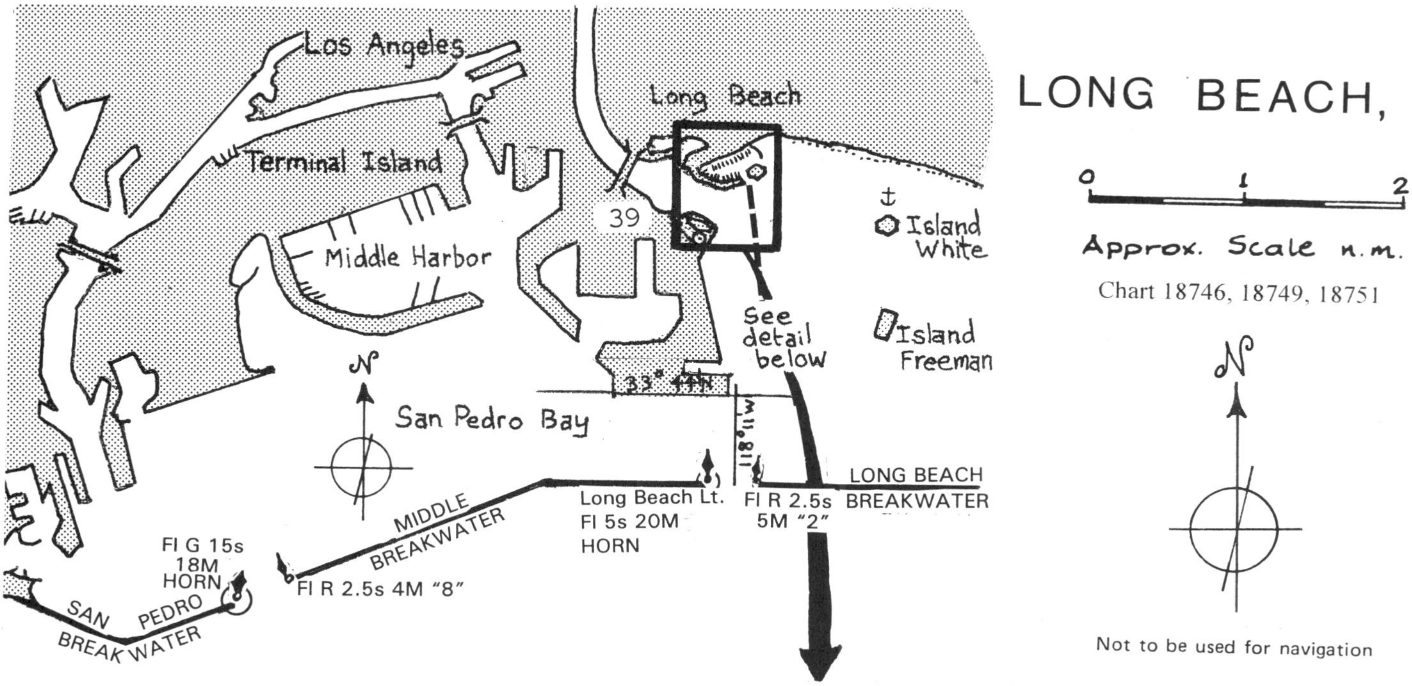

Approx. Scale n.m.

Chart 18746, 18749, 18751

Not to be used for navigation

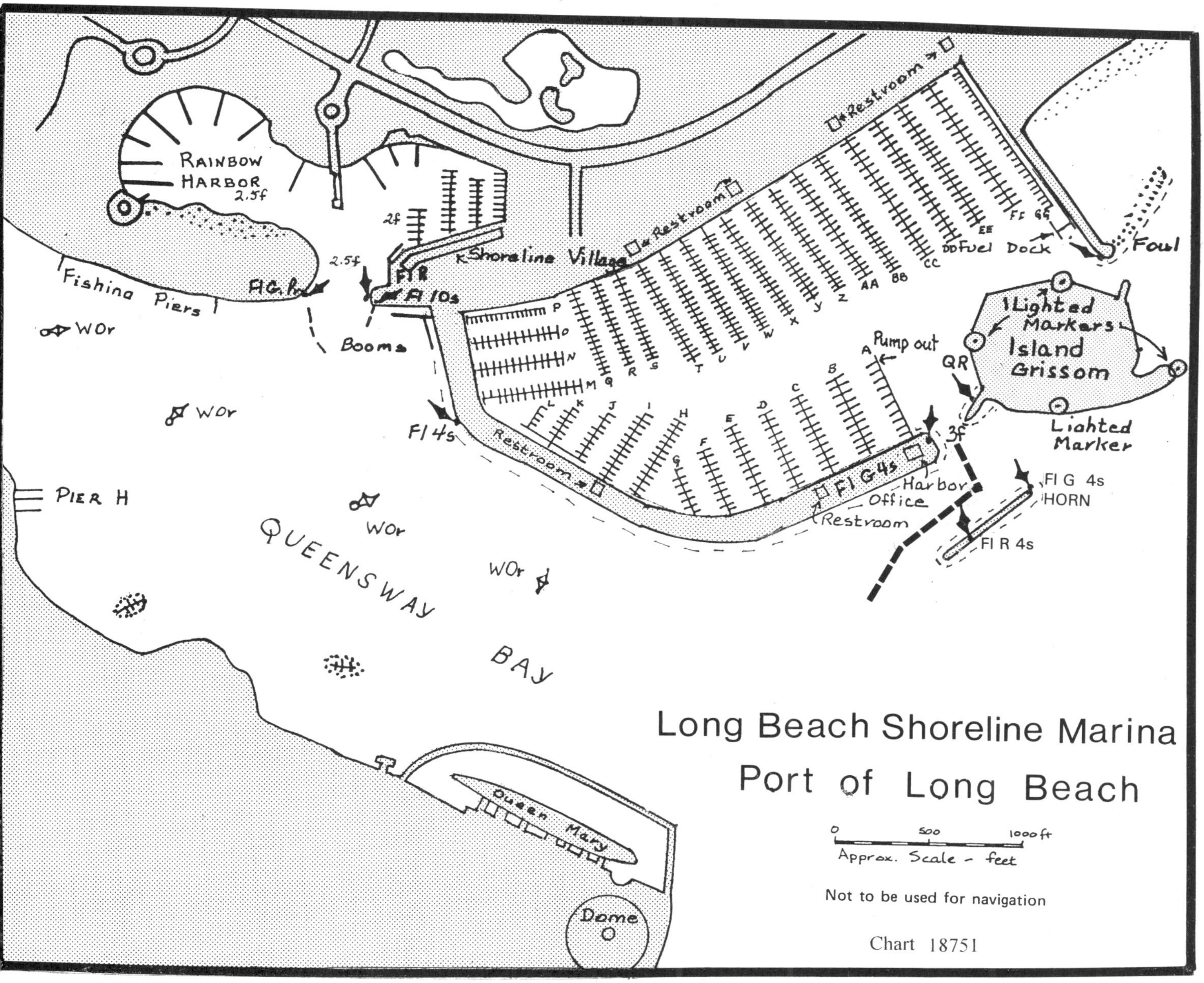

Long Beach Shoreline Marina
Port of Long Beach

Not to be used for navigation

Chart 18751

Alamitos Bay entrance channel is northeast of the eastern end of the Federal breakwater defining Long Beach Harbor. When approaching Alamitos Bay, follow the channel between the jetties and avoid the area east of the entrance where the San Gabriel River creates shoals. (Do not confuse the entrance to Alamitos Bay with that of Anaheim Bay about 1.15 miles to the southeast. Anaheim Bay is marked with a series of entrance buoys while Alamitos Bay has only the jetty lights.) The entrance is between two lit jetties that are about .5 mile north of Seal Beach pier. A fog signal is located at the end of the western jetty. The approach can be rough as swells wrap around the Federal Breakwater and refract around the Alamitos Bay and Anaheim Bay jetties. Once inside the entrance channel, these swells quickly subside.

A line of buoys within Alamitos Bay defines the lanes allocated for passage: keep the buoys to port when traveling in either direction. To counter the effects of shoaling dredging is frequent in the area. Give a wide berth to dredges and any marker buoys to avoid the submerged dredge hoses linked to the buoys. The offshore oil drilling islands in the vicinity are named after astronauts who died while on duty, Islands White, Freeman, Chaffee and Belmont.

The **Alamitos Bay Marina** is 3.25 miles ESE of the Long Beach Shoreline Marina at the east end of the Los Angeles-Long Beach harbor. Moorage for transients is provided on end-ties and reservations are recommended, especially for holiday weekends. Calling (562) 570-3215 or fax (562) 570-3247. Reservations are on a first-come, first-serve basis and you must show or fax the current Vessel Registration Card. Payment may be made using a credit card or cash. The maximum stay for visiting vessels is 15 consecutive days per month. Daily office hours are from 7:30 to 7 pm. A refundable deposit is required for washroom keys.

Other marinas in the area include:

Los Angeles Yacht Club, San Pedro - Berth 32	(310) 831-1203
Colonial Yacht Anchorage, Wilmington - Berth 204	(310) 830-1161
Shelter Point Yachting Service, San Pedro - Berth 41-A	(310) 832-7507
Fleitz Brothers, San Pedro - Berth 41-A	(310) 832-0334
Westrec Marinas, San Pedro - Berth 42-43	(310) 833-4468
Cabrillo Beach Y.C. (Cabrillo Landing), San Pedro - Berth 35	(310) 519-1694
Lighthouse Yacht, San Pedro - Berth 205-B	(310) 833-7934
Holiday Harbor/Cabrillo Marina, San Pedro - Berth 34	(310) 833-4468
Leeward Bay Marina, Wilmington - Berth 200-I	(310) 830-5621
Island Yacht No. 2, Wilmington - Berth 200-X	
Holiday Marinas, San Pedro - Berth 201	(310) 833-4468
California Yacht Marina, Wilmington - Berth 202,#36	(310) 834-7916
Cerritos Yacht Anchorage, Willmington - Berth 205-C	(310) 834-4737
Pacific Yacht Landing, Wilmington - Berth 203	(310) 830-0260
Cabrillo Beach Yacht Club, San Pedro - Berth 56-57	(310) 519-1694

Boatyards in the Long Beach/San Pedro area that cater to small craft:

1. Al Larson Boat Shop (Marina), Terminal Island - Berth 257	(310) 832-6404
2. Al Larson Boatyard, Long Beach - Berth 58	(562) 436-3322
3. Nielsen Beaumont Marine, Inc., Long Beach - Berth 75	(562) 437-1972
4. San Pedro Boat Works, San Pedro - Berth 44 and Berth 57	(310) 832-7203

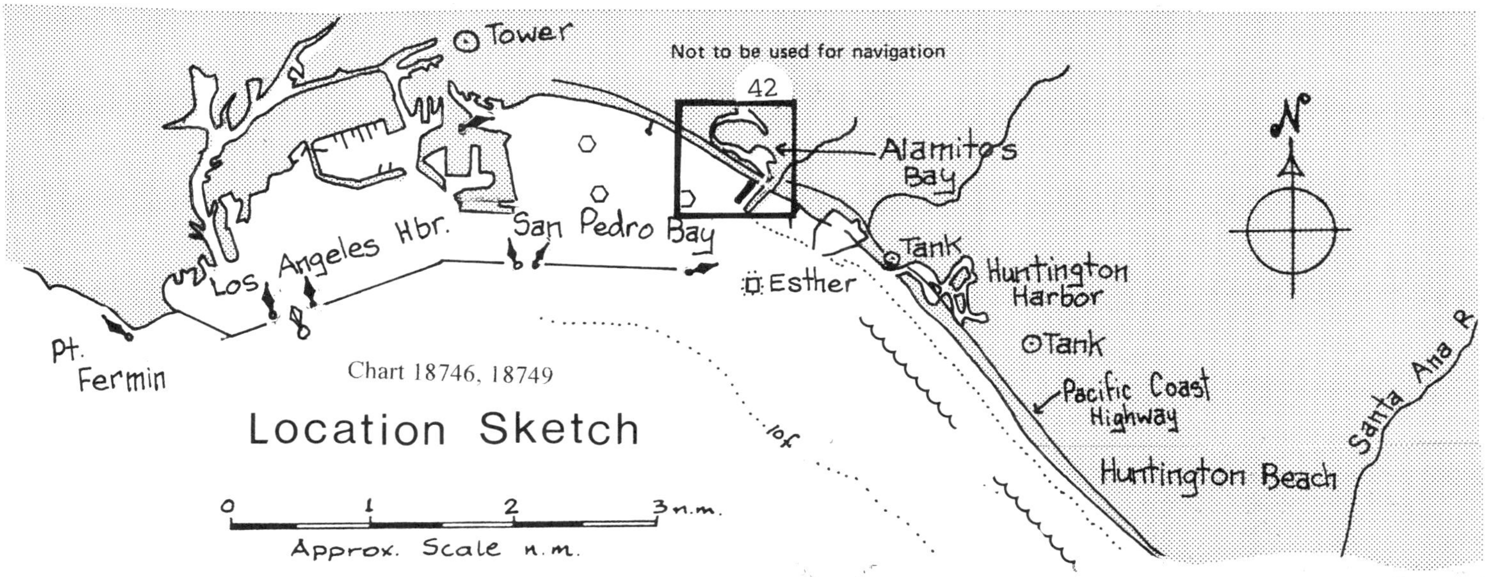

Not to be used for navigation
42
Tower
Alamitos Bay
Los Angeles Hbr.
San Pedro Bay
Esther
Tank
Huntington Harbor
Tank
Pt. Fermin
Chart 18746, 18749
Location Sketch
10f
Pacific Coast Highway
Huntington Beach
Santa Ana R.
0 1 2 3 n.m.
Approx. Scale n.m.
N

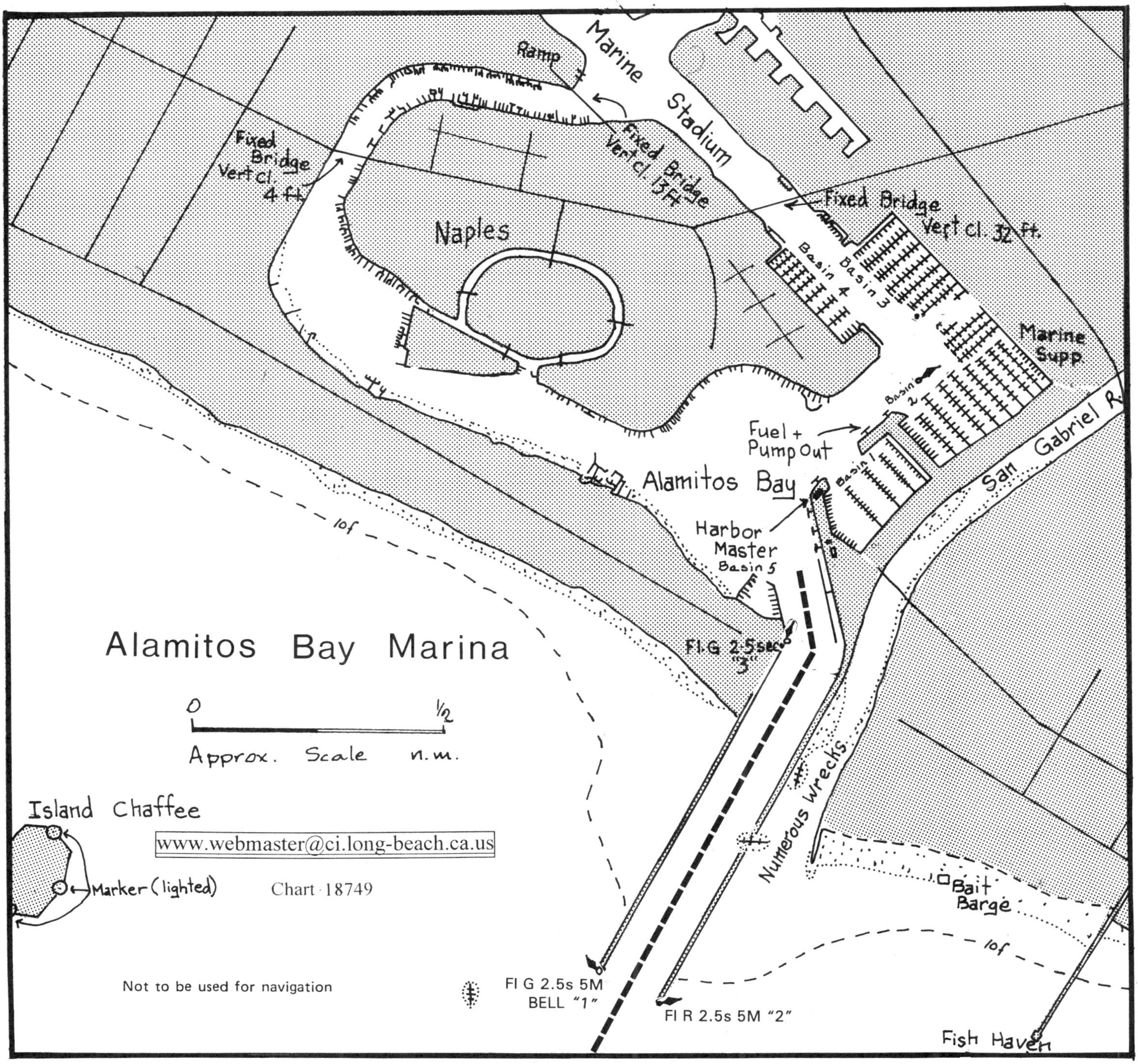

Ramp
Marine Stadium
Fixed Bridge Vert cl. 4 ft.
Fixed Bridge Vert cl. 13 Ft
Fixed Bridge Vert cl. 32 ft.
Naples
Basin 4
Basin 3
Marine Supp.
Basin 2
Fuel + Pump Out
Basin 1
San Gabriel R.
Alamitos Bay
Harbor Master
Basin 5
10f
Fl G 2.5sec "3"
Alamitos Bay Marina
0 1/2
Approx. Scale n.m.
Island Chaffee
www.webmaster@ci.long-beach.ca.us
Marker (lighted)
Chart 18749
Numerous Wrecks
Bait Barge
10f
Not to be used for navigation
Fl G 2.5s 5M BELL "1"
Fl R 2.5s 5M "2"
Fish Haven

ANAHEIM BAY (HUNTINGTON HARBOUR)

Anaheim Bay is part of the Seal Beach Naval Weapons Station and is a major ordnance depot for Naval explosives. In the past the bay was subject to a variety of restrictions. Currently, the most significant restriction with respect to pleasure craft entering Huntington Harbour is the low bridge clearance of 7m (23 ft.) where the Pacific Coast Highway crosses over the entrance. Speed limits and restrictions on the use of VHF radio within the bay are clearly posted and must be adhered to. Close to the Los Angeles/Long Beach conglomeration, Anaheim Bay is only 1.5 miles south of the entrance to Alamitos Bay, and 14 miles northwest of Newport Bay. Two yacht clubs and some 2,500 recreational boats make this their home port.

The Navy maintains large unlit buoys on both sides of the entrance channel where storage barges are moored. The channel is well marked, with buoys positioned from well outside the entrance jetties, and range markers clearly indicating the center of the channel. Follow the marked channel when traversing the entrance to the harbor, staying away from the Weapons Station vessels and moorings. Some of the shoal areas adjacent to the channel are visible at low tide making it essential that you stay in the channel. Rowboats, kayaks and smoking are prohibited and sailboats must use their engines in the inner harbor.

When approaching the harbor pick up lighted bell buoy "1" and lighted whistle buoy "2," which are a little less than a mile west of the entrance to the outer bay. Two additional markers indicate the direct course to follow to the entrance. Two long jetties protect the outer part of Anaheim Bay; lights are at the seaward ends and a fog signal is also at the end of the west jetty. The enclosed waters are within a danger zone, while an explosives anchorage is east of the channel. The channel is marked by buoys, lights and a lighted range. Anchorage is not permitted in Anaheim Bay.

After you have passed under the bridge stay well within the small channel buoys the inner harbor is subject to significant shoaling on both sides. Follow the S-curve of the channel and watch your depths. The shoaling ends in the area of the fuel dock. The harbor is dredged periodically.

Huntington Harbour is a private development composed of waterfront homes, each having a private dock. The only mooring facilities are provided by two private marinas operating in the harbor offering the usual amenities. **Peter's Landing Marina** is open 9 a.m. to 5:30 p.m. on weekdays, 10 a.m. to 5:30 p.m. on Saturday and 11 a.m. to 4 p.m. on Sunday. For information contact the office at Tel (714) 840-1387, (562)-4441 or Fax (562) 592-3816. Information is available on their website www.peterslandingmarina.com

Daily office hours for **Sunset Aquatic Marina** are 8 a.m. to 5 p.m. and they can be called at (714) 846-0179. A number of marine supply and repair businesses can satisfy most boating needs.

The Huntington Beach Pier, less than 5 mile north of the entrance to Anaheim Bay, was struck by a storm in January, 1988. While being televised, the outermost 76m (250 ft.) of the pier was completely destroyed as was the restaurant perched at the seaward end, appropriately named, "The End Cafe." Reconstruction of this well-known landmark has long since been completed.

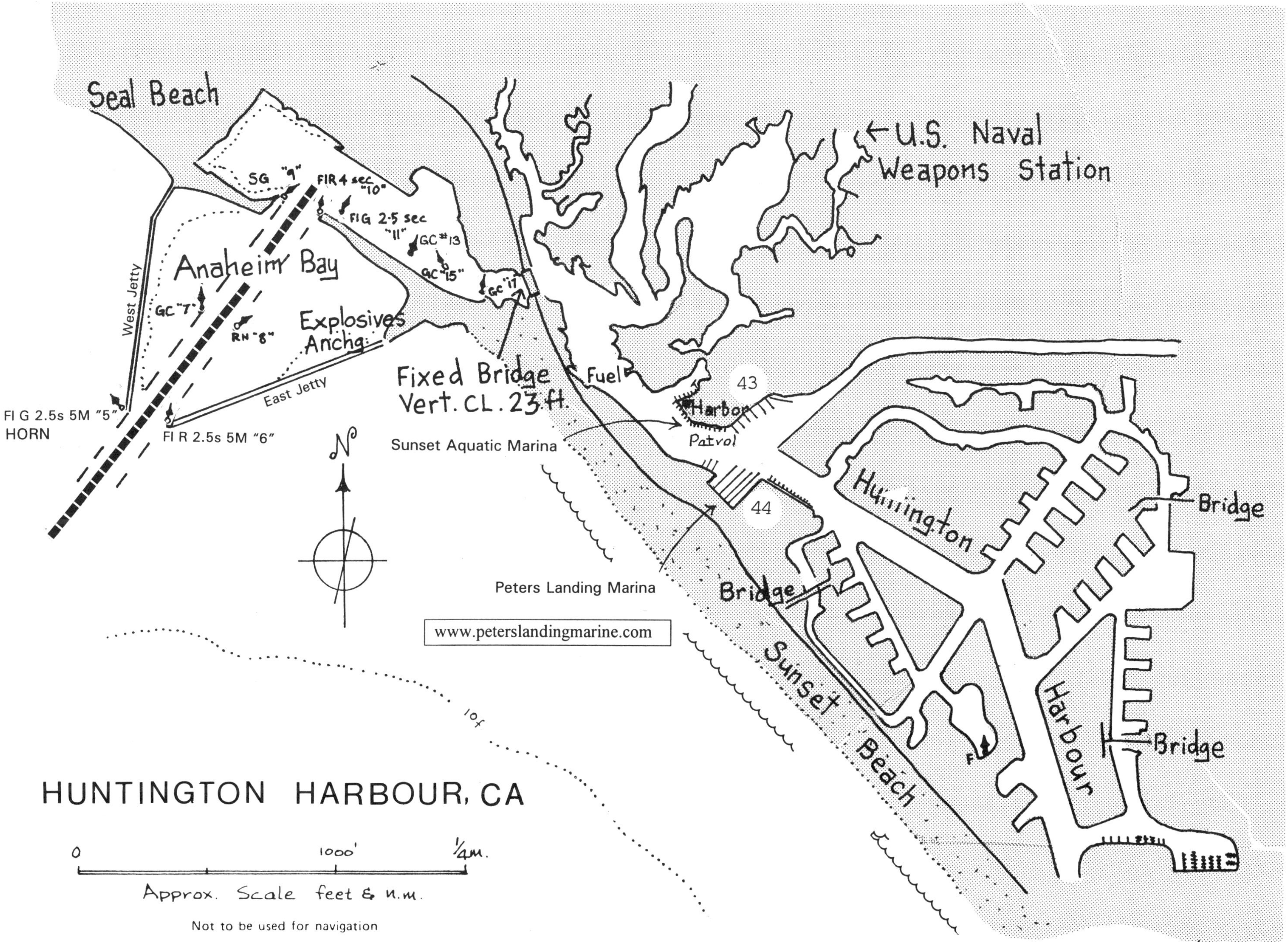

HUNTINGTON HARBOUR, CA

Not to be used for navigation

Chart 18746, 18749

NEWPORT BEACH

Until 1870 this land-locked harbor at Newport Beach was a quiet lagoon from which periodic shipments of livestock and grain were made. The entrance was first dredged in 1876, but it wasn't until 1936 that the jetties were built and the present channels and turning basins were dredged. Next door to the Los Angeles/Long Beach area, the prosperous city of Newport Beach is 14 miles southeast of Huntington Beach and 12 miles north of Dana Point. This busy yachting center is home to over 9,000 vessels, seven yacht clubs and numerous sport fishing and marine oriented businesses. A Coast Guard rescue boat is based in the harbor.

The high-rise office towers in downtown Newport—about 1.5 miles north of the entrance—make good landmarks. Balboa Pier (244m/800 ft. long) is over a mile northwest of the harbor entrance while Newport Pier (290m/950 ft.) is about 2 miles further up the coast. Two jetties, both with lights at their seaward ends protect the channel leading into the harbor. The west jetty has a fog signal at the light and shoal areas are marked by buoys.

Moorage within the harbor is under the jurisdiction of the Orange County Harbor Patrol which can be reached at (949) 723-1002. Reservations cannot be made as moorage is on a first-come, first serve basis; visiting boaters must provide the current Vessel Registration. Five berths for transient vessels (maximum length - 12m/40 ft.) are available at the Guest Dock near the Harbor Patrol Office. In contrast to most marine developments, boats are moored (fore and aft) rather than being tied to docks. When available, buoys in other parts of the harbor may be assigned to transients for up to a 20 days. Restrooms and showers are available, but the nearest laundromat is located on Balboa Island on Agate Street.

Protected anchorage in good holding mud may be taken in the area east of Lido Island for a maximum of 5 days <u>after</u> obtaining permission from the Harbor Patrol. Santa Ana winds from the northeast can make this spot rather sloppy so set the anchor with this possibility in mind. Someone must be left on board at all times; the Harbor Patrol may board the vessel to check that this rule is being followed. Take care that the vessel does not swing out of the specified area—if it does, the Orange County Police may issue a ticket. When anchoring here, it is a 1.5 mile trip by dinghy to the showers (or 3 miles overland). Since the shoreline is privately owned, landing by dinghy is difficult though you may get permission to land at the Sea Scouts base or at the Orange County Rowing Club. In an emergency anchorage can be taken in the Corona Del Mar Bend but it is adversely affected by winter storms from the south and can become quite uncomfortable.

All boating needs can be met by the many marine oriented shops in the bay except for replenishing the vessel's propane—the closest outlet for it is at Orange Coast Trailer Supply on 16th Street in Costa Mesa. Two marine supply stores of note are Minney's Yacht Exchange just a short walk up the hill and West Marine on Pacific Highway.

Two or three ferries carry up to four vehicles between the Balboa Island fuel dock and the "Fun Zone" landing on Balboa Peninsula. When transiting the channel between these points you should be on the lookout for these ferries in addition all the other traffic dashing about in the confined waters. Adjacent to the "Fun Zone" near Hills Fuel Dock are restaurants, amusement rides, gift shops and other tourist-oriented businesses. A number of restaurants within the harbor have dinghy docks for water-borne patrons. Reservations are recommended at restaurants providing this facility, particularly on week-ends.

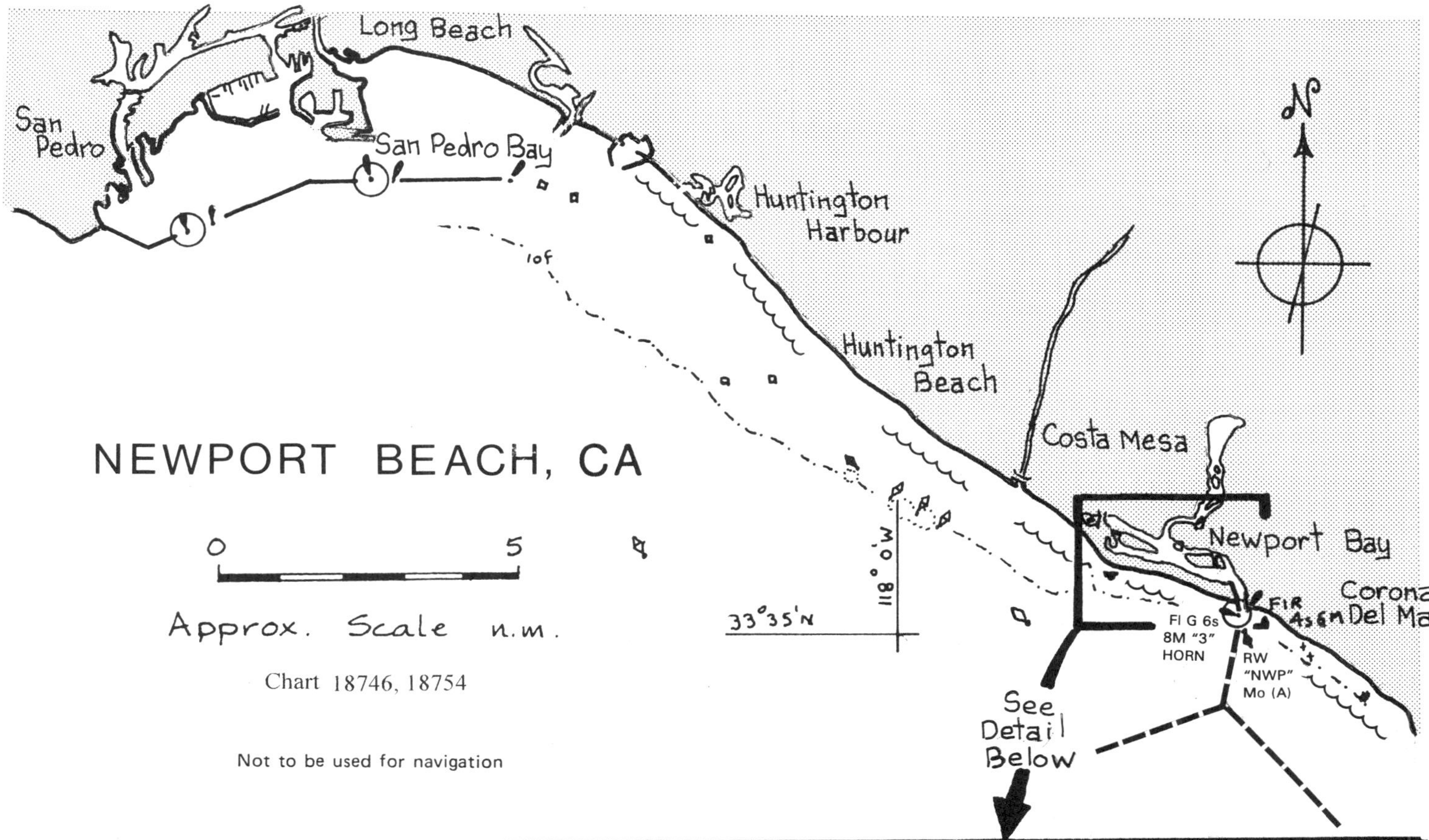
NEWPORT BEACH, CA
0
5
Approx. Scale n.m.
Chart 18746, 18754
Not to be used for navigation
Long Beach
San Pedro
San Pedro Bay
Huntington Harbour
10f
Huntington Beach
Costa Mesa
Newport Bay
Corona Del Mar
118° 0'W
33°35'N
Fl G 6s 8M "3" HORN
Fl R 4s 6M
RW "NWP" Mo (A)
N
See Detail Below

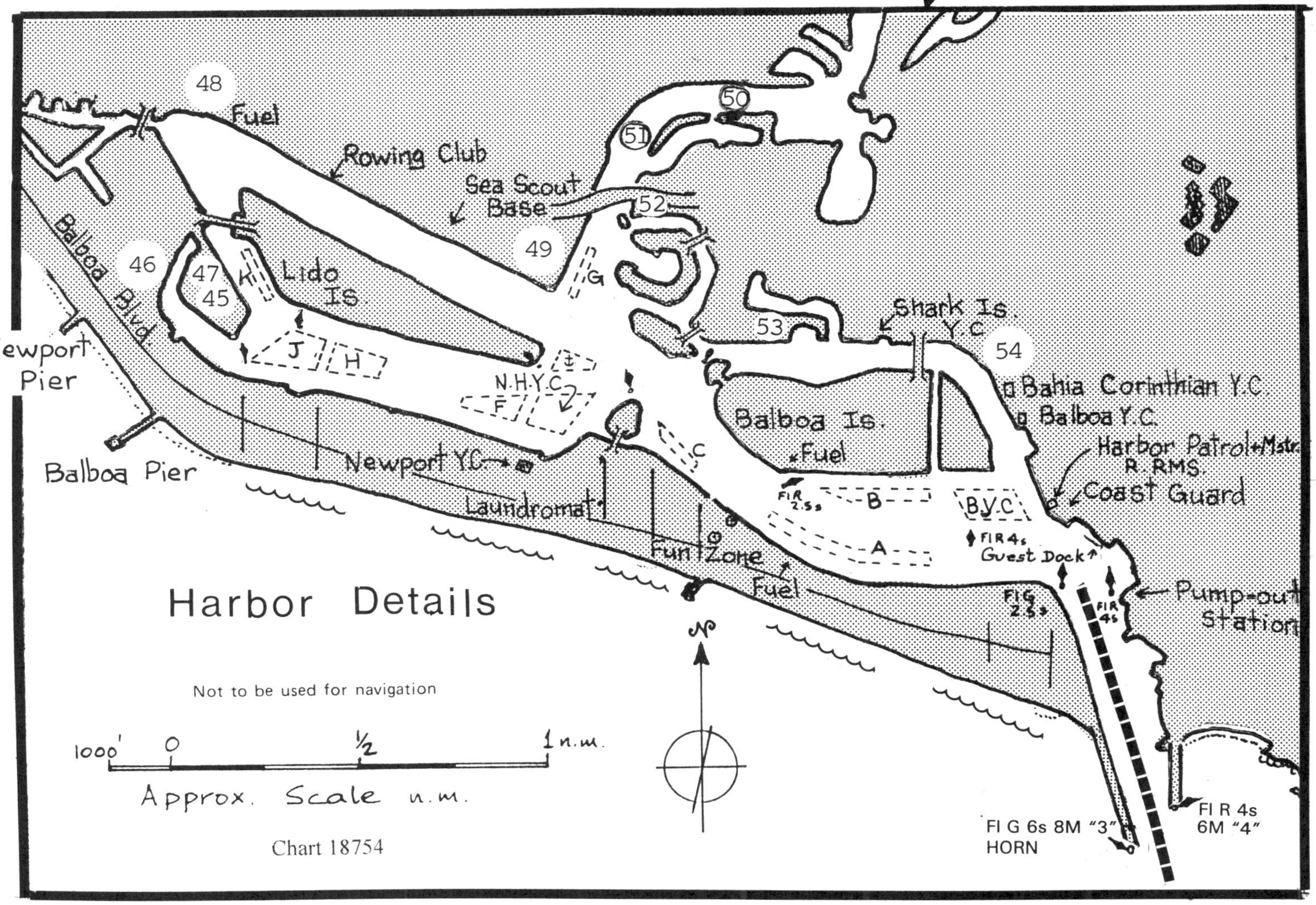
Harbor Details
Not to be used for navigation
1000'
0
1/2
1 n.m.
Approx. Scale n.m.
Chart 18754
48
Fuel
Rowing Club
Sea Scout Base
50
51
52
49
46
47
45
K
Lido Is
G
J
H
53
Shark Is. Y.C.
54
Newport Pier
Balboa Blvd
N.H.Y.C.
F
I
Bahia Corinthian Y.C.
Balboa Y.C.
Balboa Is.
Fuel
C
Harbor Patrol+Mstr. R. RMS.
Coast Guard
Balboa Pier
Newport Y.C.
Laundromat
Fl R 2.5s
B
B.Y.C.
A
Fl R 4s
Guest Dock
Fun Zone
Fuel
Fl G 2.5s
Fl R 4s
Pump-out Station
N
Fl G 6s 8M "3" HORN
Fl R 4s 6M "4"

SOUTHERN CHANNEL ISLANDS

The Southern Channel Islands comprise four islands (Santa Barbara, San Nicolas, San Clemente and Santa Catalina) scattered over a fairly large area. Normal weather consists of fairly calm conditions until about noon, when northwesterly breezes pick up and create sloppy seas (and headwinds for those crossing to the islands from southerly ports). A close eye to the weather and the vessel's position should be maintained, for the area is subject to fog, Santa Ana winds and other local disturbances.

The largest and closest island to the mainland is **Santa Catalina** (often shortened to Cat Island), less than 25 miles southwest of Los Angeles. It has numerous coves where mooring buoys are in place or anchorage may be taken during settled weather.

Santa Barbara is an isolated, wind-swept island surrounded by kelp. Less than 1 square mile in area, the saddle-shaped, treeless island has its highest point at Signal Peak 194m (635 ft.). Sutil Island is a rocky islet (91m/300 ft.) .4 miles W of the S point of Santa Barbara. Further west, about 185m (200 yds.) offshore, is a smaller islet. A dangerous underwater rock found .7 mile west of the point seldom breaks in a calm sea; in addition, kelp surrounding the island indicates foul bottom. A light 60m (195 ft.) above water is shown from a pyramidal structure on the northeast point of the island.

Santa Barbara is the only one of the southern Channel Islands included in the Channel Islands National Park. Waters surrounding the island up to 6 miles offshore are included in the Channel Islands National Marine Sanctuary and those up to 1 mile offshore are an Ecological Reserve.

During WWII Santa Barbara was the site of a U.S. Navy early warning outpost. A Quonset hut left from this period is now used as the Park ranger station and visitor center. The island is a breeding ground for sea lions and elephant seals; many different land and sea birds also nest here. In the spring the island is ablaze with color when the giant coreopsis (a relative of the sun- flower) reach 3m (10 ft.) with their bright yellow blossoms. Temporary day anchorage during calm weather only, may be taken at Landing Cove (on the northeastern shore); no special permits are necessary to go ashore.

San Nicolas and **San Clemente Islands** are military reservations, making them off-limits except with special written permission. Since most transient cruisers and local sailors have no desire to dodge military exercises, there is little interest in visiting these islands. But for those who choose, information on the times and locations for missile firing may be obtained by calling "Plead Control" on VHF Ch. 16 or the Range Officer at Point Mugu at (805) 982-8841. Aircraft with a loud hailer will order a vessel's return to port if the course observed is unsafe. The few anchorages available are described in Brian Fagan's *Cruising Guide to Southern California's Offshore Islands*.

Magnetic disturbances of up to 5° have been recorded up to 3 miles offshore on the north, east and south coasts of San Clemente Island.

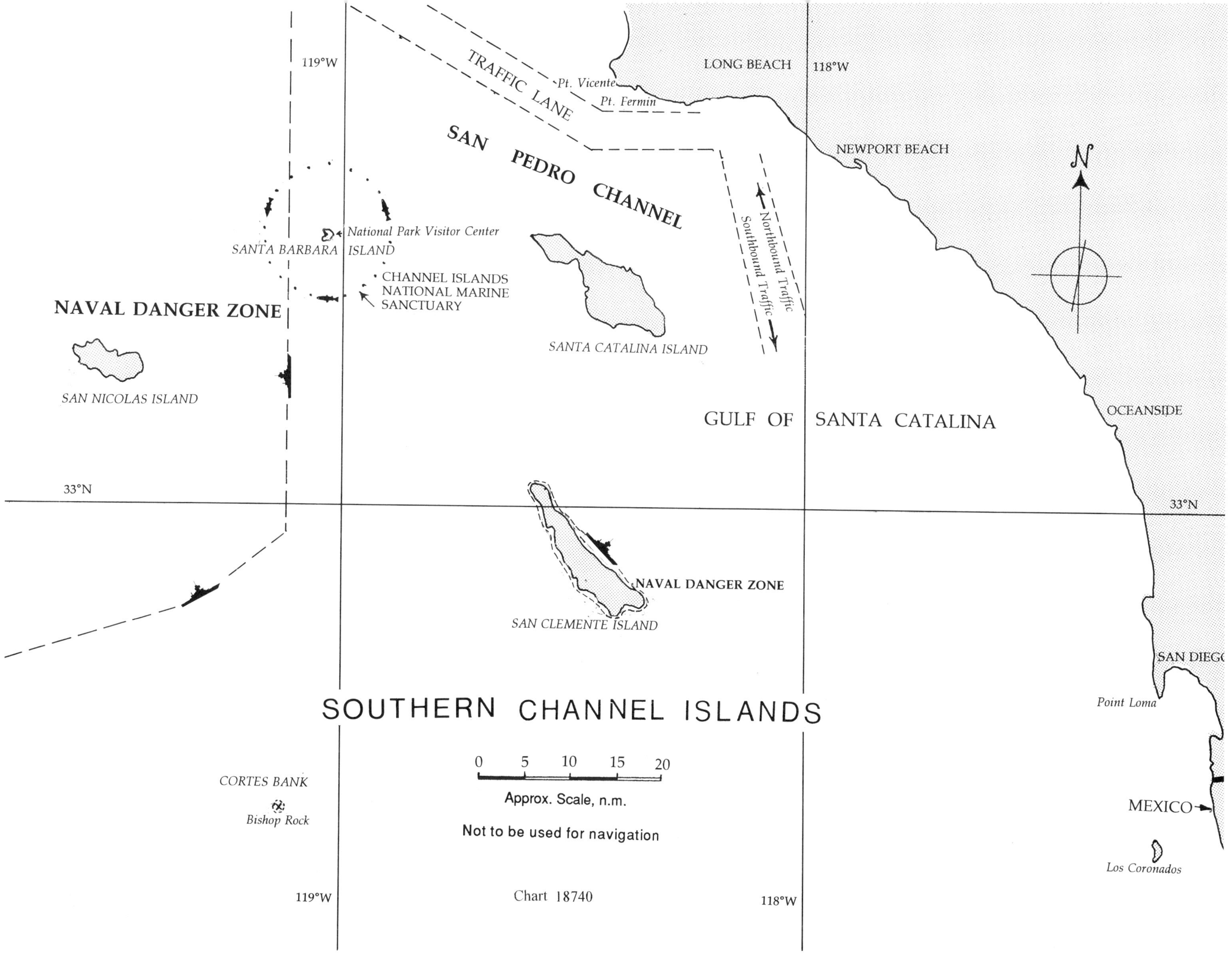

LONG BEACH
NEWPORT BEACH
OCEANSIDE
SAN DIEGO
Point Loma
MEXICO
Los Coronados
Pt. Vicente
Pt. Fermin
TRAFFIC LANE
SAN PEDRO CHANNEL
Northbound Traffic
Southbound Traffic
N
118°W
119°W
33°N
National Park Visitor Center
SANTA BARBARA ISLAND
CHANNEL ISLANDS
NATIONAL MARINE
SANCTUARY
NAVAL DANGER ZONE
SAN NICOLAS ISLAND
SANTA CATALINA ISLAND
GULF OF SANTA CATALINA
SAN CLEMENTE ISLAND
NAVAL DANGER ZONE
SOUTHERN CHANNEL ISLANDS
0 5 10 15 20
Approx. Scale, n.m.
Not to be used for navigation
CORTES BANK
Bishop Rock
Chart 18740

SANTA CATALINA ISLAND

This rugged island lies only 18 miles south of Point Fermin. The eastern shore is generally free of kelp, though a narrow band is found close to the western coast. There are no offlying dangers except in the vicinity of Isthmus Cove and a few rocks close to the shore. This is the most developed of the Channel Islands; not only does it have the little city of Avalon, but many of its coves are filled with privately-owned mooring buoys. The Santa Catalina Island Conservancy owns the land, except for the city of Avalon. Various private organizations operate summer camps at the head of many of the coves, and shore access is restricted. Landing without special permission is permitted on Catalina Island Conservancy property, Twin Harbors, Avalon, Descanso Bay and Hamilton Beach.

The best time to visit the island is prior to June 15th or after the Labor Day holiday in early September. The island's close proximity to the heavily populated urban areas of southern California makes it a poplar destination for thousands of sailors wanting to "get away from it all." During summer months the coves are filled to capacity, leaving little space for transient sailors, who will find only deep waters available for anchoring. The best time to find moorage buoys available is between Sunday and Thursday. Another reason not to visit Catalina during July and August is to avoid the noises produced by countless summer campers. Early morning renditions of "Reveille" and exuberant wake-up calls of "Rise and shine, happy campers," blasted over loudspeakers may not be your idea of a peaceful way to start the day.

In order to accommodate the maximum number of vessels, privately owned mooring buoys are located in most of the niches. If the owner is not using the buoy it is rented to sailors on a first-come, first-served basis. Mooring buoys permit fore and aft attachment by means of a spreader line, so the Harbor Patrol assigns buoys based on the vessel's length. First, pick up the floating pole so that the hawser can be raised and attached to samson post or bow cleat. (Bring the pole aboard to prevent it thumping against the hull of the vessel for the duration of your stay.) Follow the spreader line astern, until reaching the stern hawser, then attach it to a samson post or stern cleat. Though privacy is limited, many vessels are efficiently secured in a minimum of space.

Most of the coves are located on the northern shore of the island where they are exposed to northeasterly winds and thus to dangerous Santa Ana conditions. Because the island is only 18.5 miles long there is sufficient time to scoot around to protection in the lee of the island. The island may easily be circumnavigated in either direction; Avalon being such an obvious landmark, it will be used as a starting point.

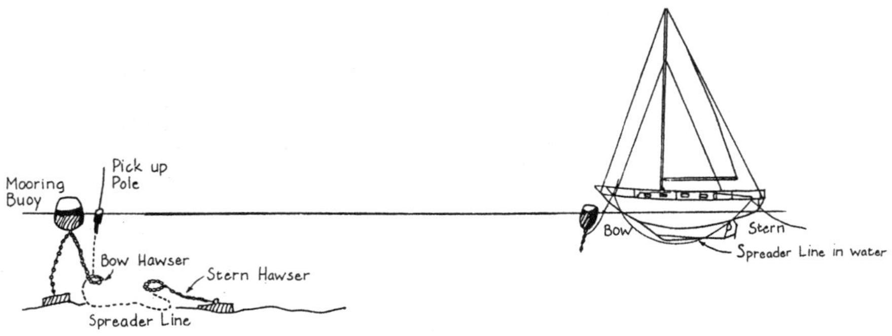

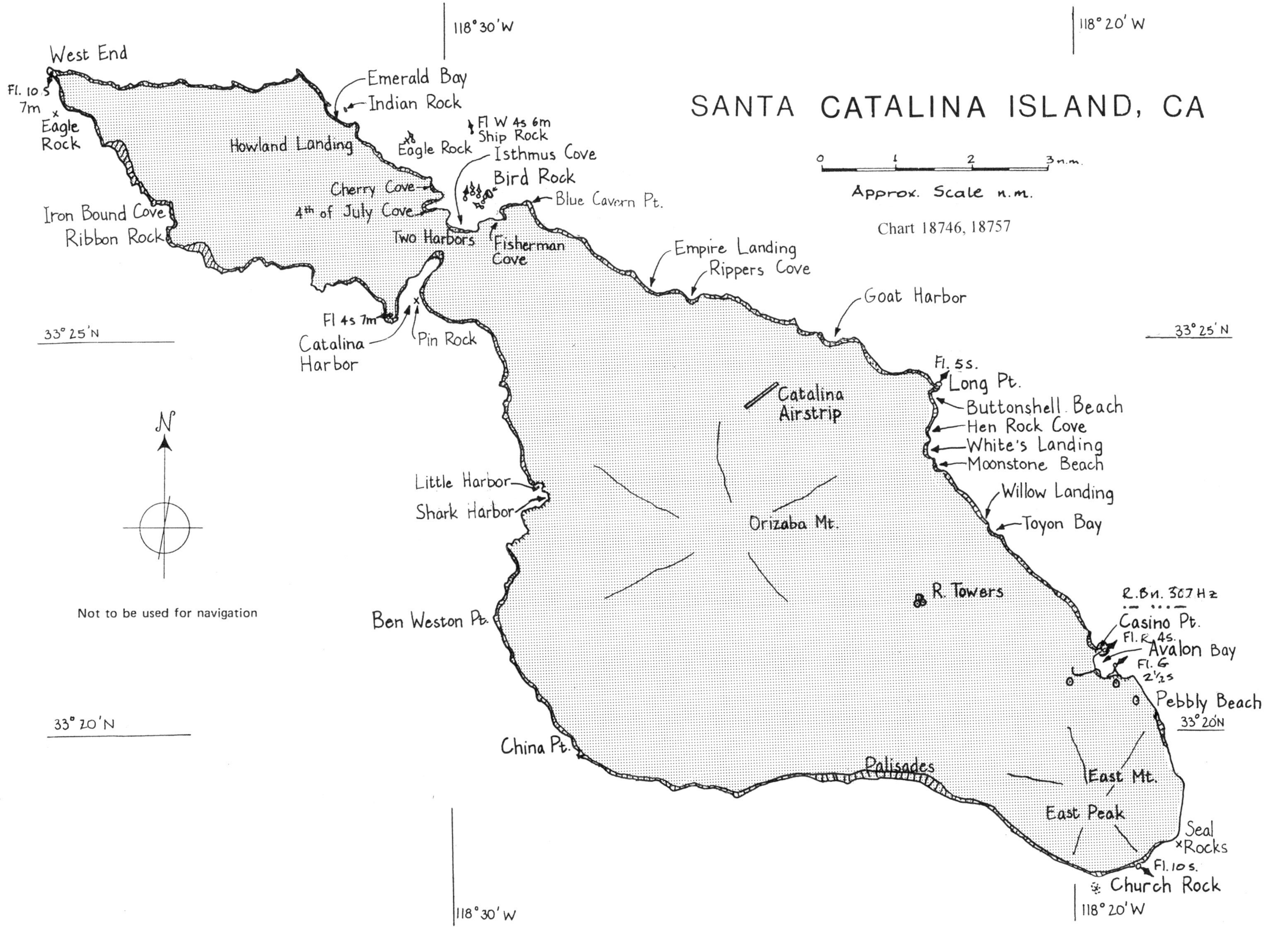
SANTA CATALINA ISLAND, CA
0 1 2 3 n.m.
Approx. Scale n.m.
Chart 18746, 18757
118°30'W
118°20'W
33°25'N
33°20'N
West End
Fl. 10 s
7m
Eagle Rock
Iron Bound Cove
Ribbon Rock
Howland Landing
Emerald Bay
Indian Rock
Eagle Rock
Fl W 4s 6m
Ship Rock
Isthmus Cove
Bird Rock
Blue Cavern Pt.
Cherry Cove
4th of July Cove
Two Harbors
Fisherman Cove
Fl 4s 7m
Catalina Harbor
Pin Rock
Empire Landing
Rippers Cove
Goat Harbor
Fl. 5s.
Long Pt.
Buttonshell Beach
Hen Rock Cove
White's Landing
Moonstone Beach
Willow Landing
Toyon Bay
Catalina Airstrip
Orizaba Mt.
R. Towers
Little Harbor
Shark Harbor
Ben Weston Pt.
China Pt.
Palisades
R.Bn. 307 Hz
Casino Pt.
Fl. R 4s.
Avalon Bay
Fl. G 2½s
Pebbly Beach
East Mt.
East Peak
Seal Rocks
Fl. 10 s.
Church Rock
N
Not to be used for navigation

AVALON HARBOR

Avalon owes its existence to tourism. Though its wintertime population is about 3,000, the summer influx of visitors swells the numbers to 10,000 or more. Tourist boats arrive several times a day, disgorging hundreds of visitors from the mainland. It has been described as a "people-watcher's Paradise, better than a zoo!"

Located on the north coast of the island, Avalon is just .5 mile from the southeastern tip. It is easy to identify as it is the only concentration of buildings on the island. Other landmarks are the Casino Point breakwater on the western side of the entrance, aptly named for the conspicuous circular building resembling a casino. In fact the building is a theater and dance hall, constructed during the big band era and is still in use. The Casino, a nearby Carillon and the huge Hamilton Cove Condominiums located high on the hill can be seen from some distance offshore.

The entrance to the harbor is between two lighted breakwaters, one off Casino Point on the north and the other extending from Cabrillo Peninsula on the south. Some protection is afforded from southwesterly weather, but Santa Ana winds can cause damage to vessels and facilities within the bay. Approach to the harbor is straightforward, the only caution relating to heavy boat and ferry traffic. The harbormaster monitors VHF Ch. 16, switching to Ch. 12 when a mooring buoy will be assigned if one is available. If the vessel has to move to another spot the skipper will be informed between 7:30 a.m. and 9:00 a.m. There is a minimum charge for vessels wishing to moor for 2 hours or less. The nearest anchorage area is west of the Casino in 25 fathoms of unprotected, rolly waters.

In order to keep the waters of Avalon Bay clean it is a "no discharge" area except for water from sinks and showers. Soon after a vessel is settled on a mooring, a Harbor Patrol officer puts a green tablet in the vessel's holding tank, which gives a tell-tale color to the water if the tank is pumped out. The pumping of holding tanks within the bay is punishable with a $500 fine and expulsion from the harbor for one year. Pump-out stations are located at the fuel dock near the casino and at a dock next to the transportation docks at the south end of the bay. Water is available at both locations, but it is free at the fuel dock when fuel is purchased.

The *Salad Bowl*, a small blue boat, picks up garbage from moored boats twice daily at no charge. A nominal fee is charged for shore boat service which can be called on VHF Ch. 9. Several dinghy docks are along the seawall; the use of locks is recommended. Showers are available near the restrooms on the Pleasure Pier and on the road to the Casino.

Propane is available at the Southern California Edison plant; a mile along the coastal road south of town. Taxis are available or you may rent a golf cart to reach the facility. Nearby Pebbly Beach Marine Center has haul-out facilities, dry storage and a work yard; they monitor VHF Ch 78. A small marine supply store is next door, where a few engine spares, paint and emergency supplies are available.

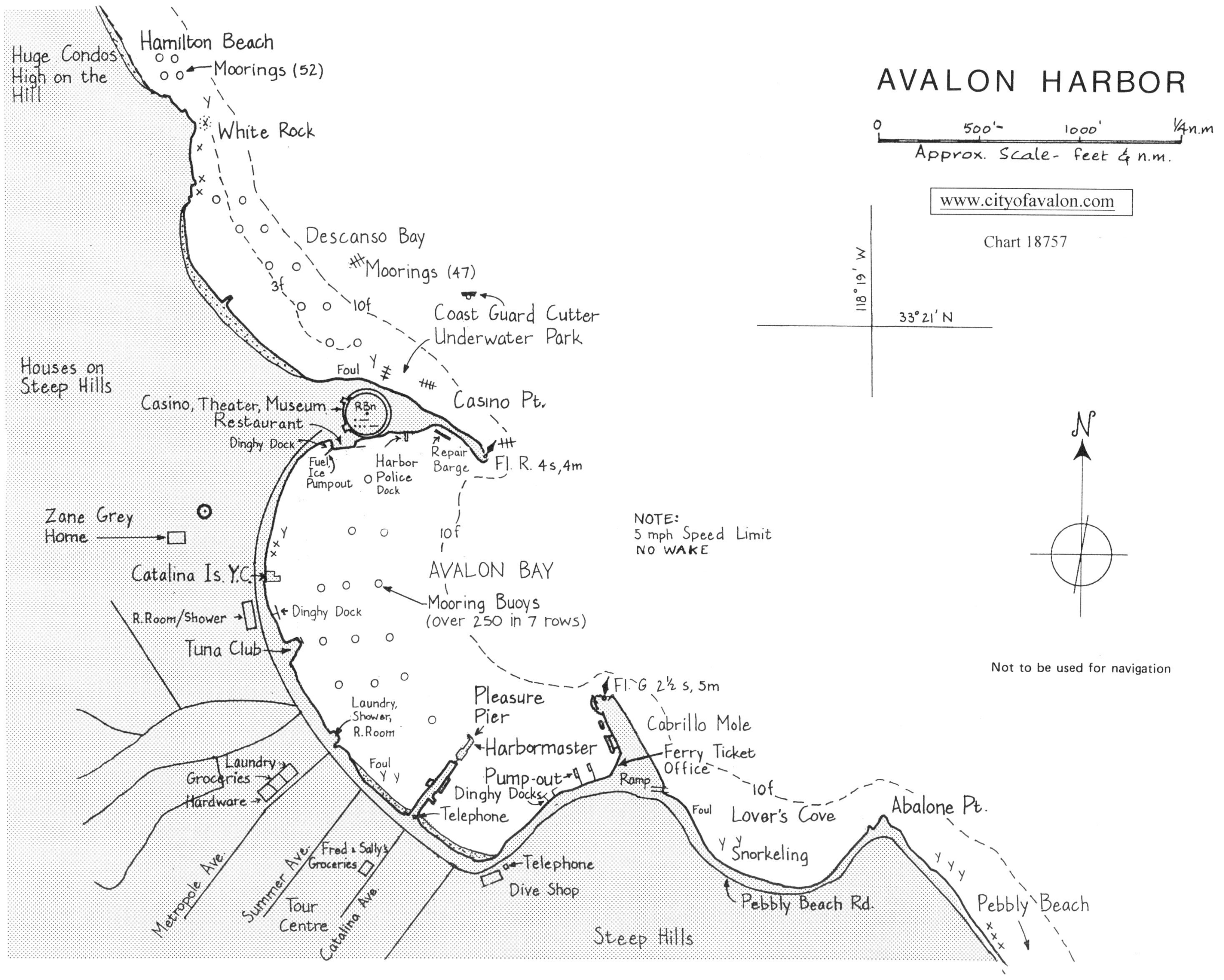
AVALON HARBOR
0
500'
1000'
1/4 n.m
Approx. Scale - feet & n.m.
www.cityofavalon.com
Chart 18757
118°19' W
33°21' N
N
Not to be used for navigation
NOTE:
5 mph Speed Limit
NO WAKE
Huge Condos High on the Hill
Hamilton Beach
Moorings (52)
White Rock
Descanso Bay
Moorings (47)
3f
10f
Coast Guard Cutter
Underwater Park
Houses on Steep Hills
Foul
Casino Pt.
Casino, Theater, Museum
Restaurant
RBn
Dinghy Dock
Fuel
Ice
Pumpout
Harbor
Police
Dock
Repair
Barge
Fl. R. 4s, 4m
10f
Zane Grey
Home
Catalina Is. Y.C.
AVALON BAY
Mooring Buoys
(over 250 in 7 rows)
R. Room/Shower
Dinghy Dock
Tuna Club
Laundry,
Shower,
R. Room
Foul
Pleasure
Pier
Harbormaster
Pump-out
Dinghy Docks
Telephone
Fl. G 2½ s, 5m
Cabrillo Mole
Ferry Ticket
Office
Ramp
10f
Foul
Lover's Cove
Abalone Pt.
Snorkeling
Laundry
Groceries
Hardware
Fred & Sally's
Groceries
Telephone
Dive Shop
Pebbly Beach Rd.
Pebbly Beach
Metropole Ave
Summer Ave
Tour
Centre
Catalina Ave
Steep Hills

SOUTH SHORE ANCHORAGES

Eagle Rock

Since Santa Ana winds are a major weather feature to be considered, it is well to be aware of havens where protection can be found so that weather reports or tell-tale signs can be acted upon. Some indicators of the approach of these winds are: a northeasterly swell or a low, brown cloud of smog to the north or northeast, which has been blown offshore. Weather warnings must be taken seriously, although sometimes the predicted Santa Ana winds bounce up and pass over the island with 50-knot winds high above, while leaving the sea level air and waters relatively calm and unaffected.

The closest refuge from Avalon is just around the southern tip of the island off the **Palisades,** 3 miles west of East End light. Anchorage may be taken in a large area below the line of cliffs in 5 - 6 fathoms, sand.

Little Harbor provides a snug refuge for about a dozen vessels about 3 miles southeast of Catalina Harbor. Caution is needed when entering to avoid the submerged reef reaching almost half way across the entrance from the N shore. Anchorage is in 2.5 - 3.5 fathoms, sand and gravel.

Iron Bound Cove is 3 miles south of West End. Anchorage may be taken southeast of Eagle Rock clear of the kelp in 6 - 8 fathoms.

Cactus Bay is .5 mile east of Eagle Rock where anchorage may be taken in 6 - 8 fathoms.

Catalina Harbor is a deep indentation on the western side of the island which is separated by a narrow neck of land from Isthmus Cove. Follow a mid-channel course when entering the deep inlet, giving good clearance to Pin Rock which is close inside the east head of the harbor. The harbor has 99 mooring buoys and there is room to anchor at least that many again in 4 - 5 fathoms, soft bottom abreast of Ballast Point. Two anchors should be set as swinging room is limited. The head of the harbor is shoal as are the waters about 137m (150 yds.) off the nearby shores. A short, pleasant walk takes you to Two Harbors where a restaurant, coffee shop and store are located. This is the social center of the island for cruisers and is the site of the annual Cruisers' Send-Off party for vessels traveling to Mexico in the fall.

The south shore coast of the island provides a variety of scenic views: gently sloping hills covered with scrub and short grass, steep cliffs and some unique rock formations such as Ribbon Rock and Eagle Rock. Generally the kelp is close inshore and quite sparse. At times, fishing is good though sighting of seals or sea lions is a fair indication that spaghetti will have to be substituted for your planned seafood feast.

At one time bison were seen roaming nearby hills but they are now confined to grazing areas in the center of the island. They were brought to the island in 1924 for the filming of "The Vanishing American" and at present about 400 remain on the island. Other animals seen include razorback boar (brought to the island to control rattlesnakes), wild goats (descendants of those left by Spanish explorers), Catalina Island fox, mule deer and white-tailed deer.

SANTA CATALINA ISLAND, CA

South Coast Anchorages

Not to be used for navigation

Chart 18746, 18757

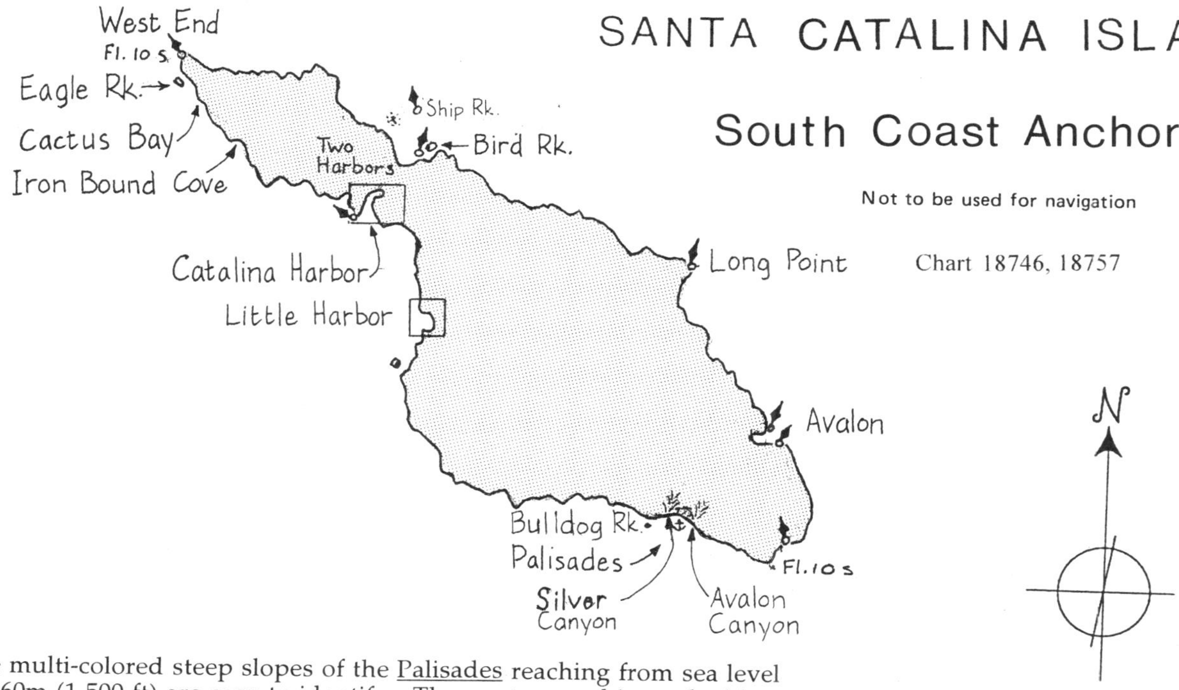

The multi-colored steep slopes of the Palisades reaching from sea level to 460m (1,500 ft) are easy to identify. The western end is marked by Silver Canyon and chunky, black, Bulldog Rock; the eastern part gives way to Avalon Canyon, a rugged, eroded, conspicuous rocky cut.

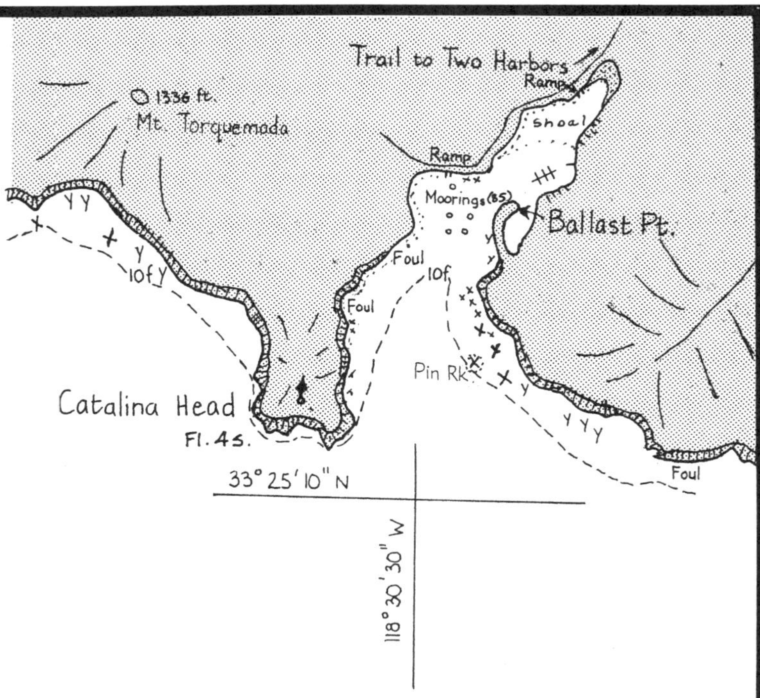

Detail – Catalina Harbor

0 1000′ ½ 1 nm

Approx. Scale - n.m. & ft.

Not to be used for navigation

Chart 18757

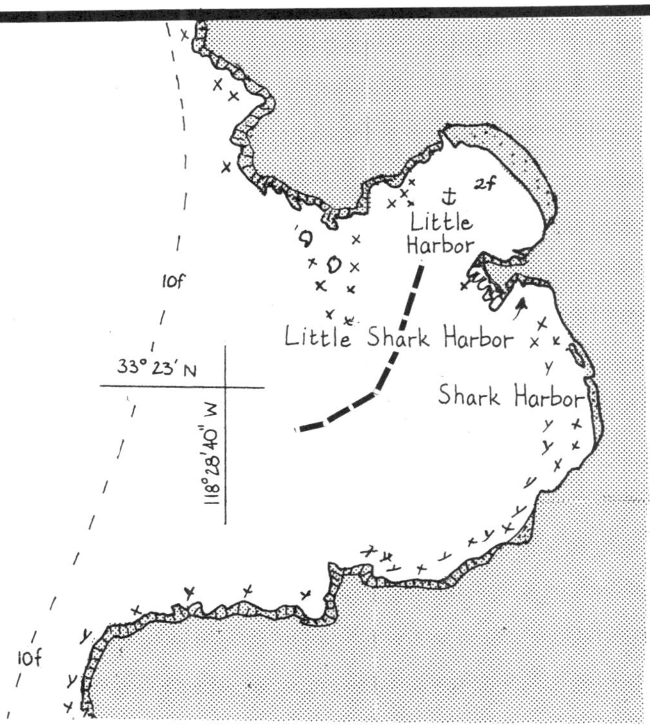

Detail – Little Harbor

0 1000′ ¼ nm

Approx. Scale n.m.

Not to be used for navigation

Chart 18757

ANCHORAGES from WEST END to ISTHMUS COVE

Emerald Cove is the prettiest anchorage on Catalina Island because of its clear, blue-green water and beautiful scenery. It is a favorite spot for local cruisers, though it is exposed to northeasterly weather and must be vacated if Santa Ana winds develop. Less than a mile southeast of Arrow Point, it is easily identified by Indian Rock, which is 270m (295 yds.) off the middle of the beach. A dangerous reef extends about 195m (210 yds.) WNW of Indian Rock, and some rocks and reefs lie off its southeastern shore.

Almost 100 mooring buoys are located before the white sand beach, leaving space for only a few boats to anchor. (For details on the use of mooring buoys see the third paragraph on page 154.) At the north end of the cove is Johnson's Landing which has two piers for the use of the Boy Scout camp scattered among the towering eucalyptus trees beyond.

Good snorkeling can be enjoyed in the vicinity of Indian Rock and its off lying reef, where you can see sand bass, opal eye and bright orange garibaldi. The water is quite cool, seldom getting above 68°, making wet suits worthwhile. A steep, short trail leads up the hill beyond the beach and joins the coastal road seen above the cove. Shore boat service to Isthmus Cove can be obtained by a call on VHF Ch 9.

Around the corner, in the cove to the east, is **Howland Landing.** This is smaller than Emerald Bay, having only 40 moorings, which leave practically no space for anchoring unless you have a small boat. You can take an out-door fresh water shower at the water spigot located at the west end of the bay. The shore facilities are leased to the Catalina Island Boys' Camp and Los Angeles Yacht Club. A conspicuous tank is east of the cove.

Big Geiger Cove is located less than .25 of a mile southeast of Howland Landing, where anchorage may be taken off the beach.

Fishing around the island can be quite productive, and with luck you can land sand bass and halibut inshore and dolphin fish offshore. Marlin tournaments are held yearly, with excellent catches recorded.

Garbaldi

SANTA CATALINA ISLAND, CA

West End to Isthmus Cove

Not to be used for navigation

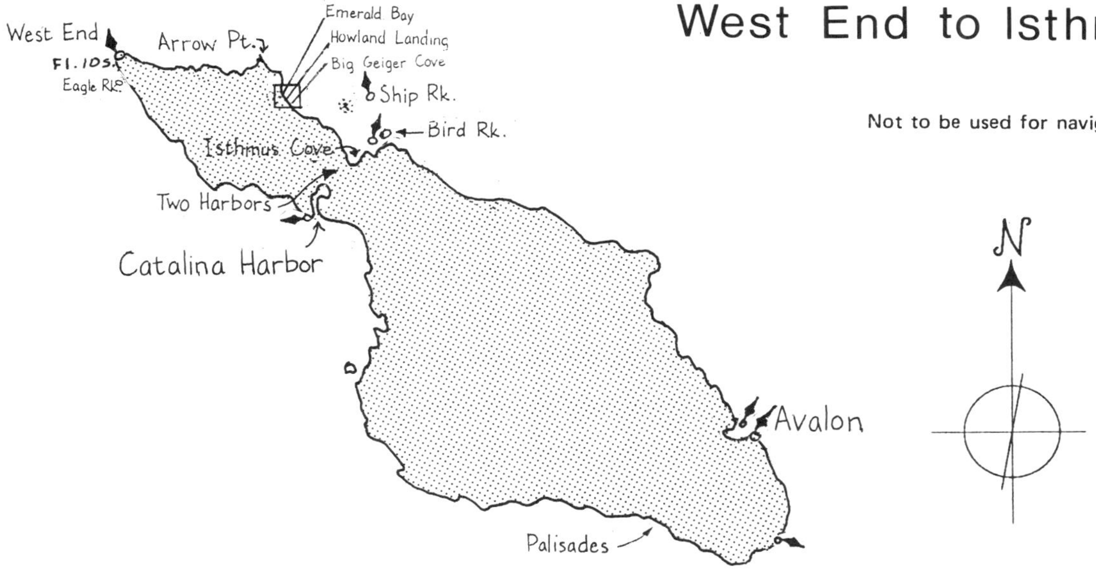

Chart 18746, 18757

Detail Sketch –

Emerald Bay
Howland Landing
Big Geiger Cove

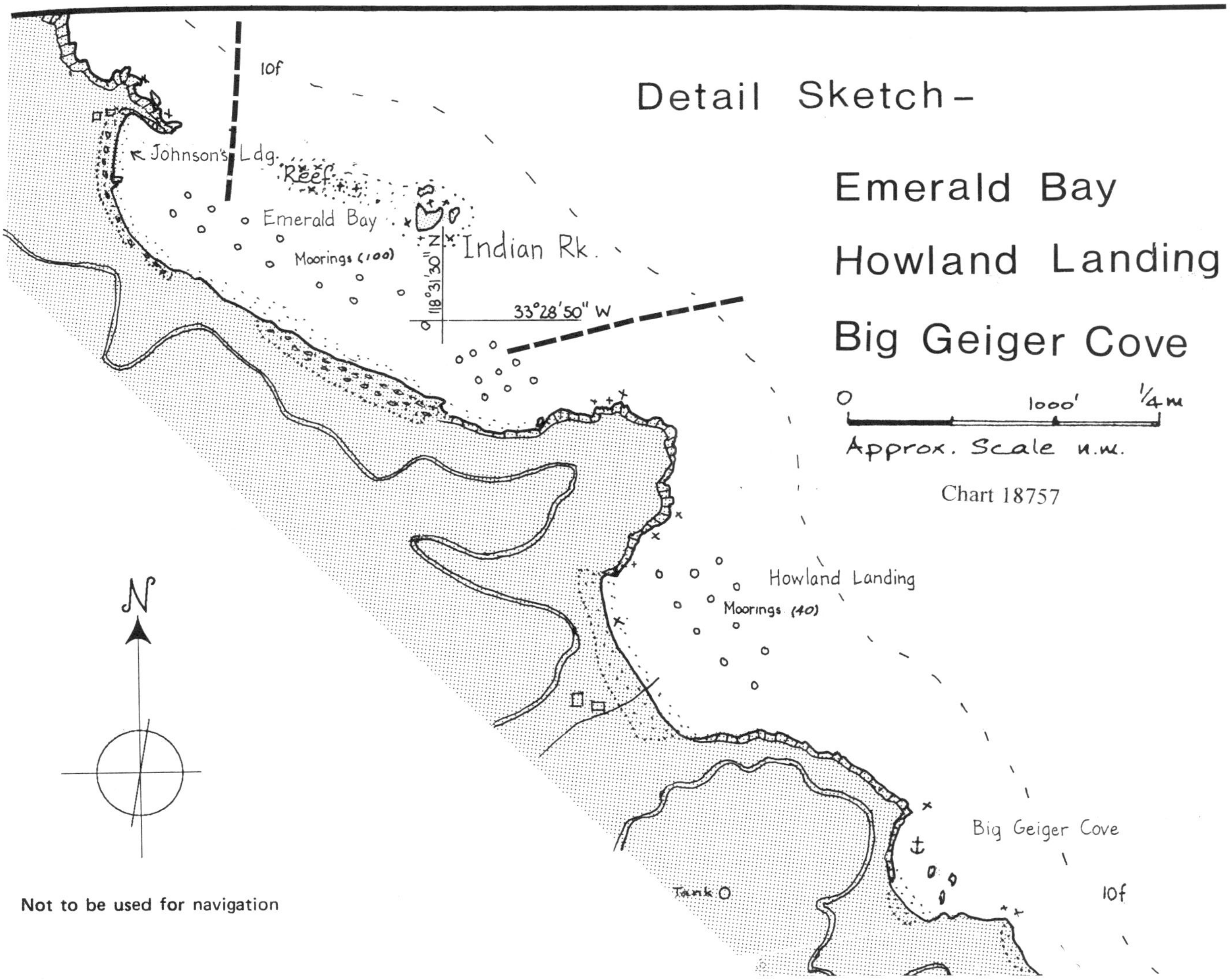

Chart 18757

Not to be used for navigation

ISTHMUS COVE and ADJACENT ANCHORAGES

Isthmus Cove is located on the north shore, six miles from the west end of the island. Several offshore rocks and reefs must be identified when approaching the large bight enclosed between Lion Head on the west and Blue Cavern Point on the east. About .6 mile WNW of Lion Head and about 550m (600 yds.) offshore is Eagle Reef, covered 1m (3 ft.) and marked by a buoy about 90m (100 yds.) to the east. Ship Rock, a guano-capped rock resembling a haystack, is about 1 mile north of the Cove; a light is shown from a pole on the rock. Closer in, about .5 mile E of Blue Cavern Point is Bird Rock, (11m/37 ft. high) a flat, sloping rock with a sharply delineated green patch of grass on its eastern end. It should not be approached closely, for reefs extend more than 90m (100 yds.) off all but its eastern side. A very dangerous area on which many vessels have gone aground is Harbor Reefs, which lie about 365m (400 yds.) southwest of Bird Rock. Measuring about 230m by 415m (250 yds. by 450 yds.), the area is well marked by kelp and buoys. A rock near the east end uncovers at about .6m (2 ft.)

Along the outer shores of Isthmus Cove are several neat niches where private buoys are set; anchorage may be taken outside the buoys: **Cherry Cove** and **Fourth of July Cove** are on the northwestern shore and **Little Fisherman Cove** and **Big Fisherman Cove** are on east part of the cove. The latter anchorage is the only place providing protection from Santa Ana winds on the north shore of the island, though its buoys are only for emergency use. Ashore is the USC Marine Science Center as well as a diver decompression chamber.

Two Harbors is a small, friendly community at the head of the cove. The harbormaster monitors VHF Ch 9 and advises regarding the availability of a mooring buoy or where anchorage may be taken (in about 10 fathoms). Boats under 20m (65 ft.) are not required to show an anchor light. Over 200 mooring buoys are available within the harbor, which has a long pier where fuel and water can be obtained. The Harbor Patrol office is at the outer end of the pier. Since a pump-out station is not available the vessel must be taken out beyond the 3-mile limit before pumping out the holding tank.

Near the head of the pier is a restaurant, coffee shop, inn and general store which sells groceries, some boat supplies and tourist items. The Cove and Camp Agency offices are located here; they collect fees for the use of moorings and sell landing permits covering all parts of the island except Avalon. During the low season you may pay for several days in advance, and by informing the officer of your next night's request, you may be able to reserve a buoy. The shore boat service monitors Ch 9; it operates from 7:00 p.m. until the bar closes around 1:00 a.m.

Emergency medical services are available from paramedics aboard LA County Lifeguard Vessel *Baywatch Isthmus*; they monitor VHF Ch 16 from 9:00 a.m. to 5:00 p.m. and can be reached 24 hours a day by calling Harbor Base Isthmus on VHF Ch 9 or Avalon on VHF Ch 12. Otherwise, medical aid can be obtained by calling Coast Guard Group Long Beach on VHF Ch 16.

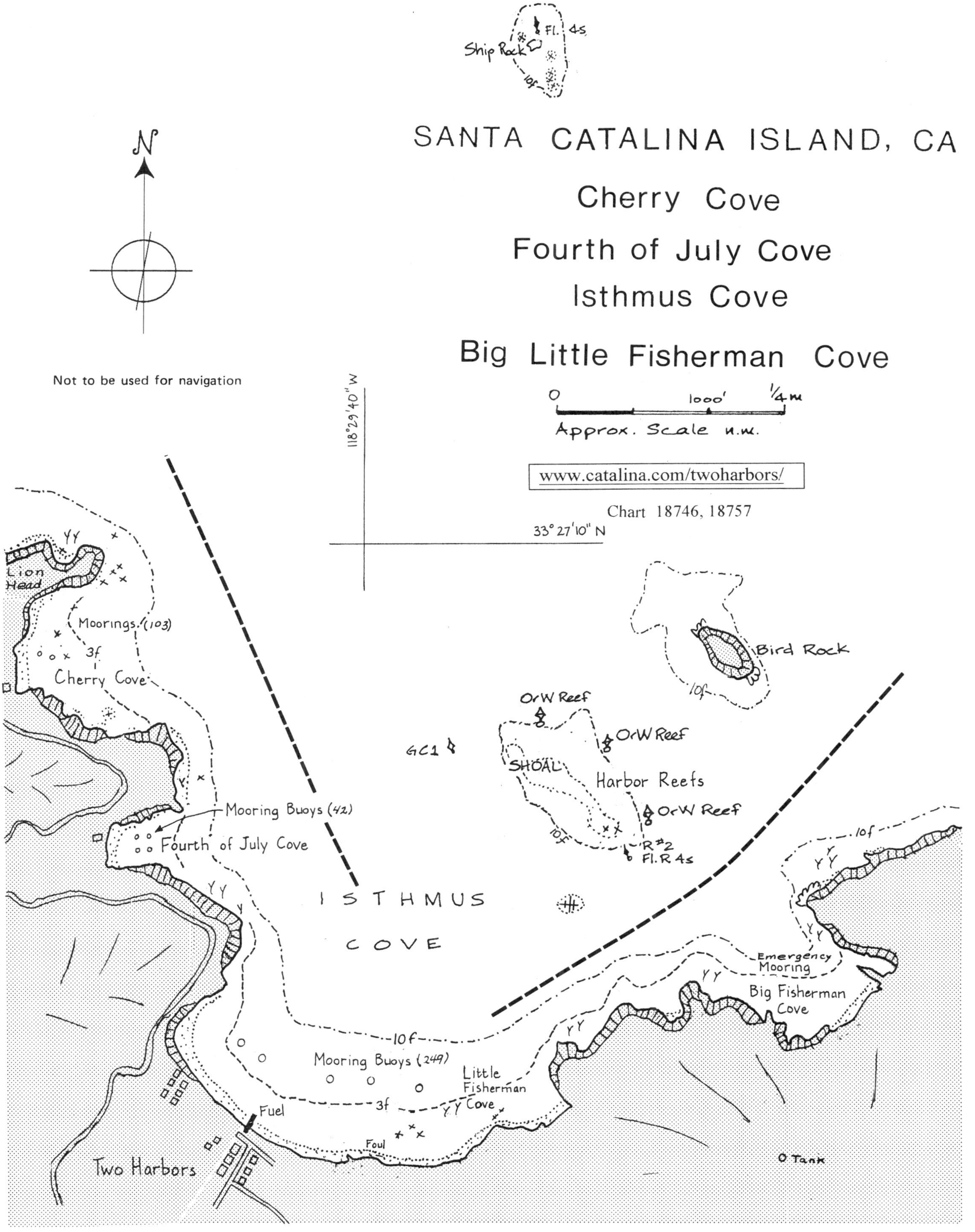
Ship Rock
Fl. 4s
10f
SANTA CATALINA ISLAND, CA
Cherry Cove
Fourth of July Cove
Isthmus Cove
Big Little Fisherman Cove
N
Not to be used for navigation
118°29'40" W
0
1000'
1/4 m
Approx. Scale n.m.
www.catalina.com/twoharbors/
Chart 18746, 18757
33° 27'10" N
Lion Head
Moorings (103)
3f
Cherry Cove
Bird Rock
10f
OrW Reef
OrW Reef
GC1
SHOAL
Harbor Reefs
OrW Reef
10f
R #2
Fl. R 4s
Mooring Buoys (42)
Fourth of July Cove
ISTHMUS
COVE
10f
Emergency Mooring
Big Fisherman Cove
10f
Mooring Buoys (249)
Little Fisherman Cove
3f
Fuel
Foul
Two Harbors
Tank

ISTHMUS COVE to AVALON

Between Isthmus Cove and Avalon the coast consists of rocky cliffs with a number of small coves, many of which are dotted with mooring buoys and backed by summer camp facilities ashore. The rugged hills beyond the coast sweeping down to the shore are covered with bunch grass and a sparse growth of shrubs. As is the practice elsewhere on the island, the buoys are rented on a first-come, first-serve basis provided the owner is not using them.

East of Blue Cavern Point, the undercut shore has a series of interesting caverns before the scar of quarry and an abandoned mine entrance become visible. The first quite exposed area used for anchoring is **Empire Landing,** which is 2 miles east of Blue Cavern Point. Just around the bulge of land to the east is **Rippers Cove,** which has a rock close to the western shore. Fore and aft anchorage may be taken before a beautiful sandy beach in good holding sand; the shore is undeveloped. About 1 mile to the southeast is **Goat Harbor,** a pretty little niche where temporary anchorage may be taken in 2 - 6 fathoms, sand.

Immediately south of Long Point is **Buttonshell Beach**, a popular anchorage which provides protection from westerly winds. Anchorage may be taken outside the moorings in 5 - 8 fathoms, sand. In the next open bight are **Hen Rock Cove**, **White's Landing (Cove)** and **Moonstone Beach.** All three have mooring buoys, with limited anchorage available outside the mooring area. The Newport Harbor Y. C. leases the Moonstone Beach shore facilities.

A little more than a mile southeast of White Cove are **Willow Cove**, and closely followed it, by **Toyon Bay**. Both spots are exposed but are used for anchorage during settled weather. Toyon Bay is a very pretty spot with tall eucalyptus, cottonwood and palm trees near the shore with the steep slopes of Swains Canyon beyond. Tanks high on the hill to the south are a remnant from the time during W.W. II when this was the site of a U. S. military training camp. A few moorings are in place, and there is room for 5 or 6 vessels to anchor. The cream-colored buildings ashore are the site of Catalina Island Marine Institute, which runs an excellent summer camp for youngsters interested in marine life.

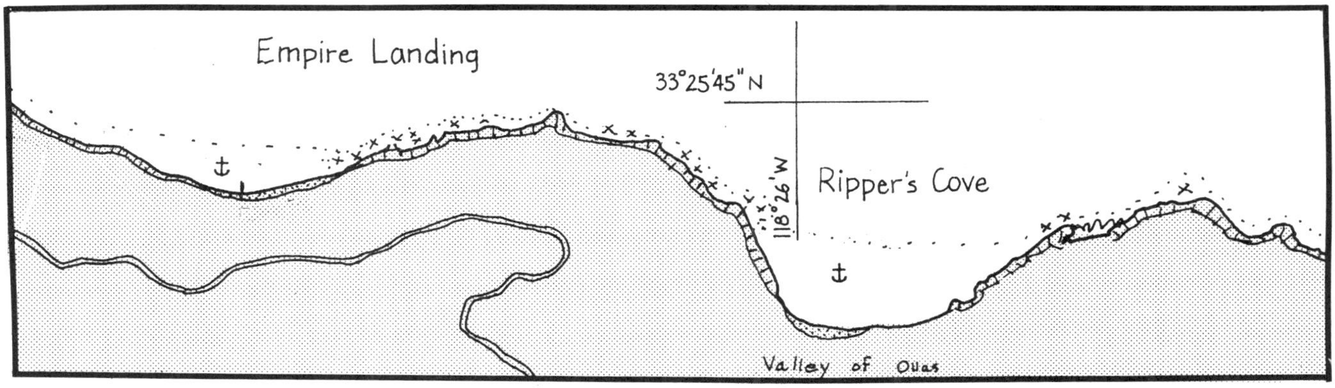

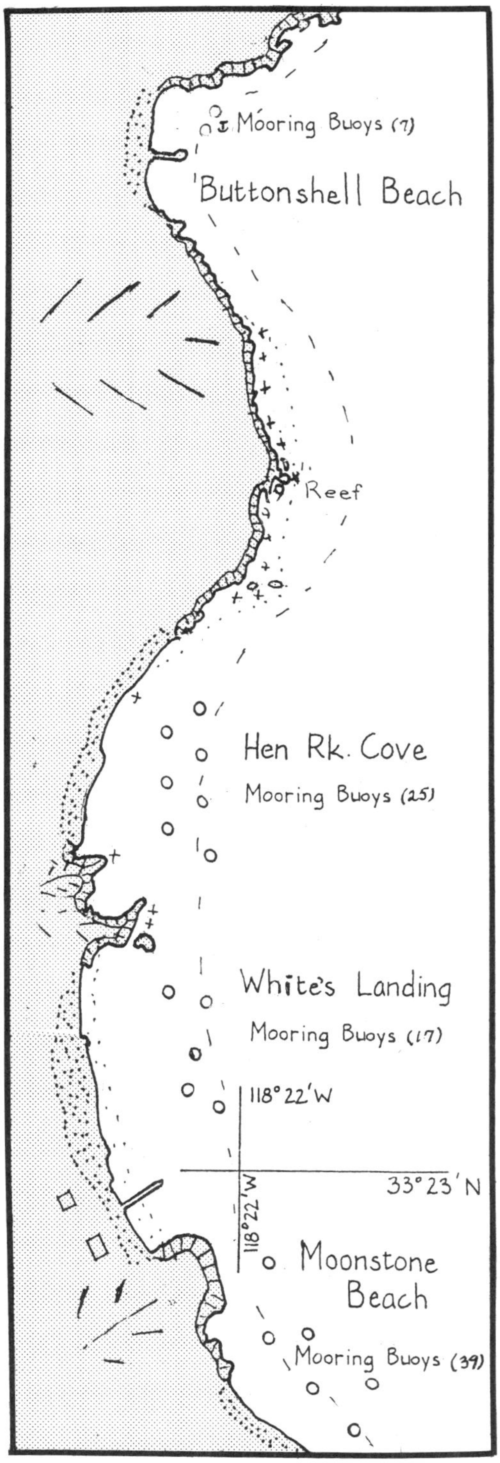

SANTA CATALINA ISLAND, CA

Empire Landing

Ripper's Cove

Buttonshell Beach

Hen Rock Cove

White's Landing

Moonstone Beach

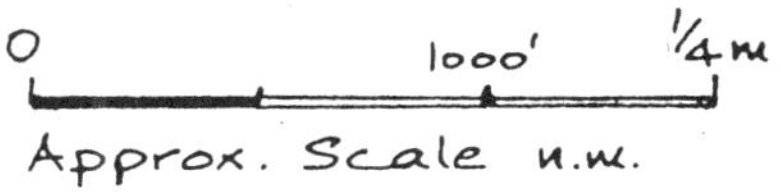

Not to be used for navigation

Chart 18746

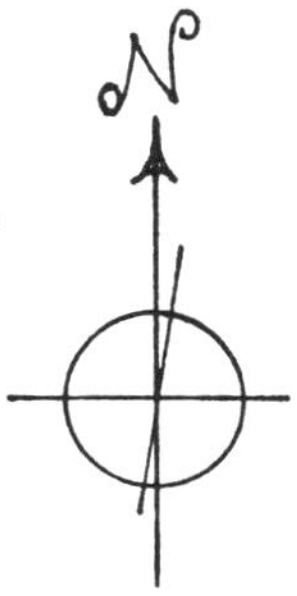

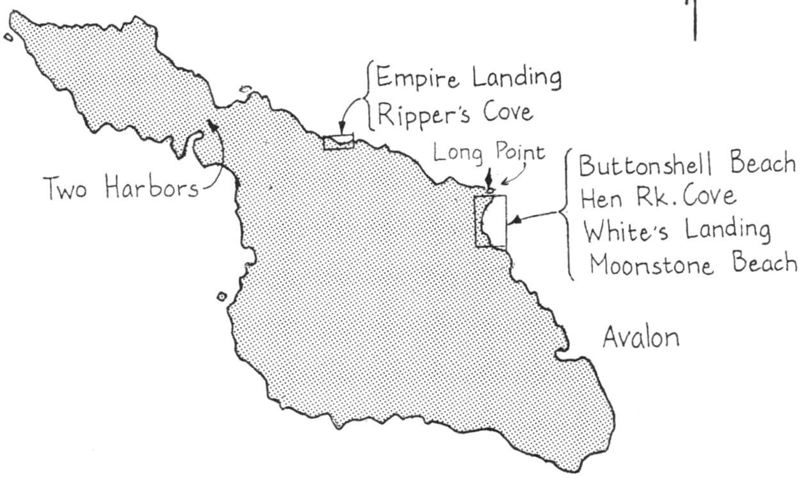

DANA POINT

In the lee of Dana Point, the harbor is 12 miles southeast of Newport Beach and about 24 miles northwest of Oceanside. Two lit breakwaters protect the entrance to the harbor; a fog signal is located at the south light. The harbor at Dana Point is a well protected, easy to enter, scientifically designed facility with about 2,500 slips. It is the first man-made harbor on the coast to be tested and redesigned while in scale model form.

When approaching from the north, pass seaward of lighted whistle buoy "2SJR" (located. about .35 mile southwest of Dana Point) to avoid reefs lining the point and some detached rocky patches. When approaching from the south avoid the buoyed rock about 275m (300 yds.) northeast of the south breakwater light. Another spot to avoid is approximately 275m (300 yds.) east of this light where a submerged sewer outfall pipe extends about .5 mile from the shore.

A multitude of lobster floats are within the kelp growing on both sides of the harbor entrance extending almost a mile offshore. To avoid running into trouble, look for red and white striped racing poles about one mile due south of the harbor entrance. From these marks proceed north to the jetty. Lobster floats frequently drift across this approach channel but are less dense than on either side of the approach. The entrance to the harbor is between two breakwaters. A bridge with a clearance of 6m (20 ft.) separates the East and West Basins. Private buoys mark the shoaling which occurs near the entrance to the West Basin.

Transient moorage is under the control of the harbormaster and you must obtain permission to use the guest slips or to anchor. The harbormaster may be contacted on VHF Chs. 12 and 16, by phone at (714) 496-2242) or by reporting to the office in person. If you arrive after the office has closed check the list of open slips posted on the office door and follow the instructions for choosing and paying for your slip (with cash or a check). When anchoring in the east basin anchorage keep to the breakwater side of the yellow cans and be prepared for lumpy wash from passing vessels. Good holding anchorage is available in the west basin which is subject to fluky west to northwest winds in the afternoon. There is a 5-day limit for use of the anchorages and someone must be aboard at all times. Vessels under 20m (65 ft.) are not required to show an anchor light. Rafting of vessels must have the harbormaster's permission.

Two marinas operate in the basin, **Dana Point Marina** (in the east basin) and **Dana West Marina**. Dana Point Marina has transient slips available for vessels under 8.5m (28 ft.) and can be reached at Tel. (949) 496-6137 or Fax (949) 496-0788. When berths are available, Dana West offers "hot-berthing" privileges to boaters from the two "sister" marinas at Ventura West and Harbor Island West in San Diego. For information call (949) 493-6222 or Fax (949) 493-7531. Slips in Dana Point tend to be narrow relative to their length and generous overhang allowances tend to impinge on the fairways. To assure adequate maneuvering room and slip width, particularly for beamy boats, ask for a slip assignment one size larger than your vessel's length.

Mariners Village and Dana Wharf feature a variety of interesting shops and restaurants and a $1 shuttle service makes it easy to reach the commercial establishments in the city or at the top of the hill on Pacific Coast Highway. The Orange County Marine Institute, an education facility, is located at the west end of the harbor. Anchored nearby is a full-sized replica of Richard Henry Dana's square-rigger that is used as a stage for concerts and various other functions. If time permits, a beautiful side trip can be taken to the Mission at San Juan Capistrano which is only 4 miles distant.

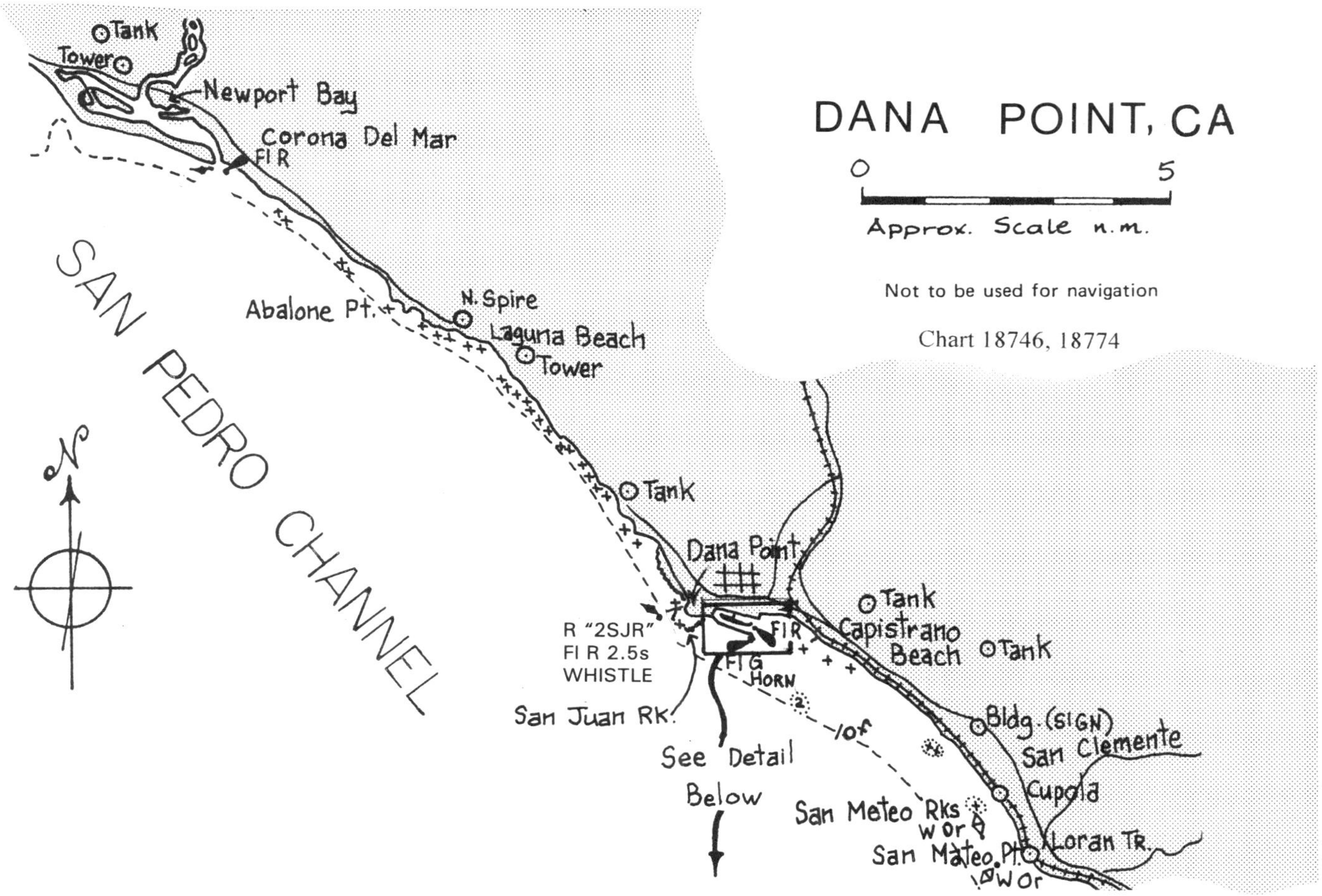
DANA POINT, CA
0
5
Approx. Scale n.m.
Not to be used for navigation
Chart 18746, 18774
Tank
Tower
Newport Bay
Corona Del Mar
Fl R
SAN PEDRO CHANNEL
Abalone Pt.
N. Spire
Laguna Beach
Tower
Tank
Dana Point
Tank
Capistrano Beach
Tank
R "2SJR"
Fl R 2.5s
WHISTLE
Fl R
Fl G
HORN
San Juan Rk.
See Detail Below
10f
Bldg. (SIGN)
San Clemente
Cupola
San Meteo Rks
W Or
San Mateo Pt
W Or
Loran Tr.

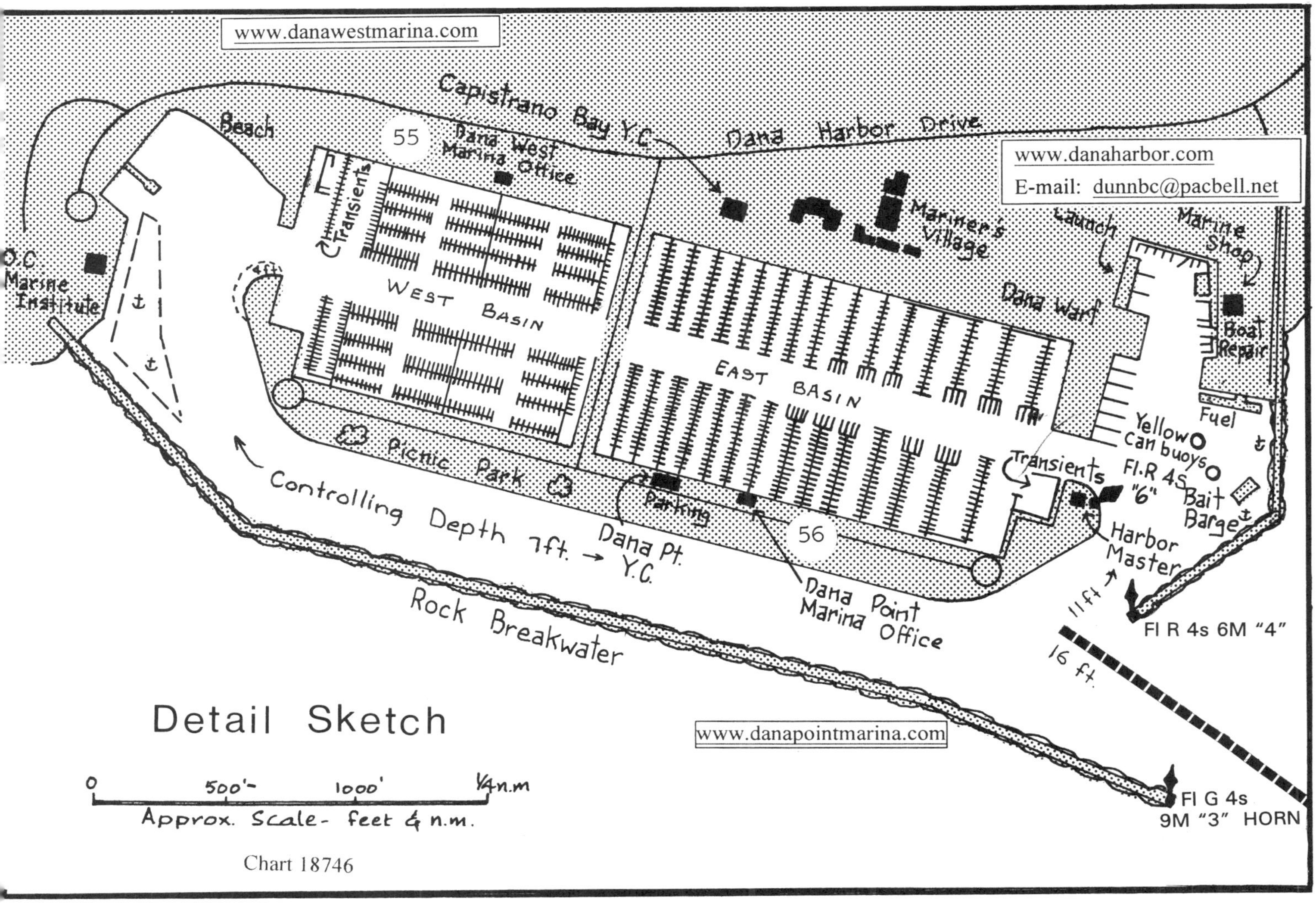
www.danawestmarina.com
Capistrano Bay Y.C.
Dana Harbor Drive
www.danaharbor.com
E-mail: dunnbc@pacbell.net
Beach
55
Dana West Marina Office
Transients
Mariner's Village
Launch
Marine Shop
O.C. Marine Institute
West Basin
Dana Warf
Boat Repair
East Basin
Fuel
Yellow Can buoys
Picnic Park
Transients
Fl.R 4s "6"
Bait Barge
Parking
56
Harbor Master
Controlling Depth 7ft.
Dana Pt. Y.C.
Dana Point Marina Office
11ft
Fl R 4s 6M "4"
Rock Breakwater
16 ft.
Detail Sketch
www.danapointmarina.com
0
500'
1000'
1/4n.m
Approx. Scale - feet & n.m.
Fl G 4s
9M "3" HORN
Chart 18746

OCEANSIDE

There are three bays in the coast on the north side of the city of Oceanside. The northernmost bay is the Del Mar Boat Basin which is restricted for use by the Camp Pendleton Marine Corps, while the two to the south provide space for over 900 slips in the Oceanside Small Craft Harbor. Located 24 miles south of Dana Point, Oceanside is 30 miles north of Mission Bay. Various landmarks are shown on the sketch but the most obvious is the large lighted sign, "OCEANSIDE," on a hill over-looking the harbor. From a distance, another significant landmark is a 12-story building above and beyond the southernmost bay.

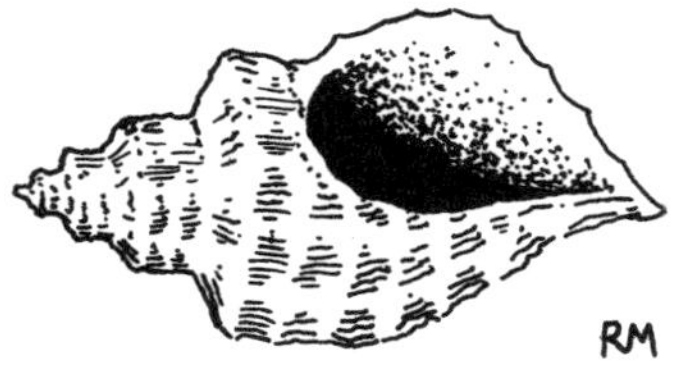

Sculptured rock whelk (Ocenebra interfossa)

When approaching the entrance, pick up lighted bell buoy "OC" that is about 915m (1,000 yds.) southwest of the end of the east jetty. The entrance to all three basins is protected by two jetties—the longer western jetty which is almost parallel to the beach also protects the channel to the Del Mar Basin and the shorter, eastern jetty which angles out from the beach. Both have a light at their seaward ends. The east jetty has a fog signal at its tip and another light is about 60m (200 ft.) further along where the jetty bends in a more easterly direction. Buoys, lights and a day beacon mark the entrance to the harbor. Once inside the breakwater, stay within the marked channel to avoid shifting shoal areas and a submerged hazard on the north side.

Be sure to use the most recent edition of NOAA chart 18774 since it corrects several errors in light characteristics found on a previous edition; corrections were reported in a *Notice to Mariners*.

Caution is advised as this can be a difficult entrance when a heavy southerly swell is running and the tide is ebbing. In these conditions vessels attempting to transit the entrance and stay within the marked channel will find themselves broadside to the seas and at risk of a broach or of being driven onto the breakwater. Though dredging of the channel is done regularly, shoaling continues unabated. Contact the Harbor Patrol (VHF Ch 16) for advice on channel conditions. A Coast Guard cutter is stationed at the dock on the south side of the entrance

Berths for 50 transient vessels are available at the Harbor Headquarters Dock on the east side of the harbor on a first-come, first-serve basis. Call the Harbor Police on VHF Ch 16 to make berthing arrangements or tie to a dock in front of the Harbor Office before arranging for moorage. The usual amenities are available.

A short walk around the harbor takes you to a collection of shorefront shops, fast food outlets and restaurants clustered around the southern basin.

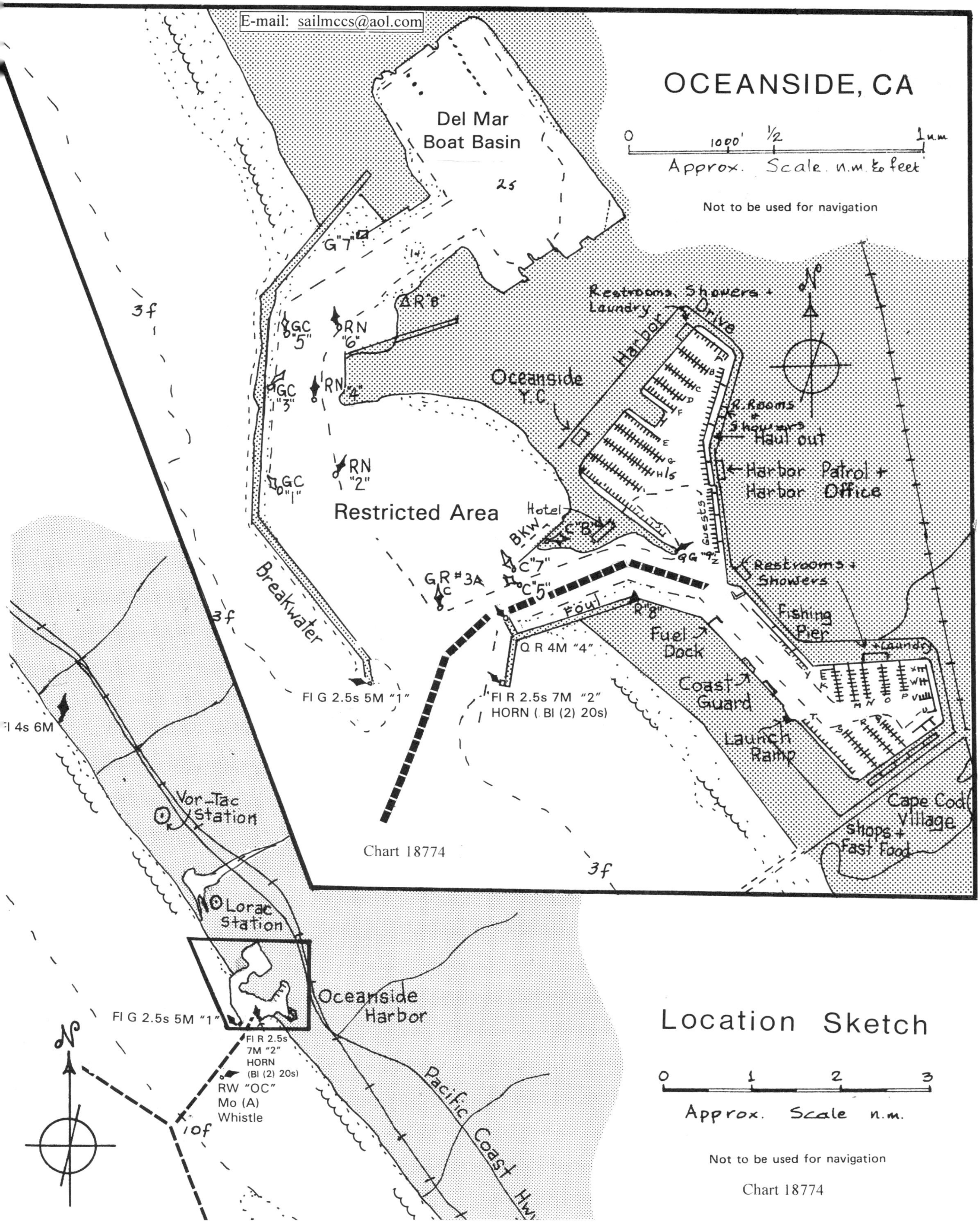

E-mail: sailmccs@aol.com
OCEANSIDE, CA
0 1000' 1/2 1nm
Approx. Scale n.m. to feet
Not to be used for navigation
Del Mar Boat Basin
25
G"7"
14
R"8"
GC "5"
RN "6"
GC "3"
RN "4"
RN "2"
GC "1"
Restricted Area
3f
Breakwater
Oceanside Y.C.
Harbor Drive
Restrooms, Showers + Laundry
R.Rooms + Showers
Haul out
Harbor Patrol + Harbor Office
Guests
Hotel
BKW
C"8"
C"7"
C"5"
QG "9"
GR #3A C
Foul
R"8"
Restrooms + Showers
Fishing Pier
Fuel Dock
Coast Guard
Launch Ramp
Laundry
Q R 4M "4"
Fl G 2.5s 5M "1"
Fl R 2.5s 7M "2"
HORN (Bl (2) 20s)
Cape Cod Village
Shops + Fast Food
Chart 18774
Fl 4s 6M
Vor-Tac Station
Lorac Station
Oceanside Harbor
Fl G 2.5s 5M "1"
Fl R 2.5s 7M "2"
HORN (Bl (2) 20s)
RW "OC" Mo (A) Whistle
10f
Pacific Coast Hwy
Location Sketch
0 1 2 3
Approx. Scale n.m.
Not to be used for navigation
Chart 18774

MISSION BAY

Most of the waterfront and all of the waters and islands within Mission Bay are a 4,600-acre park administered by the city of San Diego. It is surrounded on all sides by the residential areas of Pacific Beach, Morena, Loma Portal and Ocean Beach which are all within San Diego city limits. Though it covers a large area, most of its inner waters are shallow 1.8m/6 ft. and much of it is blocked off by the Mission Bay Bridge (Clearance 11.5m/ 38 ft.) which joins two lobes of land in the southwest corner of the park. Two other bridges having clearances of 11.5m/38ft. and 9.5m/31 ft. further restrict bay traffic.

Located 30 miles south of Oceanside, Mission Bay is easily identified by the high Sea World Tower and the Hyatt Islandia Hotel. A fishing pier at Ocean Beach is about .25 mile south of the entrance to Mission Bay.

Two jetties protect the entrance channel leading to the small craft facilities in Quivera Basin, Mariners Basin and beyond. Current charts are incorrect with respect to the lights at the entrance. Although the chart indicates a flashing green light on a buoy and a flashing red light on the jetty, the actual configuration is a flashing green light on a tall tower at the end of the northern jetty. The flashing red light is on a buoy which is some distance west of the southern jetty. **Caution**: a 365m (1,200 ft.) weir separating the San Diego River from the entrance channel must not be crossed. When tides are above 2m (6 ft.) this section appears to be open water when in fact, less than .5m (2 ft.) covers the weir.

There is no problem entering the bay under normal conditions, but this can be a dangerous entrance which must be avoided if a moderate to heavy swell is running. At such times breakers occur at the entrance and roll in along the south jetty. Favoring the northern side of the channel is advisable when seas are choppy.

The only anchorage in Mission Bay is in Mariners Basin where the holding is good in sand and mud and there is no charge. When setting the anchor be aware that the bottom shoals rapidly on the eastern side of the basin. Dinghies can be beached anywhere and restrooms are located on the beach.

Several large private marinas line the waterfront in Quivera Basin: **Islandia Marina**, **Seaforth Marina**, and **Mission Bay Marina**. Transient berthing is limited and depends on boat size and there is a maximum stay of three months. All amenities are within a short walking distance from the docks. Fuel and pump-out facilities are at Islandia Marina and Marina Village Marina.

Beyond Mission Bay Bridge (Clearance: 11.6m/38 ft.) are several small boat marinas that are filled with local boats and no transient berths are available. Fuel is sold at Dana Landing where there is a launch ramp. Launch ramps are also found around the bay at De Anza Cove, Ski Beach, Santa Clara Cove, San Juan Cove, South Shores and La Jolla Shores.

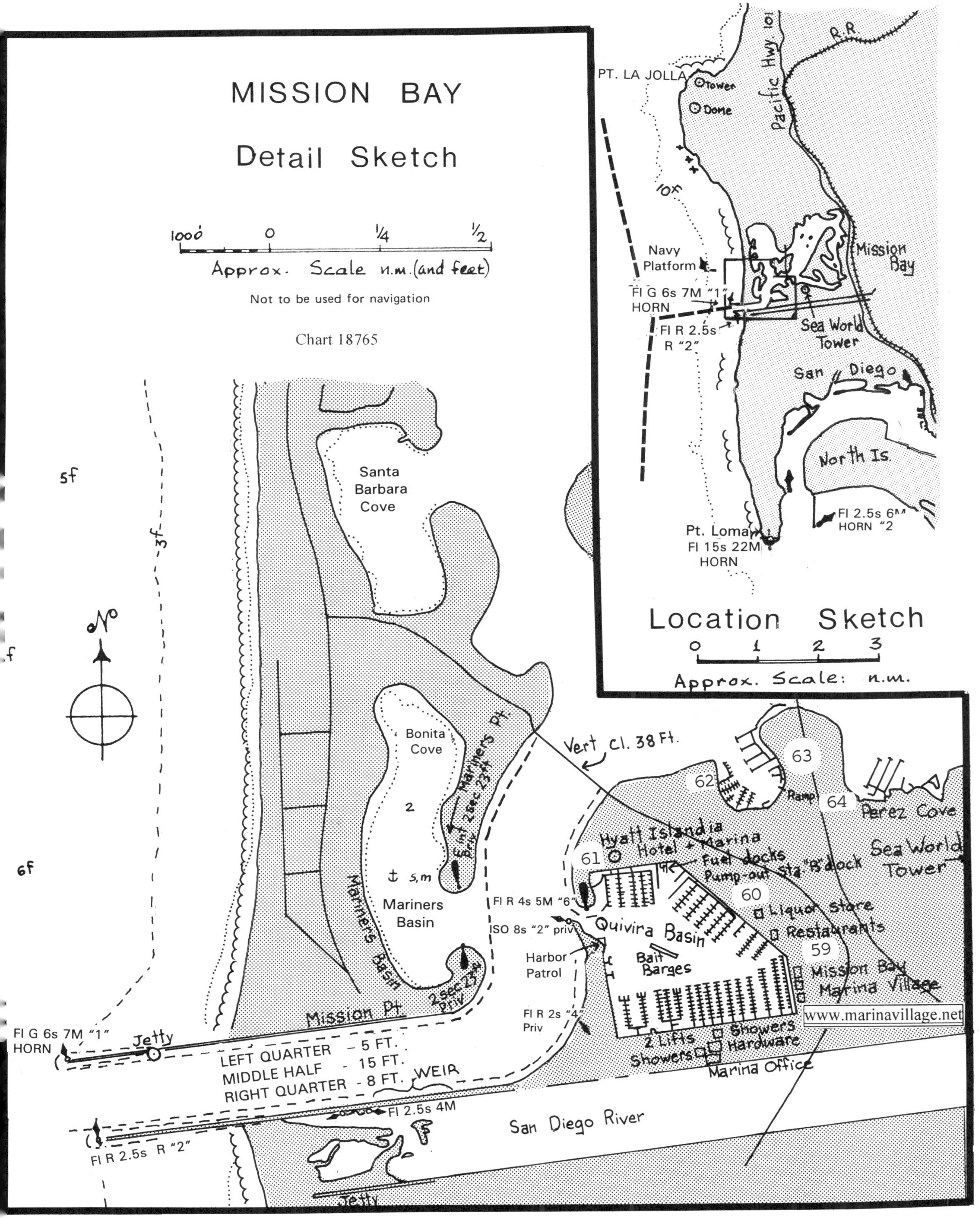
MISSION BAY
Detail Sketch
1000' 0 1/4 1/2
Approx. Scale n.m. (and feet)
Not to be used for navigation
Chart 18765
Santa Barbara Cove
Bonita Cove
Mariners Pt.
2
Mariners Basin
Mariners Basin
Mission Pt.
Jetty
Fl G 6s 7M "1" HORN
LEFT QUARTER - 5 FT.
MIDDLE HALF - 15 FT.
RIGHT QUARTER - 8 FT.
WEIR
Fl 2.5s 4M
Fl R 2.5s R "2"
San Diego River
Jetty
5f
3f
6f
Vert Cl. 38 Ft.
Fl R 4s 5M "6"
ISO 8s "2" priv
Harbor Patrol
Fl R 2s "4" Priv
Quivira Basin
Bait Barges
Hyatt Islandia Hotel + Marina
Fuel docks
Pump-out sta "B" dock
61
62
63
64
60
59
Ramp
Perez Cove
Sea World Tower
Liquor store
Restaurants
Mission Bay Marina Village
www.marinavillage.net
Showers
Hardware
2 Lifts
Showers
Marina Office
PT. LA JOLLA
Tower
Dome
Pacific Hwy 101
R.R.
10f
Navy Platform
Mission Bay
Fl G 6s 7M "1" HORN
Fl R 2.5s R "2"
Sea World Tower
San Diego
North Is.
Pt. Loma
Fl 15s 22M HORN
Fl 2.5s 6M HORN "2"
Location Sketch
0 1 2 3
Approx. Scale: n.m.

SAN DIEGO BAY

Less than 10 miles north of the Mexican border, this area was first visited by a European in 1542 by Juan Rodriquez Cabrillo. It was the site of the first of the 21 missions built by Father Junipero Sera linking Catholic outposts to present-day San Francisco. Following the defeat of Mexico in the Mexican-American war in 1847, it became part of the United States. "Old Town" is well worth visiting to see reminders of its colorful past.

San Diego Bay, a natural harbor, is home port to over 7,000 boats and 15 yacht clubs. It is also a major stop for vessels entering and leaving the United States enroute to the cruising ground of the Pacific Northwest, Mexico, the islands of the South Pacific and the Panama Canal. Temporary slips and moorings for transient vessels are sometimes limited but there is always anchorage space available though at the busiest time you may have to move around because of time limitations. While San Diego offers all the services and provisioning resources a cruising vessel could ask for, finding a slip or mooring for other than a brief stop may require considerable advance planning.

There is very heavy commercial, naval and recreational traffic in the entrance and within the harbor and boaters must be extremely careful to obey the rules of the road and avoid collisions. Pleasure craft must take care not to impede the passage of large vessels.

When approaching pick up sea buoy "SD" which is about 2.5 miles from the tip of Point Loma. Proceed 1 mile NNE to buoy "3" and a further mile to buoy "5" which marks the beginning of the series of buoys marking the channel to the narrows and the entrance to San Diego Bay. When approaching from the north keep well offshore (west of 117°17.5'W) to avoid thick kelp beds extending some distance off the coast from Mission Bay to Point Loma. Detached kelp may be seen scattered in the area, particularly after a storm and should be avoided.

The high, steep-sided peninsula ending in Point Loma protects the entrance from northwesterly winds. The Point Loma Light and fog horn are located at the south end of the point below abandoned lighthouse located high on the ridge. If it is too foggy to see either lighthouse, extensive kelp beds offshore are a good sign that Point Loma is close by. When approaching from the south you can see the Silver Strand beaches with a backdrop of office towers and tall condominiums of the downtown area. Fifteen miles south of Point Loma and 7 miles offshore are the wild and uninhabited Coronado Islands (Los Coronados) which are Mexican territory.

A jetty extends for 1 mile south of the southwestern tip of North Island from Zuniga Point. This is a dangerous area since the outer two-thirds is submerged at high water and is marked by only three lights along its length and a light and fog signal at its end. Within the three-mile space south of Point Loma three buoys are located, the last of which marks the beginning of the channel leading into the harbor. Another light and fog signal are off Ballast Point which is almost .5 mile northeast of the east side of Point Loma. The large U.S. Navy submarine base with its many imposing structures and tenders is just beyond. A submerged jetty extends outward from North Island toward Ballast Point. This hazard is identified by striped red and white poles, a small caution buoy (white with red) and a day mark on a piling. By staying well within the marked channel this danger is avoided.

Vessels of foreign registry or those arriving from a foreign port must go directly to the Harbor Police Dock at the western end of Shelter Island and make a formal entry through Customs for San Diego is a **Port of Entry**. When a vessel is returning from Mexican waters (even if landfall has not been made) it must check in with Customs at the Harbor Dock.

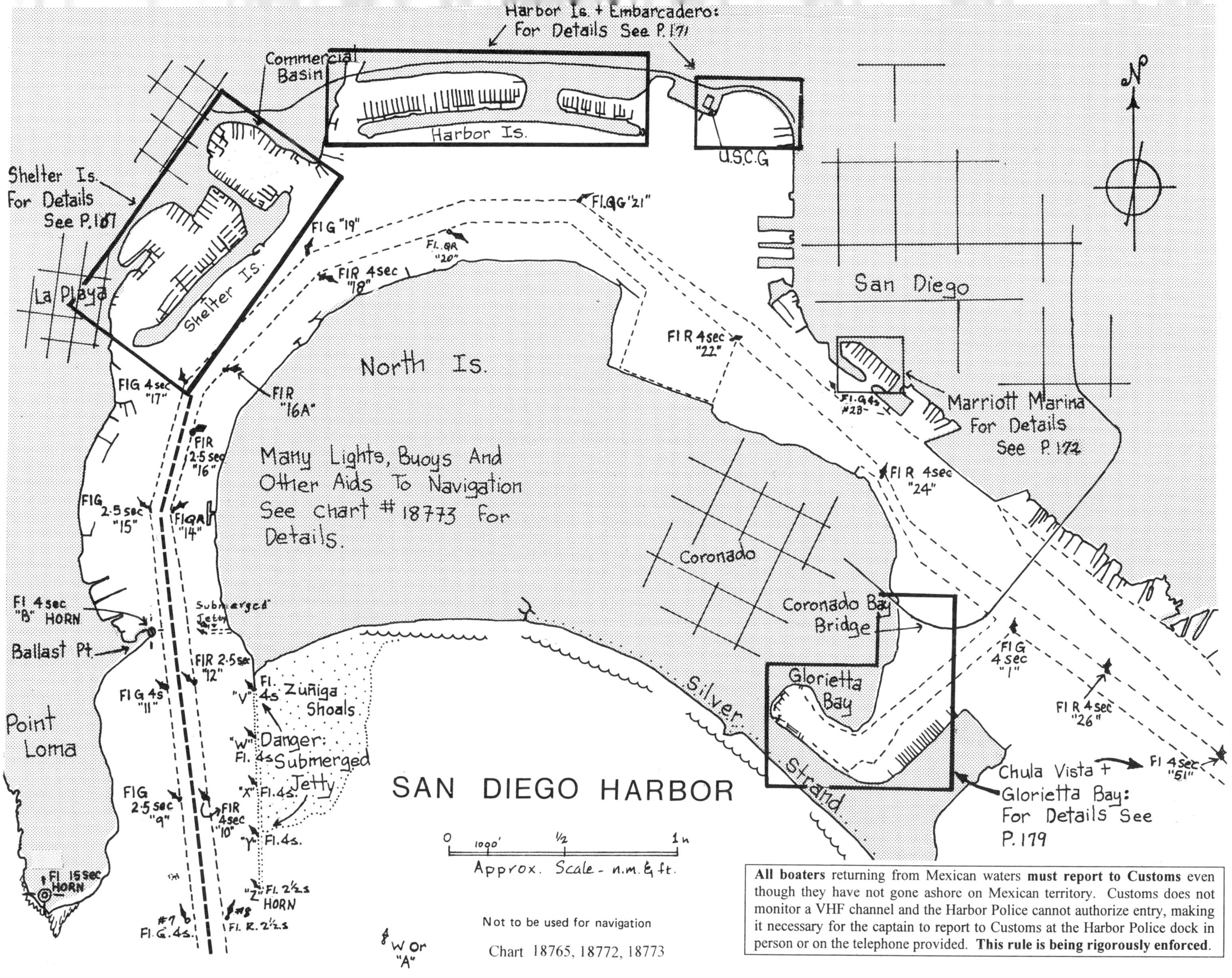

SAN DIEGO HARBOR

Not to be used for navigation

Chart 18765, 18772, 18773

All boaters returning from Mexican waters **must report to Customs** even though they have not gone ashore on Mexican territory. Customs does not monitor a VHF channel and the Harbor Police cannot authorize entry, making it necessary for the captain to report to Customs at the Harbor Police dock in person or on the telephone provided. **This rule is being rigorously enforced**.

Hours of operation are from 8 a.m. to 5 p.m., Monday to Saturday. You may stay at the Municipal Dock (next to the Police Dock) until all formalities have been completed.

Pumping sewage into San Diego Bay is prohibited. Pump-out stations operate at Harbor Police Dock and at the fuel docks on Harbor Island. When entering the bay cruisers should check with the Harbor Police to determine **current regulations** for transient sailors. The Harbor Police monitor moorage and anchorage areas with clearly marked (*Harbor Police)* inboard and inflatable vessels and shore facilities are checked by a bicycle patrol.

SHELTER ISLAND

Shelter Island is the most convenient area for transient sailors to stay while visiting San Diego since all cruising needs can be met in the nearby boatyards, marine supply stores and marinas. Use caution when rounding either end of Shelter Island to avoid shoal areas marked by buoys and pilings that are clearly noted on nautical charts. Nevertheless, even small boats cutting corners regularly run aground.

The moorage closest to the entrance of San Diego Bay is the **roadstead anchorage** on the south side of Shelter Island, about 1 mile east of Point Loma. There is a waiting list for rental of the mooring buoys and anchorage can only be taken in a designated spot after obtaining a Permit. Though a rather rough spot because of the wake of vessels passing up and down the channel, anchorage here gives access to Shelter Island facilities and landing is easy on the beach or at the dinghy landing near the launching ramp. Vessels must show a proper anchor light at night and display an anchor ball during the day.

The forest of masts on the north side of Shelter Island is the location of 3 yacht clubs, 3 fuel docks, 5 marinas and 1 anchorage. The **La Playa Cove anchorage** is between the San Diego Yacht Club (a large brown clubhouse) and Southwestern Yacht Club. Anchoring is permitted from Friday night to Monday morning, by Permit from the Harbor Police.

The area northeast of Shelter Island is called **America's Cup Harbor** (formerly Commercial Basin) and was the center of activity when the race was hosted in San Diego. The two marinas seldom have transient berths and buoys are rented on a yearly basis.

The closest large grocery store (Von's) is several long blocks east of Shelter Island Drive on Rosecrans Street. For a major replenishment of groceries it is worth visiting Smart & Final, a restaurant supply outlet which sells in case lots. Membership in Costco gives access to a discount food warehouse which is an excellent place to stock up. Fresh produce and "health" foods are available at Henry's Market on Rosecrans Street at Midway, close to Smart & Final. Convenience stores and restaurants are in the immediate area and a laundromat is across from the Post Office.

One of the best features of shopping for boat gear and nautical items on Shelter Island is that all marine supplies and services can be found within walking distance of the Basin. Downwind Marine on Canon Street and West Marine on Rosecrans are the two most important marine supply stores and there are many other specialty shops featuring a variety of nautical merchandise and services. A block south of West Marine on Scott street is Seabreeze Ltd., a nautical gift and bookstore which, in addition to navigational charts, has the largest selection of nautical books on the west coast. Shelter Island also has many excellent restaurants and seafood markets serving light lunches and first-rate meals where you can find the best clam chowder and smoked fish burgers on the west coast. Map Centre on Sports Arena Blvd. features an exceptional selection of globes in addition to nautical books and charts.

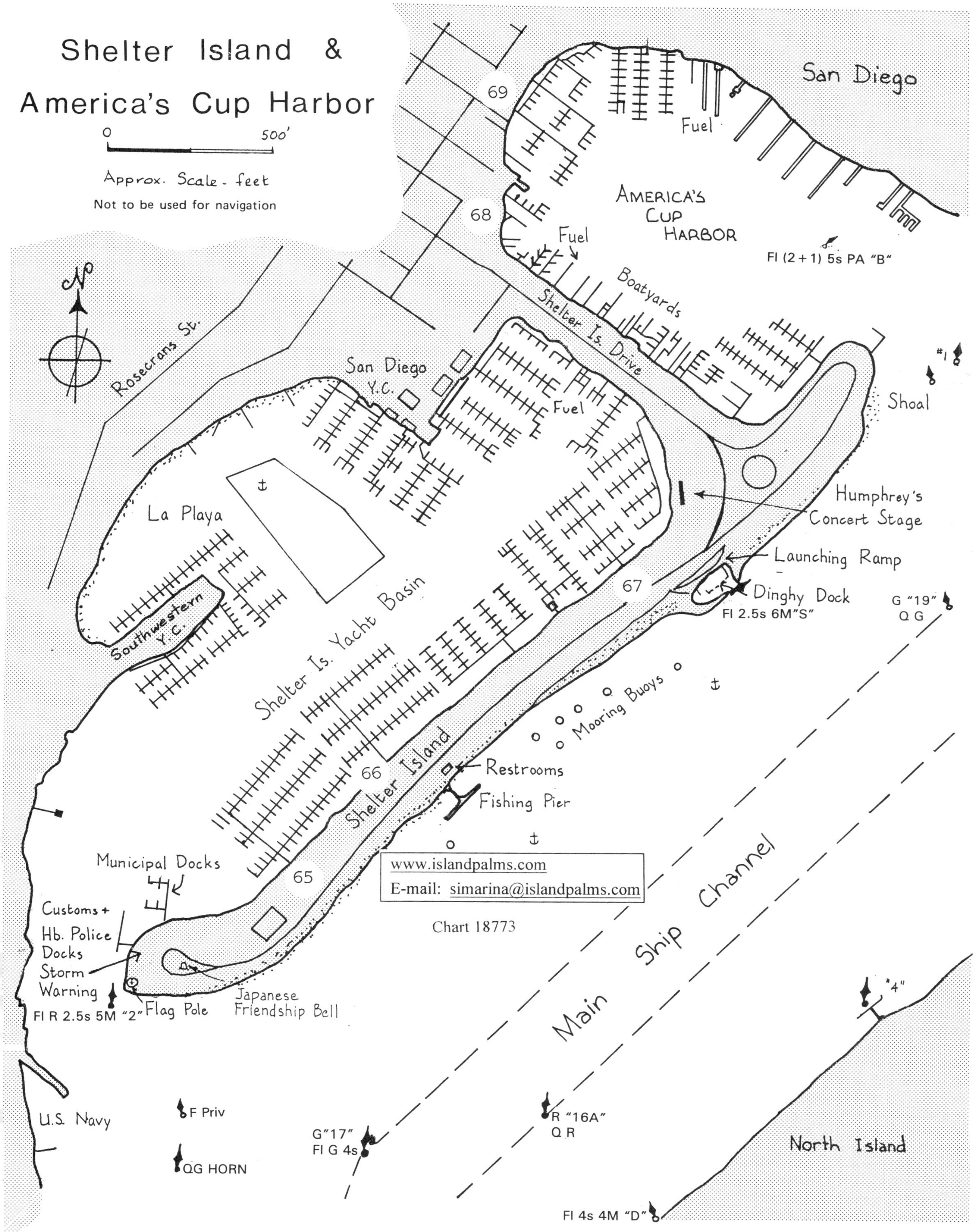

Shelter Island & America's Cup Harbor
0
500'
Approx. Scale - feet
Not to be used for navigation
N
San Diego
69
68
Fuel
AMERICA'S CUP HARBOR
Fuel
Fl (2+1) 5s PA "B"
Boatyards
Shelter Is. Drive
Rosecrans St.
San Diego Y.C.
Fuel
#1
Shoal
Humphrey's Concert Stage
Launching Ramp
Dinghy Dock
Fl 2.5s 6M"S"
G "19"
Q G
67
La Playa
Southwestern Y.C.
Shelter Is. Yacht Basin
Mooring Buoys
Shelter Island
66
Restrooms
Fishing Pier
Municipal Docks
65
www.islandpalms.com
E-mail: simarina@islandpalms.com
Chart 18773
Main Ship Channel
Customs + Hb. Police Docks
Storm Warning
Flag Pole
Japanese Friendship Bell
Fl R 2.5s 5M "2"
"4"
U.S. Navy
F Priv
QG HORN
G"17"
Fl G 4s
R "16A"
Q R
North Island
Fl 4s 4M "D"

HARBOR ISLAND

As you travel further into San Diego Bay, the next pleasure boat facilities are located at the massive development at Harbor Island. The three marinas in the west basin (**Harbor Island West, Marina Cortez** and **Cabrillo Isle Marina**) have over 1,600 slips, and in the east basin **Sunroad Marina** has 540 slips. Transient berths may be available. Facilities and charges vary, making it advisable to compare charges for electricity, use of the pool, whether or not the deposit on the key to showers is 100% refundable, and if there is a live-aboard charge.

Two fuel docks are located in the west basin, and there are several small marine shops on the Island. The nearest repair facilities are on Shelter Island. Though there are several restaurants ashore there are no grocery stores, so you must go to Point Loma or downtown. It is well to call ahead for reservations at Boathouse Restaurant (619) 291-8081, Tom Hams Lighthouse Restaurant (619) 291-9110 and Sheraton Harbor Island Hotel (619) 291-1900.

Just a short distance away is a large Coast Guard Air Station, and to the south of it is the **Embarcadero**. Mooring buoys where vessels are tied Mediterranean style are available for a maximum of 30 days. Fees are payable in advance and vary with the length of stay. To determine if any buoys are available call the Wharfinger's Office at (619) 291-3900, Ext. 340, 8 a.m. to 5 p.m. on weekdays or after hours call (619) 291-1799. Dinghy docks along the Embarcadero provide access to restroom facilities ashore. Noise and air turbulence from the Coast Guard Station can be a nuisance, and jet traffic at the International Airport on Harbor Island add to the sound effects.

Southeast of the Coast Guard Station is the **Laurel Street Anchorage,** where the maximum stay is three months. The Harbor Patrol has jurisdiction over this spot but there is seldom space for transients as it is normally filled with local vessels.

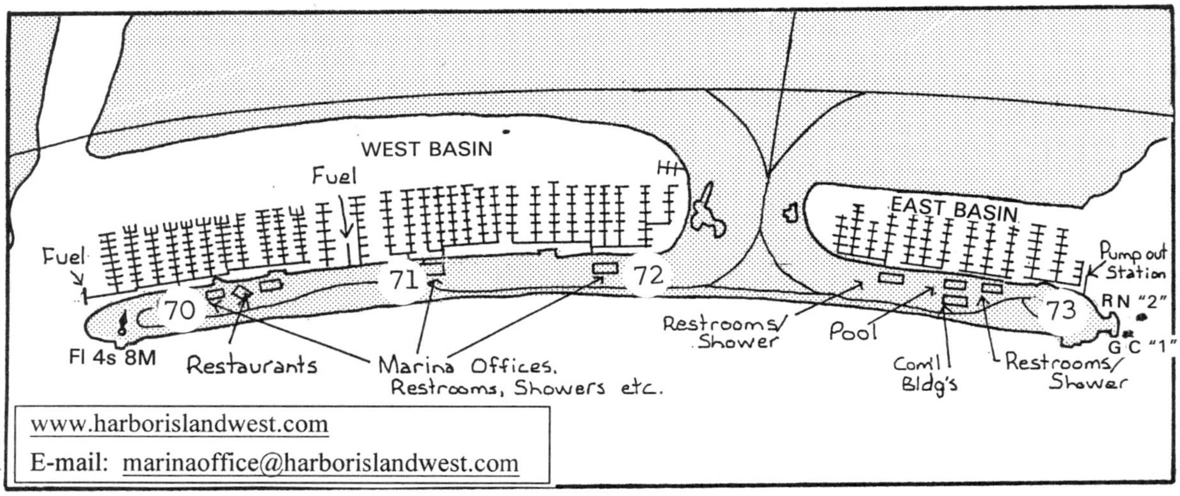

Harbor Island

Chart 18773

Two events of interest to cruisers are held on Shelter Island. The first is a series of concerts held on the open-air stage at Humphry's Half Moon Marina, featuring a variety of musical groups and soloists. Details regarding the concert schedule can be obtained from the marina and most Shelter Island marine stores. Boaters congregate in their dinghies and inflatables, drifting around in a tight cluster (like islands of kelp) since anchoring or tying to docks, pilings or vessels at the marina is prohibited. These concerts offer an opportunity to meet fellow cruisers with the only rule being to keep the fire lane clear.

The second event is held at Downwind Marine, 2804 Canon Street, where an annual "Cruisers' Kick-off Party" is held, usually on the last Sunday of October. This provides an excellent opportunity for cruisers to Mexico to meet fellow yachtsmen who are heading south after the Baja-Ha-Ha crowd has departed.

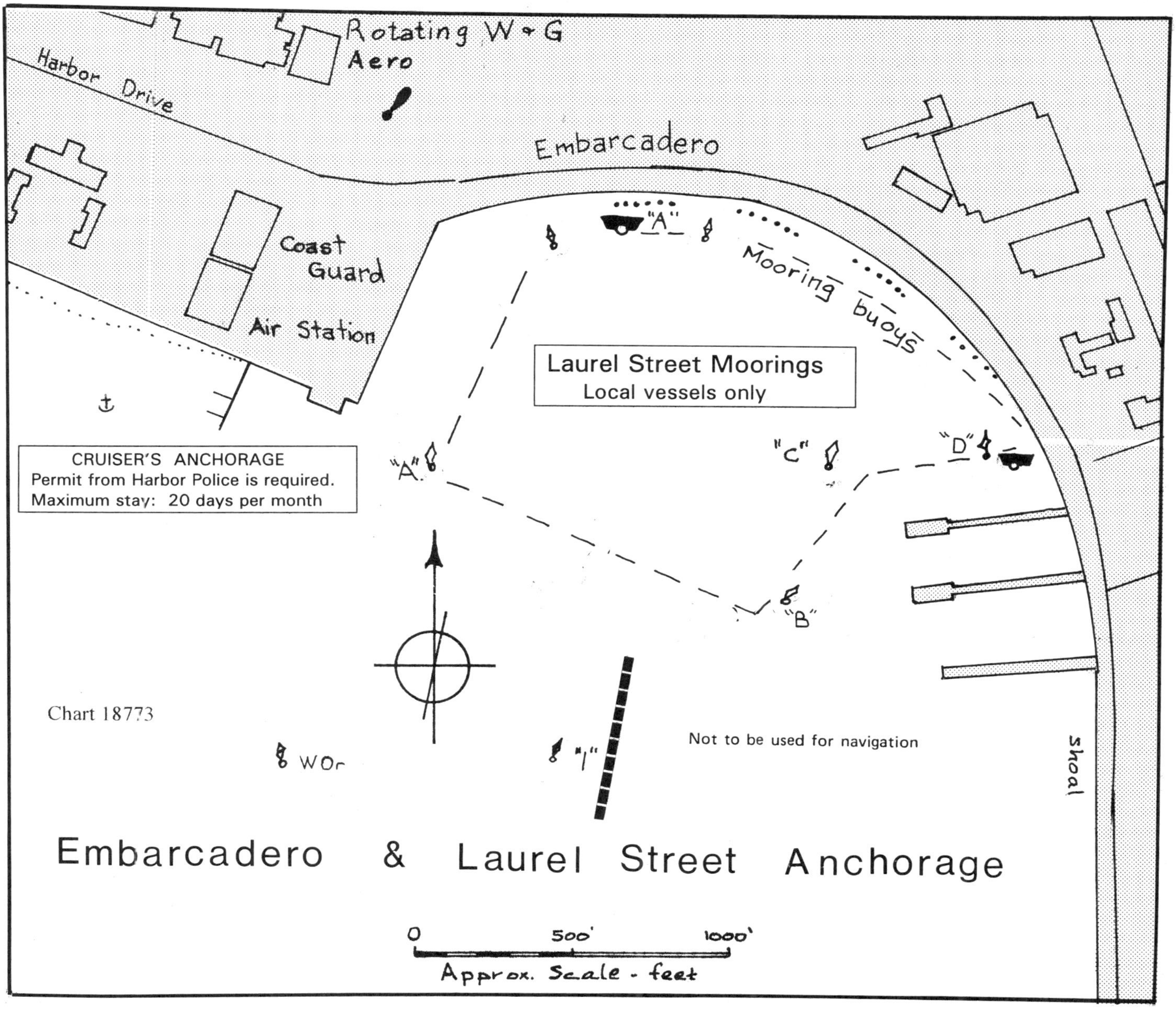

Marriott Marina is a fully modern 445-slip facility formerly known as the Intercontinental Marina. It is advisable to call ahead for rates and availability of space (619-230-8955). There is a $100 refundable deposit on each key and you have the use of the luxurious swimming pools, Jacuzzi and a first-class restaurant is close by. Next door is a very interesting collection of shops and restaurants in the rambling bayside development of Seaport Village. It's worth a visit to Cost Plus, a nearby market which has a selection of interesting foods at competitive prices.

Watch your depths when proceeding south in the bay, and keep within the delineated channels. Two anchorage areas are available: the **Bay Bridge roadstead anchorage** south of the Coronado Bay Bridge near the Coronado end. and the **Glorietta Bay anchorage** across from the boat launching ramp where your stay is limited to 72 hours. Mooring buoys in the area north of the Coronado Bay Bridge are controlled by the Harbor Police. The **Glorietta Bay Marina** has transient berths, though no fuel. Another **anchorage** further south in the bay is at Crown Cove. The anchorage and dock near Silver Strand Park are for the use of military personnel only.

There are a number of marinas in the southern part of San Diego Bay offering a variety of services and prices. Coronado Cays Marina has a fuel dock but no laundry facilities. **Loews Crown Isle Marina** is a luxury resort across the bay from Chula Vista. **Anchorage** may be taken in the South Bay (32 39N 117 07.3W) west of the 24th Street Marine Terminal in National City.

Transient slips are often available at the two marinas in Chula Vista Harbor: **Chula Vista Yacht Harbor** (619-691-1860) on the north side and **California Yacht Marina** (619-422-2595) on the south. Call ahead to verify availability of slips as it is a long trip down the bay from other moorage areas. Take care to follow the buoyed channels leading to the marinas as numerous shoal areas exist, some immediately adjacent to the dredged channel. It is essential that a current chart is used and care be taken when traveling in this area.

Although it is some distance from the activity of Shelter Island and downtown, Chula Vista is worth investigating if you plan on being in the San Diego area for an extended period. There is a boatyard immediately to the north with haul-out and maintenance facilities for small to medium-size boats, and there is a yard capable of servicing large vessels further north in National City.

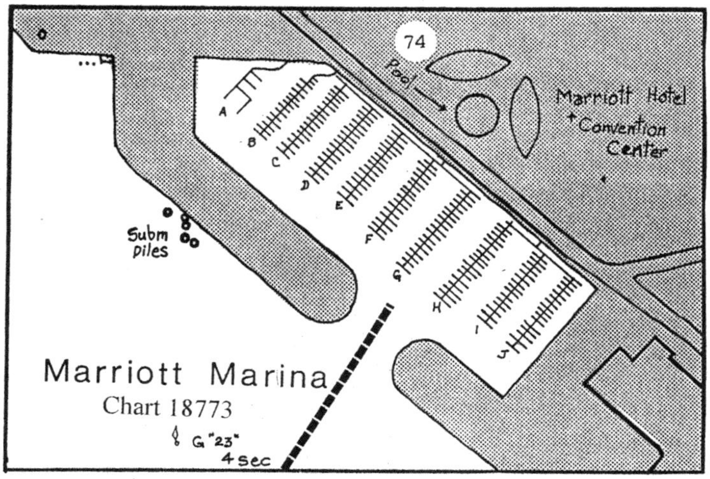

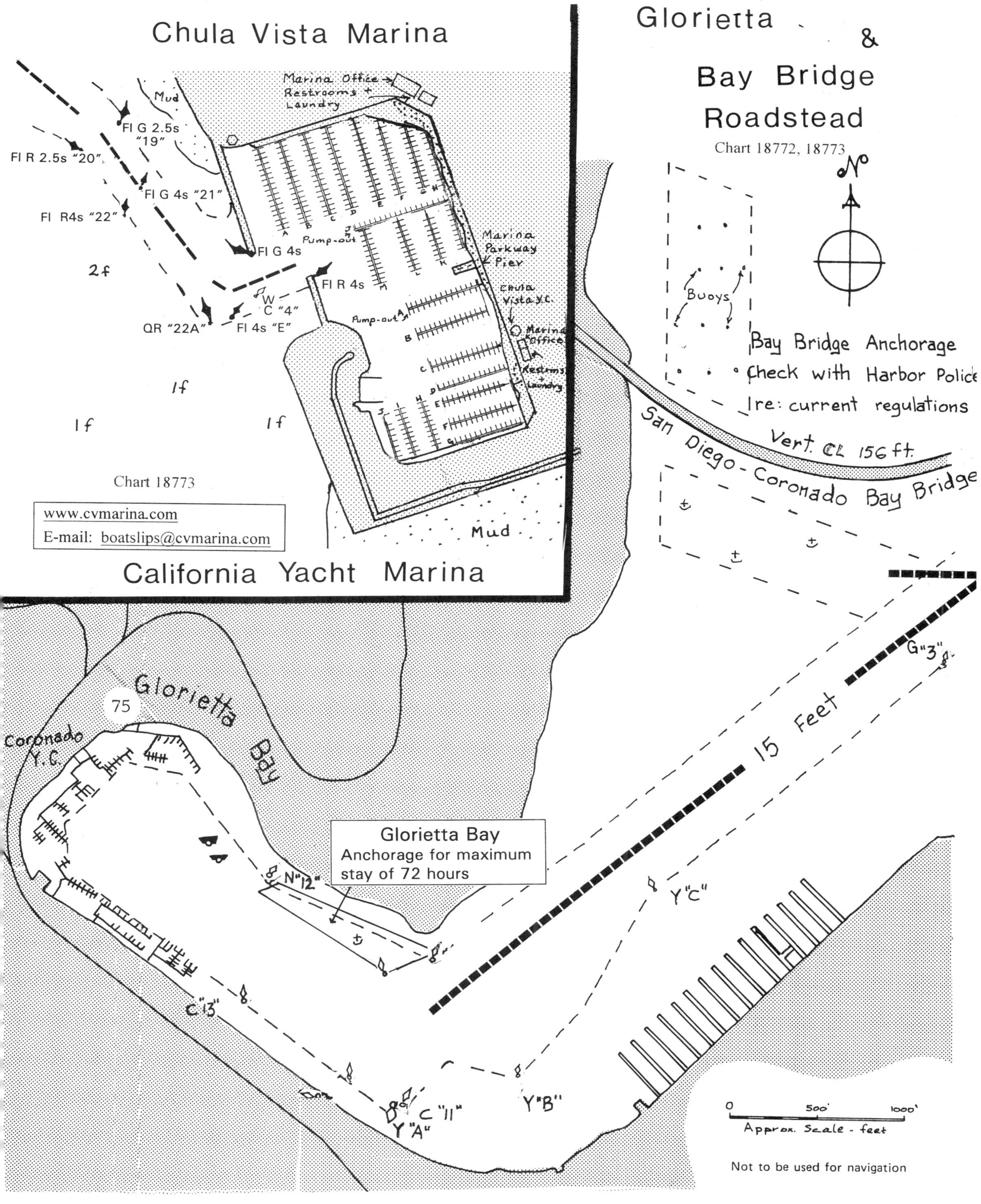

Chula Vista Marina
Marina Office
Restrooms + Laundry
Mud
Fl G 2.5s "19"
Fl R 2.5s "20"
Fl G 4s "21"
Fl R4s "22"
Fl G 4s
Pump-out
Marina Parkway Pier
Fl R 4s
2f
W
C "4"
QR "22A"
Fl 4s "E"
Chula Vista Y.C.
Marina Office
Restrms + Laundry
1f
1f
1f
Chart 18773
www.cvmarina.com
E-mail: boatslips@cvmarina.com
Mud
California Yacht Marina
Glorietta & Bay Bridge Roadstead
Chart 18772, 18773
N
Buoys
Bay Bridge Anchorage
Check with Harbor Police
re: current regulations
Vert. CL 156 ft.
San Diego-Coronado Bay Bridge
G "3"
15 Feet
75
Glorietta Bay
Coronado Y.C.
Glorietta Bay
Anchorage for maximum
stay of 72 hours
N "12"
Y "C"
C "13"
C "11"
Y "A"
Y "B"
0 500' 1000'
Approx. Scale - feet
Not to be used for navigation

APPENDIX I: CHARTS, PUBLICATIONS and REFERENCES

Coastal Charts

18473 Oak Bay to Shilshole Bay
18471 Approaches to Admiralty Inlet
18427 Anacortes to Skagit Bay
18465 Strait of Juan de Fuca - Eastern Part
18480 Approaches to Strait of Juan de Fuca (includes Destruction Is.)
18500 Columbia River to Destruction Island
18520 Yaquina Head to Columbia River
18580 Cape Blanco to Yaquina Head
18600 Trinidad Head to Cape Blanco
18620 Point Arena to Trinidad Head
18640 San Francisco to Point Arena
18680 Point Sur to San Francisco
18700 Point Conception to Point Sur
18720 Point Dume to Purisima Point
18721 Santa Cruz Island to Purisima Point
18725 Port Hueneme to Santa Barbara
18740 San Diego to Santa Rosa Island
18022 San Diego to San Francisco

Detail Charts

18450 Seattle Harbor, Elliott Bay, Duwamish Waterway
18443 Approaches to Everett
18468 Port Angeles
18484 Neah Bay
18502 Grays Harbor
18521 Columbia River
18558 Tillamook Bay
18561 Approaches to Yaquina Bay
18581 Yaquina Bay and River
18583 Siuslau River
18584 Umpqua River
18587 Coos Bay
18588 Coquille River Entrance
18589 Port Orford to Cape Blanco
18601 Cape Sebastian to Humbug Mountain
18602 Pyramid Point to Cape Sebastian
18603 St. George Reef and Crescent City Harbor
18605 Trinidad Harbor
18622 Humboldt Bay
18626 Elk to Fort Bragg
18643 Bodega and Tomales Bays
18645 Gulf of the Farallones

NOTE: This list is presented to assist boaters in choosing charts which cover the areas of their interest. Since NOAA charts and chart numbers are constantly being revised, some data will be outdated soon after publication of this guide.

APPENDIX I: CHARTS AND PUBLICATIONS Cont'd.

18647 Drakes Bay
18649 San Francisco Entrance
18682 Half Moon Bay
18685 Monterey Bay
18703 Estero Bay
18704 San Luis Obispo Bay
18744 Santa Monica Bay
18749 San Pedro Bay
18754 Newport Bay
18765 Approaches to San Diego (includes detail of Mission Bay)
18774 Gulf of Santa Catalina (includes detail of Oceanside)
18773 San Diego Bay
18727 San Miguel Passage
18728 Santa Cruz Passage
18729 Anacapa Passage
18757 Santa Catalina Island

U.S. Coast Pilot 7 (latest edition): Pacific Coast (Calif., Ore., Wash. and Hawaii)
Tide Tables, West Coast, North and South America
Tidal Current Tables, Pacific Coast of North America and Asia
Light List, Vol. VI (latest edition) - *Pacific Coast and Pacific Islands*
Reed's Nautical Almanac - West Coast.

BOOKS and REFERENCES

Bailey, Jo. and Nyberg, Carl: *Gunkholing in the San Juans,* Norwesting, Inc.
Calhoun, Bruce, *Cruising the San Juan Islands*, Weatherly Press
California Boater's Guide to the Harbors and Marinas of San Francisco Bay, Delta and Hawaii. Bald Eagle Enterprises..
Coffey, D. J. *Dolphins, Whales and Porpoises and Encyclopedia of Sea Mammals,* Macmillan Publishing Co. Inc.
Daily, Marla. *California's Channel Islands: 1001 Questions Answered.*
Ehorn, Nancy (principal writer). *Channel Islands National Park,* Friends of Channel Islands National Park.
Fagan, Brian M. *Cruising Guide to Southern California's Offshore Islands*, Caractacus Corp., Santa Barbara, CA
Haig-Brown, Roderick. *Captain of the Discovery.* Macmillan of Canada
Haley, Delphine (Ed.). *Marine Mammals,* 2nd Ed. Pacific Search Press, Seattle, WA
Herzog, Carl (Ed.). *Reed's Nautical Almanac: North American West Coast.* Current edition., Courier Corp. Westford, MA
Hinshaw-Patent, Dorothy. *Seals, Sea Lions & Walruses*, Holiday House, N.Y.,
Papastavrou, Vassilli. *Seals and Sea Lions.* Wayland Publishers Ltd.
Santa Monica Bay Restoration Project. *Southern California Boaters' Guide.* Los Angeles, CA
Schell, Hal. *Hal Schell's Guide to Cruising California's Delta*, Schell Books, Stockton, CA
Scott, James W. and Reuling, Melly A. Washington Public Shore Guide. University of Washington Press
Snively, Gloria. *Exploring the Sea Shore in British Columbia, Washington and Oregon.*
Wagner, Henry, Editor. *Spanish Voyages to the Northwest Coast of America*: 16th Century.
Wicklund, Bruce. *Boating and Diving Catalina Island.*

APPENDIX II: GPS WAYPOINT LIST

This waypoint list is intended to assist a cruiser to locate offshore approach buoys and avoid reefs, points and capes by a safe margin Waypoints are not given following the channel into marinas because facilities, river channels and other areas subject to shoaling are constantly changing. Consequently it is necessary for the skipper to use the latest navigational information available and follow channel markers and other aids visually.

Since GPS receivers have their individual finger-prints it follows that a navigator is advised to use the list with caution, determine the pattern of variance from the description and remarks and make adjustments as required.

ST	CITY/AREA	HARBOR/BAY	DESCRIPTION	LAT(North)	LON(West)	REMARKS
WA	Seattle	Shilshole Bay	Outer marker	47°40.9'	122°25.1'	Bell Buoy
	Seattle	Elliott Bay	Marina	47°37.7'	122°23.7'	Entrance
		Puget Sound	Point No Point	47°55.0'	122°31.0'	Avoid shipping lanes
		Puget Sound	Bell Buoy R"2"	47°57.0'	122°35.5'	.5 n.m. E of buoy
		Admiralty Inlet	Nodule Point	48°02.0'	122°39.5'	.5 n.m. E of point
		Admiralty Inlet	Marrowstone Pt.	48°06.2'	122°40.5'	.5 n.m. E of point
	Port Townsend	Admiralty Inlet	Point Wilson	48°09.4'	122°44.7'	.5 n.m. NE of pt, traffic
	Port Townsend	Admiralty Inlet	Point Hudson Hbr.	48°07.0'	122°44.0'	.5 n.m. E of entrance
	Port Townsend	Admiralty Inlet	Boat Haven	48°06.4'	122°46.1'	.1 n.m. off entrance
		Juan de Fuca Str	Dungeness Spit	48°11.7'	123°05.4'	.2 n.m. NE of buoy #2
	Port Ludlow	Puget Sound	S. end of Colvos R.	47°56.7'	122 39.5	Red nun
	Everett	Possession Sound		47°59.0'	122°15.6'	.5 n.m. SW bell buoy #1
		Possession Sound	Everett Harbor	47°58.9'	122°14.1'	Entrance
	Semiahmoo Bay		Shoal marker	48°59.7'	122 46.8	Green buoy #1
	Anacortes	Fidalgo Bay	Cap Sante Pt.	48°30.8'	122°35.6'	.1 n.m. N buoy #A
		Deception Pass	Hoypus Pt.	48°24.8'	122°36.5'	1.5 n.m. ENE of pass
		Deception Pass		48°24.3'	122°39.2'	.5 n.m. W of pass
		Northwest Pass		48°25.0'	122°40.5'	.5 n.m. NNW of Deception I.
		Swinomish Channel	Skagit Bay	48°21.7'	122°33.4'	Entrance to channel
	Bellingham	Bellingham Bay	Squalicum Harbor	48°45.0'	122°31.5'	.2 n.m. W of entrance
	Friday Harbor	Marina	Entrance lights	48°32.4'	123° 00.9	Fl G 2.5s
	Sequim	Juan de Fuca Str.	John Wayne Marina	48°05.1'	123°01.5'	.2 n.m. NE of buoy #2
	Port Angeles	Juan de Fuca Str.	Ediz Hook	48°10.4'	123°22.5'	Approach
	Port Angeles	Juan de Fuca Str.	Ediz Hook	48°08.6'	123°23.4'	Turning point
	Port Angeles		PA Boat Haven	48°07.7'	123°26.8'	Buoy #3
		Juan de Fuca Str.	Pillar Point	48°14.5'	124°05.0'	1.5 n.m. N of Pillar Pt.
	Sekiu	Clallam Bay	Approach	48°15.9'	124°15.5'	N of entrance
	Neah Bay	Waadah Island	Entrance	48 23.1'	124 36.0'	ISO white, 6s
	Neah Bay	Neah Bay	Entrance	48°22.7'	124°35.0'	NW of bell buoy #1
		Koitlah Point	Enroute	48°25.0'	124°40.0'	2 n.m. WNW Koitlah Point
	Cape Flattery		Turning point	48°24.0'	124°45.0'	.5 n.m. NW of Tatoosh I.
	Cape Alava		Enroute	48°10.0'	124°50.0'	Buoy #2UR
	La Push	Quillayute River	Offshore buoy	47°53.1'	124°40.4'	Whistle Buoy #Q
	Grays Harbor		Offshore buoy	46°51.9'	124°14.4'	Lighted Wh. buoy #GH
OR	Columbia River		Approach	46 11.1'	124 11.1'	Lighted horn buoy #CR
	Columbia River		Entrance	46 13.4'	124 11.0'	Bell buoy #1 Fl G
	Ilwaco	Baker Bay	Entrance	46°18.0'	124°02.4'	Mooring Basin light #2
	Astoria	Columbia River	Tansy Pt. Turn	46°11.6'	123°54.6'	Change of direction
	Garibaldi	Tillamook Bay	Offshore buoy	45°34.3'	123°59.6'	Whistle buoy #T on range
	Depoe Bay	Depoe Bay	Offshore buoy	44°48.5'	124°05.2'	Whistle buoy #DB
	Newport	Yaquina Bay	Offshore buoy	44°35.9'	124°06.7'	Whistle buoy #Y
	Florence	Siuslaw River	Offshore buoy	44°01.1'	124°09.7'	Whistle buoy #S
	Salmon Harbor	Umpqua River	Offshore buoy	43°40.0'	124°14.3'	Whistle buoy #U
	Coos Bay	Coos Bay	Offshore buoy	43°22.3'	124°23.0'	Whistle buoy #K
	Bandon	Coquille River	Offshore buoy	43°08.2'	124°27.8'	Whistle buoy #2
		Cape Blanco	Reef avoidance	42°45.2'	124°39.3'	Whistle buoy #20R

CHARLIE'S CHARTS assumes no liability of any kind for consequences of the use of this list. The use of these waypoint is at the discretion of the captain of each vessel and the safe navigation of his/her vessel is entirely his/her responsibility.

ST	CITY/AREA	HARBOR/BAY	DESCRIPTION	LAT(North)	LON(West)	REMARKS
OR	Port Orford	Port Orford	Approach anch	42°43.3'	124°30.6'	Bell buoy #1
	Rogue River Reef		Reef avoidance	42°28.0'	124°35.0'	Reef avoidance
	Gold Beach	Rogue River	Offshore buoy	42°23.6'	124°28.6'	Whistle buoy #R
	Gold Beach	Rogue River	Entrance	42°24.8'	124°26.7'	Green can buoy #1
CA	Brookings	Chetco River	Offshore buoy	42°01.7'	124°17.0'	Whistle buoy #CR
		St George Reef	Reef avoidance	41°50.5'	124°23.5'	Lighted horn buoy #2SG
		St George Reef	Reef avoidance	41°45.0'	124°23.5'	Clearance at SW
	Crescent City	Crescent Cty Hbr.	Offshore buoy	41°43.1'	124°11.8'	Lighted whistle buoy #2
	Crescent City	Crescent Cty Hbr.	Approach	41°44.2'	124°11.5'	Entrance light, horn
	Trinidad	Trinidad Head	Offshore buoy	41°03.1'	124°10.4'	Whistle buoy #26
	Trinidad	Trinidad Head	Approach buoy	41°03.0'	124°09.6'	Gong buoy #2
	Eureka	Humboldt Bay	Offshore buoy	40°46.4'	124°16.2'	Whistle buoy #HB
	Eureka	Mid Channel	Breakwater entr.	40°46.0'	124°14.5'	On range #4
	Shelter Cove	Entrance	Entrance buoy	40 00.6'	124 03.6'	Bell buoy #1
	Cape Mendocino	Blunts Reef	Reef avoidance	40°27.0'	124°32.0'	4.m. NW of horn buoy #2B
	Punta Gorda	Punta Gorda Pt.	Point clearance	40°13.0'	124°26.5'	4.5 n.m. SW of point
		Shelter Cove	Offshore buoy	40°00.2'	124°04.8'	Whistle buoy #20
		Shelter Cove	Approach	40°01.0'	124°03.7'	Anchorage
	Fort Bragg	Noyo anchorage	Offshore buoy	39°25.9'	123°49.9'	Whistle buoy #NA
	Fort Bragg	Noyo anchorage	Approach	39°25.7'	123°49.6'	.3 n.m. from breakwater
		Punta Arena	Point clearance	38°57.4'	123°48.0'	1 n.m. W of point
	Bodega Bay	Bodega Head	Offshore buoy	38° 17.1'	123 04.2'	Lighted whistle buoy #12
	Bodega Bay	Bodega Harbor	Approach buoy	38°17.2'	123°02.4'	Lighted gong buoy #BA
	Bodega Bay	Bodega Harbor	Entrance light	38°18.3'	123°02.9'	Entrance light, horn 10s
		Point Reyes	Point clearance	37°59.7'	123°03.4'	Pt. Reyes light, Fl W 5s
		Drakes Bay	Approach anch'g.	38°59.0'	122°57.3'	Whistle buoy #1 Fl G 6s
		Duxbury Reef	Reef avoidance	37°51.5'	122°41.7'	Lighted whistle buoy #IDR
	Marin Penin.	Bonita Channel	Channel entrance	37°50.0'	122°34.0'	Buoyed channel #2 Fl R 6s
	San Francisco	Northern approach	Point clearance	37°48.6'	122°31.7'	.5 n.m. S of Pt Bonita
	San Francisco	Offshore approach	Lighted horn buoy	37°45.0'	122°41.6'	4 n.m. ENE to channel
	San Francisco	Southern approach	Clear Mile Rocks	37°47.7'	122°30.7'	Close N of Mile Rocks Light
	San Francisco	South Channel	Seal Rocks	37°46.8'	122°30.4'	Clear coastal rocks
	San Francisco	Farallon Islands	NW end	37°48.0'	123°11.0'	15-mile island chain
	San Francisco	Farallon Islands	Clear SE end of Is.	37°41.0'	123°02.0'	25 n.m. W of Golden Gate
		San Pedro Rocks	Rocks avoidance	37°36.0'	122°31.4'	Clear coastal rocks
		Ano Nuevo Isl.	Point clearance	37°00.0'	122° 20.3'	4.8 n.m S of island
	San Francisco	South Traffic Lane	Approaching buoy	37°39.0'	122°41.4'	Yellow Bell buoy S
	Half Moon Bay	Pillar Point Hbr.	S. end of SE Reef	37°27.9'	122°28.2'	Lighted gong buoy #1S
	Half Moon Bay	Pillar Point Hbr.	Approach	37°28.3	122°30.9'	Lighted whistle buoy #PP
	Half Moon Bay	Pillar Point Hbr.	Entrance	37°29.2'	122°30.4'	Lighted gong buoy #1, FLG
	Half Moon Bay	Pillar Point Hbr.	Inner Harbor light	37°30.0'	122°29.1'	Triangular dayb'd on b'water
		Bolsa Point	Point clearance	37°11.0'	122°32.5'	4 n.m. W of point
		Needle Rock Pt.	Point clearance	36°54.0'	122°08.0'	3.5 n.m. SW of point
	Santa Cruz	Monterey Bay	Offshore buoy	36°56.3'	122°00.6'	Whistle buoy #SC
	Santa Cruz	Monterey Bay	Breakwater entr.	36°57.6'	122°00.1'	West breakwater light G 4s
	Moss Landing	Monterey Bay	Entrance	36°48.0'	121°48.1'	Lighted bell buoy #MLA
	Monterey	Monterey Harbor	Approach buoy	36°36.5'	121°53.3'	Junct'n. buoy R & G nun
		Point Cabrillo	Point clearance	36°37.7'	121°53.5'	.2 n.m. NE bell buoy #4
		Point Pinos	Point clearance	36°39.1'	121°56.6'	.2 n.m. N bell buoy #2
		Cypress Point	Point clearance	36°35.2'	121°59.4'	.3 n.m. W of point
		Point Sur	Point clearance	36°17.0'	122°05.0'	10 n.m. W of point
		Pt PiedrasBlancas	Point clearance	35°38.0'	121°19.0'	2.5 n.m. W of Point
	San Simeon	San Simeon Bay	S of SanSimeon Pt.	35°37.5'	121°11.3'	S of bell buoy #1
		Von Helm Rock	Rock avoidance	35°31.3'	121°08.2'	Rock avoidance
		Point Estero	Point clearance	35°26.0'	121°01.7'	2 n.m. SW of point
	Morro Bay	Estero Bay	Approach	35°21.7'	120°52.6'	Lighted whistle buoy #MB
		Point Buchon	Point clearance	35°15.0'	120°56.3'	2 n.m. W of point
		Westdahl Rock	Rock avoidance	35°08.6'	120°47.2'	Bell buoy #1

CHARLIE'S CHARTS assumes no liability of any kind for consequences of the use of this list. The use of these waypoint is at the discretion of the captain of each vessel and the safe navigation of his/her vessel is entirely his/her responsibility.

ST	CITY/AREA	HARBOR/BAY	DESCRIPTION	LAT(North)	LON(West)	REMARKS
CA	SanLuis Obispo	San Luis O. Bay	Approach buoy	35°09.4'	120°43.8'	.3 n.m. SE of buoy #2
		Point Arguello	Point clearance	34°31.0'	120°49.0'	10 n.m. W of point
		Point Conception	Point clearance	34°22.0'	120°29.0'	5 n.m. S of point
	Santa Barbara	Santa Barbara Pt.	Point clearance	34°23.2'	119°42.4'	.5 n.m. S of point
	Santa Barbara	Santa Barbara	Hbr. approach	34°24.1'	119°40.8'	Lighted bell buoy #SB
	Santa Barbara	Santa Barbara	Approach	34°24.3'	119° 41.0'	Lighted buoy #3
	Santa Barbara	Santa Barbara	Entrance	34°24.3'	119° 41.3	Breakwater extension light
	Ventura	Pierpoint Bay	Approach buoy	Variable	Variable	Lighted whistle buoy #2
	Ventura	Pierpoint bay	Entrance	34°14.7'	119°16.3'	Enter S of breakwater
	Oxnard	Channel Is. Hbr.	Approach	34°09.3'	119°16.2'	Breakwater S light #1
	Oxnard	Channel Is. Hbr.	Entrance	34°09.4'	119°13.5'	South jetty light #2
	Marina del Rey	MarinaDelRey Hbr.	Approach	33°57.5'	118°27.6'	Breakwater S light #1, Fl G
	Marina del Rey	MarinaDelRey Hbr.	Approach	33°57 .8'	118 27.9'	Breakwater N light #2, Fl R
	Redondo Beach	King Harbor	Approach	33°50.4'	118°23.7'	Lighted bell buoy #RB
	Redondo Beach	King Harbor	Entrance	33° 50.5'	118 23.7	West jetty light #3
	Los Angeles	San Pedro Bay	Approach	33°41.4'	118°14.5'	.1 nm off whistle buoy #3
	Los Angeles	San Pedro Bay	Approach	33°41.9'	118 14.7'	Approach lighted buoy #5
	Los Angeles	San Pedro Bay	Approach	33.42.5'	118 15.1	White cylindrical tower #8
	Los Angeles	San Pedro Bay	Main channel	33° 42.8'	118 15.5'	Lighted buoy #10
	Long Beach	San Pedro Bay	Approach	33°41.4'	118°11.0'	Lighted whistle buoy #LB
	Long Beach	San Pedro Bay	Approach	33° 41.4'	118 08.2'	East end of breakwater #1
	Long Beach	Alamitos Bay	Approach	33°44.2'	118°07.3'	.1 n.m. W of breakwater
	Long Beach	Alamitos Bay	Approach	33° 44.7'	118°07.0'	Channel light #3
	Long Beach	L.B. ShorelineMarina	Approach	33° 45.6'	118°10.9'	Jetty light QW on post
	Long Beach	L.B. ShorelineMarina	Approach	33° 45.4'	118°10.9'	West end of breakwater
	Long Beach	L.B. ShorelineMarina	Entrance	33° 45.6'	118°11.5'	East entrance light
	Huntington Hbr.	Anaheim Bay	Approach	33° 43.0'	118°06.7'	Entrance bell buoy #1
	Huntington Hbr.	Anaheim Bay	Approach	33° 43.7'	118°06.1'	West jetty light #5
	Newport Beach	Newport Bay	Approach	33°35.1'	117°52.6'	Lighted buoy #NWP
	Dana Point	San Juan Rocks	350 yds. off D.Pt.	33° 27.3'	117° 43.3'	Lighted whistle buoy 2SJR
	Dana Point	Dana Point	Approach	33°26.0'	117°43.5'	S of point
	Dana Point	Dana Point	Approach	33°27.2'	117°41.9'	Lighted buoy #1 G
	Dana Point	Dana Point	Entrance	33 27.3'	117 41.5'	Breakwater light #1
	Oceanside	Oceanside Harbor	Approach	33°12.1'	117°24.0'	Off entrance
	Oceanside	Oceanside Harbor	Entrance	33° 12.3'	117 24.1'	Breakwater light #3
	Mission Bay	Mission Bay	Approach	32° 45.0'	117°16.5'	Off entrance breakwater
	Mission Bay	Mission Bay	Approach	32° 45.5	117 15.8'	North jetty light #1
	San Diego	Point Loma	Point clearance	32°40.0'	117°17.0'	2 n.m W of point
	San Diego	Point Loma	Point clearance	32° 37.3'	117°14.0'	2.5 n.m S of point
	San Diego	San Diego Hbr.	Entrance channel	32° 39.0'	117°13.5'	Off entrance
	San Diego	Danger	Entrance channel	32° 40.1'	117 13.5	Zuniga submerged jetty lt.
	Channel Islands					
	San Miguel Is.	Cuyler Harbor	Approach	34°04.5'	120°20.5'	NE of harbor
	Santa Rosa Is.	Bechers Bay	Approach	34°00.0'	120°01.0'	NE of bay
	Santa Cruz Is.	Lady's Harbor	Approach	34°03.6'	119°47.3'	N of anchorage
	Santa Cruz Is.	Fry's Harbor	Approach	34°03.4'	119°45.2'	N of anchorage
	Santa Cruz Is.	Prisoner's Is.	Approach	34°01.5'	119°41.0'	N of anchorage
	Anacapa Island	East end of island	Approach	34° 01.1'	119 21.6'	Fl 2W, horn
	Anacapa Island	Frenchy's Cove	Approach	34°00.5'	119°24.4'	N of anchorage
	Santa Catalina	Long Point Light	Approach	33 24.4'	118.21.9'	Long Point Light
	Santa Catalina	Avalon Harbor	Approach	33°20.9'	118°19.0'	.3 n.m. E of Avalon #2
	Santa Catalina	Avalon Harbor	Approach	33 20.7'	118°19.3'	Light #1, Fl G
	Santa Catalina	Catalina Harbor	Approach	33 25.4'	118.30.8'	Fl W 4s.
	Santa Catalina	Isthmus Cove	Offshore approach	33°28.1'	118°29.0'	.5 n.m. NE of Ship Rock
	Santa Catalina	Isthmus Cove	Danger	33° 27.8'	118°29.4'	Ship Rock light
	Santa Catalina	Isthmus Cove	Danger	33° 26.9'	118°29.4'	Harbor Reefs light
	Santa Catalina	Isthmus Cove	Entrance	33° 26.9'	118°29.6'	North entrance buoy #1

APPENDIX III: TABLE of MILEAGES – COASTWISE DISTANCES

Entrance buoy to San Diego ....8.3 miles
Entrance buoy to Long Beach ....4.3 miles
Entrance buoy to Los Angeles ....3.8 miles
Entrance buoy to San Francisco................................. 15.0 miles
Humboldt Bay entrance buoy to Eureka ....5.5 miles
Entrance buoy to Coos Bay City .... 13.3 miles
Yaquina Bay entrance buoy to Newport ....3.3 miles
Columbia River Large Navigation Buoy to Astoria .... 17.8 miles

	San Diego, CA	Newport Beach, CA	Long Beach, CA	Los Angeles, CA	Santa Barbara, CA	Port San Luis, CA	Monterey, CA	San Francisco, CA	Eureka, CA	Crescent City, CA	Coos Bay, OR	Florence, OR	Newport, OR	Depoe Bay, OR	Astoria, OR	Cape Flattery, WA
Newport Beach, CA	78															
Long Beach, CA	94	25														
Los Angeles, CA	95	27	3													
Santa Barbara, CA	174	108	94	90												
Port San Luis, CA	259	193	179	175	91											
Monterey, CA	370	304	290	286	203	121										
San Francisco, CA	455	389	374	371	287	205	96									
Eureka, CA	653	587	572	569	485	403	294	232								
Crescent City, CA	704	638	624	620	537	455	346	283	64							
Coos Bay, OR	817	751	736	733	649	567	459	396	180	125						
Florence, OR	848	782	768	764	681	599	490	427	212	156	59					
Newport, OR	881	815	800	797	713	631	522	459	244	188	92	43				
Depoe Bay, OR	891	825	810	807	723	641	532	469	254	199	101	54	16			
Astoria, OR	989	922	908	904	821	739	630	567	352	296	201	153	115	101		
Cape Flattery, WA	1104	1038	1024	1020	937	754	746	683	468	411	321	273	235	222	153	
Seattle, WA	1228	1162	1148	1144	1061	978	870	807	592	535	445	397	359	346	277	124

The intersection of columns between two places is the approximate distance in nautical miles. For example: Santa Barbara is 713 nautical miles from Newport, Oregon.

APPENDIX IV: WEATHER FORECASTS

Weather information originates from a variety of sources such as radio, radiotelephone, fax, SSB frequencies and websites with an infinite number of links. The following list provides a summary and practical starting point.

NOAA Weather Radio VHF Weather Channels and Frequencies

WX1	162.550 MHZ	WX5	162.450MHZ
WX2	162.400MHZ	WX6	162.500MHZ
WX3	162.475MHZ	WX 7	162.525MHZ
WX4	162.425MHZ		

For local programming within about 40 miles of a transmitter one can use a special **NOAA Weather Radio Receiver** and thereby maintain a simultaneous watch on NOAA Weather Radio and marine VHF channels. A warning alarm feature is available for specially designed receivers. NOAA Weather stations are found at: http//205.156.54.206/nwr/nwrtrans.htm

High Frequency Voice Broadcast (HF) is an upper sideband (USB) radio-telephone voice synthesized broadcast of the High Seas and Offshore Forecasts:

OFFPZ6 Pt. Reyes (NMC) Times (UTC): 0430, 1030, 1030 Frequencies (kHZ): 4426.0, 8764.0, 13089.0

High Frequency SITOR Broadcast also known as Narrow Band Direct Printing (NBDP) is a radiotelex broadcast of the High Seas Forecasts in a text form:

HSFEP1 Pt. Reyes (NMC) Times: 0005, 1800 Frequencies: 8416.5, 16806.5

NAVTEX is a radiotelex broadcast of coastal and offshore forecasts in text form for near shore users at 518.0 kHZ. USCG Station broadcast times (UTC):

OFFPZ5 Astoria, OR 0130, 0530, 0930, 1330, 1730, 2130

OFFPZ6 Cambria, CA 0045, 0445, 0845, 1245, 1645, 2045

OFFPZ6 Pt. Reyes, CA 0000, 0400, 0800, 1200, 1600, 2000

Medium Range Broadcast is a medium frequency radiotelephone voice broadcast of urgent marine information, including offshore forecasts at 2670.0 kHZ after first being announced on the distress, safety and calling frequency 2182.0 kHZ.

OFFPZ5 Port Angeles, WA 0615, 1815

OFFPZ5 Astoria, OR 0533, 1733

OFFPZ5 North Bend, OR 0603, 1803

OFFPZ6 Humbolt Bay, CA 0303, 1503

OFFPZ6 San Francisco, CA 0203, 1403

OFFPZ6 Los Angeles/Long Beach, CA 0503, 1301, 2103

Numerous USCG stations make multiple VHF radiotelephone voice broadcasts at 156.8 mHZ (Ch. 16 and 157.1 mHZ VHF FM (Ch 22A). Typical coverage of these broadcasts is up to 25 nm offshore.

Website coverage of coastal waters from Seattle to San Diego (up to 60 nm offshore): http://www.nws.noaa.gov/om/marine/zone/west/westmz.htm

A map of the US Pacific coast appears that is divided into zones depicting waters 20 nm offshore and 20 to 60 nm offshore. Click on the part of the coast where a weather forecast is desired. A more detailed map appears and you can access information for a specific area (coastal waters or 20 to 60 nm offshore) by clicking on a spot e.g. between Florence and Cape Arago, OR. A detailed forecast then appears on the monitor.

APPENDIX V: MARINA TELEPHONE NUMBERS

CALIFORNIA

Dana Point Harbor

55	Dana West Marina	(949) 493-6222
56	Dana Point Marina	(949) 496-6137
	Embarcadero Marina	(949) 496-6177

Eureka

	Eureka Public Marina	(707) 268-1973
	Woodley Island Marina	(707) 443-0801

Huntington Harbor

	Sunset Aquatic Marina	(714) 846-0179
	Peter's Landing	(714) 840-1387
	Huntington Harbor	(714) 840-5545

Long Beach Harbor

	Alamitos Bay Marina	(562) 570-3215
	Cerritos Yacht Anch'g.	(562) 431-6575
	Downtown Shoreline Marina	(562) 570-4950

Los Angeles

33	Al Larson Marina	(310) 832-0526
25	Cabrillo Marina	(310) 732-2252
	Cabrillo Way Marina	(310) 514-4985
36	Cerritos Yacht Anchorage	(310) 834-4737
34	Colonial Yacht Anchorage	(310) 830-1161
89	California Yacht Marina	(310) 834-7113
26	Flietz Bros. Watchorn Hbr.	(310) 832-0334
30	Holiday Harbor Marina	(310) 835-3952
37	Island Yacht Anchorage #1	(310) 830-1111
37	Island Yacht Anchorage #2	(310) 830-3382
29	Leeward Bay Marina	(310) 830-5621
35	Lighthouse Yacht Landing	(310) 834-9595
90	New Marks Yacht Center	(310) 834-5016
32	Pacific Yacht Landing	(310) 830-0260
28	San Pedro Marina	(310) 519-8177
38	Terminal Island Marina	(562) 432-3387
31	Yacht Haven	(310) 834-6892

Marina del Rey

15	Bar Harbor Marina	(310) 827-3687
	Bay Club Marina	(310) 821-8881
21	Catalina Yacht Anchorage	(310) 823-3684
13	Deauville Marina	(310) 823-4655
	Dock 77	(310) 823-4595
16	Dolphin Marina	(310) 587-0566
18	Holiday Harbor Marina	(310) 578-0566
12	Neptune Marina	(310) 823-4555
20	Marina City Y.C. (Ext. 313)	(310) 822-0611
22	Marina del Rey Hotel	(310) 823-4593
9	Marina Harbor #1 and #2	(310) 822-1659
19	Mariners Bay	(310) 822-2001
	SMYC Marina	(310) 823-2252
10	Tahiti Marina	(310) 823-4504
17	Panay Marina	(310) 823-2026
14	Villa del Mar Marina	(310) 823-4644

Mission Bay

79	Bahia Resort Hotel Marina	(858) 539-7695
78	Campland on the Bay	(858) 581-4224
62	Dana Inn and Marina	(619) 222-6440
63	Dana Landing	(619) 224-2513
61	Islandia Marina	(619) 221-4858
59	Driscoll Mission Bay Marina	(619) 223-5191
58	Marina Village Marina	(619) 224-3125
60	Seaforth Marina	(619) 224-6807
64	Sea World Marina	(619) 226-3910

Monterey Bay

22	Breakwater Cove Marina	(831) 373-7857
	Monterey Municipal Marina	(831) 646-3950

Morro Bay

23	Morro Bay Fuel Dock Marina	(805) 772-8617
24	Morro Bay Marina	(805) 772-8085
25	Morro State Park Marina	(805) 772-8796

Newport Harbor

52	Harbor Marina	(949) 644-9730
53	Balboa Yacht Basin	(949) 673-1761
54	Carlsberg Marina	(949) 723-8485
51	De Anza Bayside Marina	(949) 673-1331
	Lido Marina Village	(949) 675-8662
45	Lido Dry Stack Anchorage	(949) 673-9330
	Port Calypso Marina	(949) 645-6900
50	Newport Dunes Resort Marina	(949) 729-1100
	Swales Anchorage	(949) 548-1501
	Villa Cove Marina	(949) 644-9730

Oceanside

	Del Mar Marina	(760) 725-7245
57	Oceanside Harbor	(760) 435-4000

Oxnard (Channel Islands Harbor

Anacapa Island Marina	(805) 985-6035
Bahia Cabrillo Marina	(805) 985-6400
Channel Islands Marina	(805) 985-7558
Channel Islands Landing	(805) 985-6059
Peninsula Yacht Anchorage	(805) 985-6400
Vintage Marina	(805) 984-3366

Redondo Beach (King Harbor)

King Harbor Marina	(310) 376-6926
Portofino	(310) 798-5869
Port Royal Marina	(310) 376-0431
Redondo Beach Marina	(310) 374-3481

San Diego

80	Bay Club Marina	(619) 222-0314
72	Cabrillo Isle Marina	(619) 297-6222
82	California Yacht Marina	(619) 422-2595
77	Chula Vista Marina	(619) 691-1860
	Driscoll's Wharf	(619) 222-4930
	Fiddler's Cove Marina	(619) 522-8680
81	Gold Coast Anchorage	(619) 225-0588
75	Glorietta Bay Marina	(619) 435-5203
67	Half Moon Anchorage	(619) 224-3401
70	Harbor Island West Marina	(619) 291-6440
65	Kona Kai Marina	(619) 224-7547
83	Loew's Crown Island Marina	(619) 424-4455
71	Marina Cortez	(619) 291-5985
74	Marriott Marina	(619) 230-8955
	Montego Bay Marina	(619) 224-2471
68	Shelter Cove Marina	(619) 224-2471
66	Shelter Island Marina	(619) 223-0301
84	Sheraton Harbor Island	(619) 692-2249
69	Sun Harbor Marina	(619) 222-1167
73	Sun Road Resort	(619) 574-0736

San Francisco Bay

	Alameda Marina	(510) 521-1133
	Bair Island Marina	(650) 701-0382
1	Berkeley Marina	(510) 644-6376
	Berkeley Marine Center	(510) 843-8195
	Ballena Isle Marina	(510) 523-5528
	Benica Marina	(707)745-2628
	Brickyard Cove Marina	(510) 236-1933
2	Brisbane Marina	(650) 583-6975
	Clipper Yacht Harbor	(415) 332-3500
	Coyote Point Marina	(650) 573-2594
	Embarcadero Cove Marina	(510) 834-4591
3	Emery Cove Marina	(510) 428-0505
4	Emeryville City Marina	(510) 654-3716
	Fifth Avenue Marina	(510) 834-9815
6	Glen Cove Marina	(707) 552-3236
7	Grand Marina	(510) 865-1200
	Jack London Marina	(510) 834-4591
	Liberty Ship Pier	(415) 331-5020
9	Marina Village Yacht Harbor	(510) 521-0905
	Marina Plaza Harbor	(415) 332-4723
10	Oyster Cove Marina	(650) 952-5540
11	Oyster Point Marina	(650) 952-0808
	Pete's Harbor	(650) 366-0922
12	Pier 39	(415) 705-5556
	Pacific Marina (Oakland Y.C.)	(510) 522-6868
	Port of Oakland Marina	(510) 834-4591
	Redwood City Marina	(650) 363-1390
8	Richardson Bay Marina	(415) 332-5510
15	Richmond Marina Bay	(510) 236-1013
16	San Francisco Marina	(415) 292-2013
17	San Leandro Marina	(510) 357-7447
18	Schoonmaker Point Marina	(415) 331-5550
19	South Beach Harbor	(415) 495-4911
	Treasure Island Marina	(415) 981-2416
	Union Basin	(510) 834-4591
	Vallejo Marina	(707) 648-4370

Santa Barbara

26	Santa Barbara Harbor	(805) 564-5530

Santa Cruz

	Capitola Bay Moorings	(831) 462-2208
21	Santa Cruz Harbor	(831) 475-6161

Ventura

	Harbor Town Marina	(805) 650-9273
1	Ventura Isle Marina	(805) 644-5858
2	Ventura West Marina	(805) 644- 1402

OREGON

Astoria Mooring Basin	(503) 325-8279
Bandon, Port of	(541) 347-3206
Brookings, Port of	(541) 469-2218
Charleston Marina Coos Bay	(541).368-5745
Gold Beach, Port of	(541) 247-6269
Newport Marina, Port of	(541) 867-3321
Port Orford, Port of	(541) 332-712
Salmon Harbor Marina, Winchester Bay	(206) 282-5555

WASHINGTON

Anacortes Marina	(360) 293-4543
Blaine Harbor	(360) 647-6176
Cap Sante Boathaven,Anacortes	(360) 293-0694
Dagmar's Marina, Everett	(425) 259-6124
Davidson's Marina, Seattle	(206) 486-7141
Elliott Bay Marina, Seattle	(206) 285-4817
Harbor Village Marina, Seattle	(425) 485-7557
John Wayne Marina, Sequim	(360) 417-3440
La Conner Marina, La Conner	(360) 466-3118
Leschi Yacht Basin, Seattle	(206) 328-6777
Makah Marina, Neah Bay	(360)645-3012
Nautical Landing, Seattle	(206) 284-2308
Olson's Resort, Sekiu	(360) 963-2311
Point Roberts Marina	(360) 945-2255
Port Angeles Marina	(360) 876-5535
Port of Friday Harbor	(360) 378-2688
Port Hudson Hbr., Pt Townsend	(360) 385-2355
Port of Port Townsend	(360) 385-2355
Port Ludlow Marina	(360) 437-0513
Quileute Marina, La Push	(360) 374-5392
Roche Harbor Resort	(360) 378-2155
Semiahmoo Marina, Blaine	(360) 371-5700
Shilshole Marina, Seattle	(206) 728-3006
Shipyard Cove Marina, Fri.Hbr.	(360) 378-5101
Skyline Marina, Anacortes	(360) 293-5134
Squalicum Harbour, Bellingham	(360) 676-2542

APPENDIX VI: PINNIPED IDENTIFICATION

Pinniped means "fin-footed." Seals, sea lions and walruses belong to the group of mammals known as pinnipeds. Of the 34 species only 6 are seen along this coast and San Miguel Island is unique for it is the only place where all 6 species can be found.

Pinnipeds have stream-lined bodies, tapered at both ends, making them graceful, efficient swimmers. They are insulated from the cold not only by the layer of fat surrounding their body but also by the automatic control of blood circulation to the outer parts of the body. Because they spend most of the time under water the nostrils are naturally closed and they must consciously open their nostrils to breathe.

Harbor seals and elephant seals are known as "true seals" or earless seals since they have only a tiny hole for their ears. Their front flippers are used for steering as they are propelled by the rear flippers; on land, they move like caterpillars. Sea lions and fur seals are known as eared seals and their ear lobes can be easily seen. They can turn their hind flippers forward to function like feet when on land; when swimming the front flippers are used to pull them through the water while the rear flippers act as a rudder for steering.

Though natural predators such as killer whale and sharks take their toll, the worst enemy of pinnipeds has been man. Unrelenting slaughter during the 1800's brought several species to the brink of extinction and now the use of driftnets, and dumping chemical pollutants into rivers and oceans jeopardizes the long-term survival of the species.

GUADALUPE FUR SEAL

Little is known about this rare seal which was almost annihilated during the 19th century. Before the wholesale killing began large numbers of the species lived on rocky offshore islands of the pacific coast from California to Chile. In the middle 1920s it was believed that the Guadeloupe fur seal was extinct but in the early 1950s a small colony was discovered and has slowly increased to about 500. It is on the endangered species list.

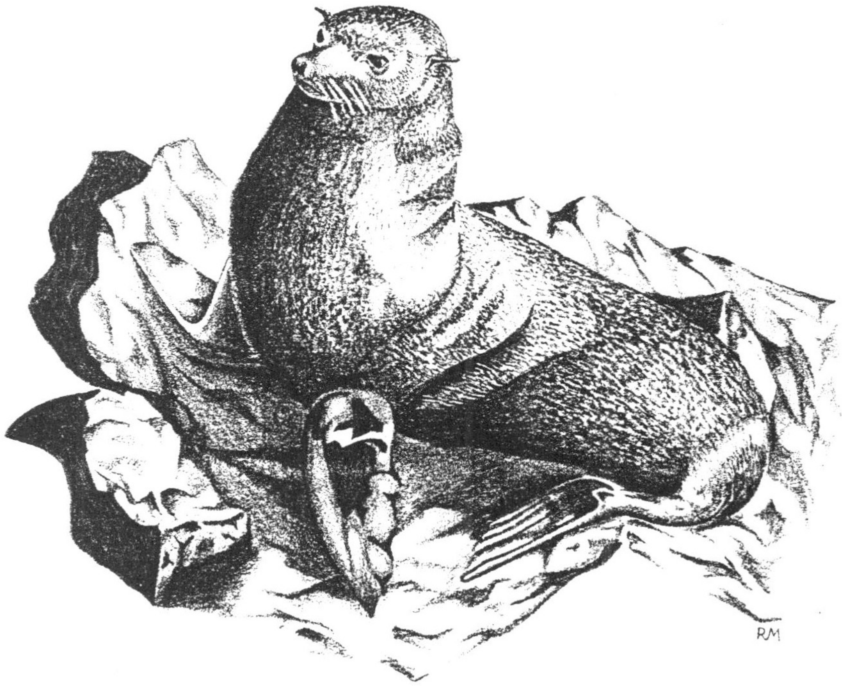

Length: Males may reach 1.8m (6 ft.); females are smaller.

Weight: Males may weigh up to 160 kg. (350 lbs.)

Range: From Guadaloupe Island in Mexico to San Miguel Island in the Channel Islands where they haul out in caves.

NORTHERN FUR SEAL or ALASKAN FUR SEAL

The high quality of its pelt almost led to the demise of this animal by Russian and American seal hunters. In addition, it has a very high mortality rate while young; almost 70% die by the age of three. Causes of death are bad mothering, harsh weather conditions, and parasitic worm infestations prevalent in areas regularly frequented by them.

Length: Males grow to 2.1m (15 ft.), females, 1.5m (5 ft.)
Weight: Males reach 270 kg. (600 lbs.), females, 70 kg. (155 lbs.)
Color: Males are dark brown with a greyish brown neck, females, grey with a tan belly.
Range: Females and immature animals spend the winter in the Channel Islands while the bulls winter in the Aleutian Islands.
Food: Lamprey, squid and small amounts of fish
Pups: Length at birth: 60 cm. (2 ft.) and weigh 4.5 kg. (10 lbs.) Born early in July, they remain on land, by August they are swimming. They nurse for about 3 months. For the first week the mother is very attentive then she goes fishing, returning once a week to nurse the pup. During her absence the pups gather in groups until answering the call of the mother.
Enemies: Killer whale and man (as a result of hunting and the spread of plastic pollutants)

NORTHERN ELEPHANT SEAL

These are the largest of the pinnipeds. The distinctive fleshy proboscis is used as a resonating chamber to produce a variety of sounds when bulls are establishing their territorial rights before the cows arrive. They swim by means of body undulations rather than using their forward flippers for power. During the annual molt large sheets of skin and hair are shed aided by much scratching and rubbing against rocks. Molting lasts about a month, during which period no food is eaten.

By 1860 sealers had reduced their numbers so drastically they were no longer considered worth hunting and in 1884 no more could be found. In 1892 a group of 8 were found on a remote island off Baja California by a collector for the Smithsonian Institution who killed 7 of them. Fortunately, less than 100 had survived elsewhere and it is from those animals the present population grew. In 1922 Mexico was the first nation to give protection to elephant seals, the U.S. followed later.

Length:	Males exceed 6m (18 ft.), females average 3.7m (12 ft.)
Weight:	Males exceed 3,600 kg. (7,900 lbs.), females, 1,125 kg. (2,500 lbs.)
Color:	Males are grayish brown, females are a rich, dark brown.
Range:	Found off the western coast of North America. Breeding grounds are restricted to San Nicolas and San Miguel in the Channel Islands and San Benito and Guadalupe Islands in Mexico.
Population:	They now number about 20,000.
Pups:	Newborn pups are covered with black woolly fur which changes to gray within two months when the mothers return to the sea. Meanwhile, the pups eat small invertebrates and ingest some stones which are believed to aid in grinding food.

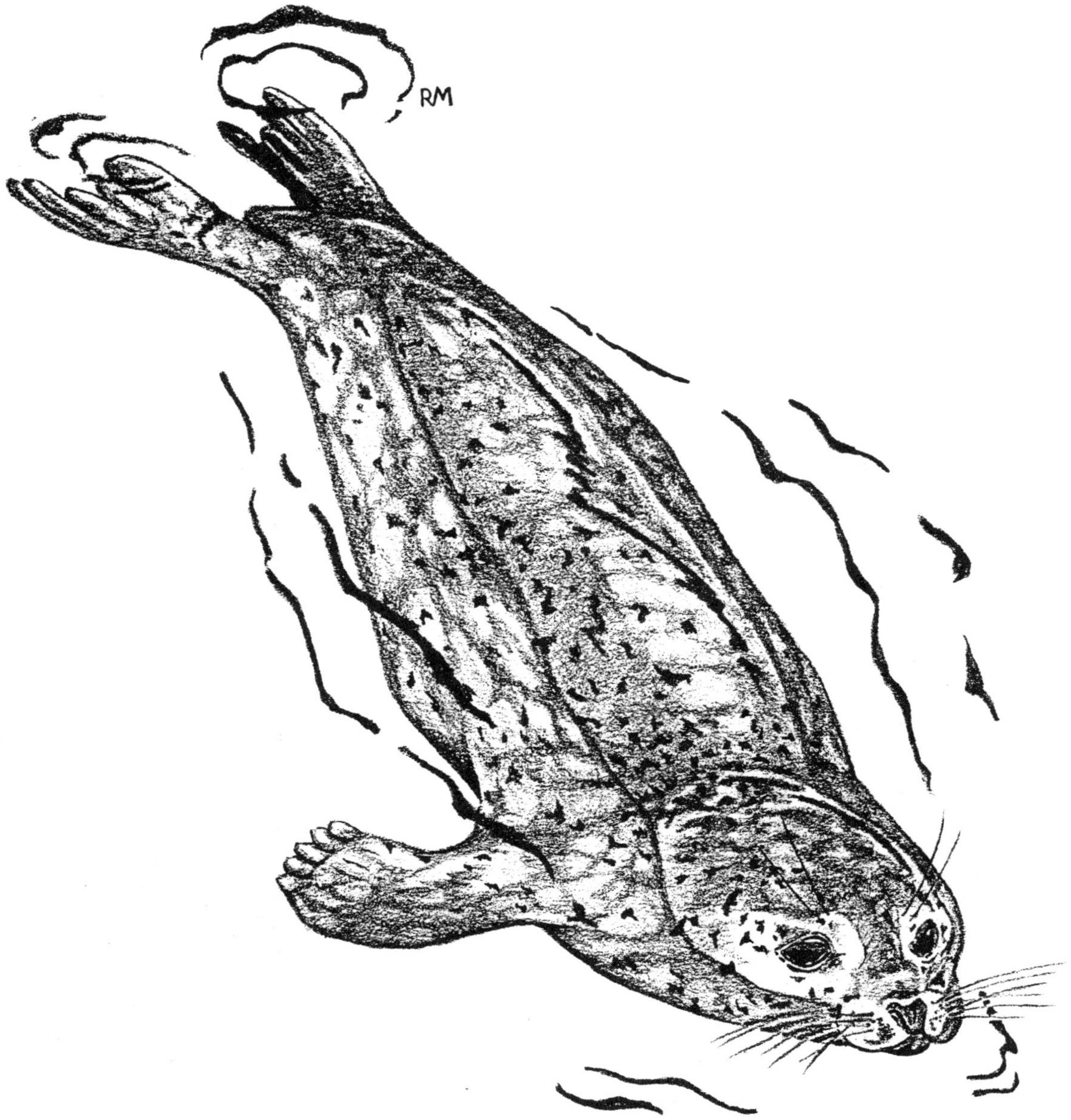

Range:	Shallow sand banks, river estuaries and coastal waters.
Region:	Widely distributed in both the Pacific and Atlantic Oceans. The Pacific subspecies is found from Alaska to Guadalupe Island off the Mexican coast.
Population:	Total world population of 5 subspecies: 200,000 to 400,000
Food:	Fish, crustacea and mollusks
Pups:	Length at birth: 90 cm (1 yd.), Weight: 9 - 14 kg. (20 - 30 lbs.) Nurse for about 3 weeks, then are abandoned to fend for themselves. They are the only one of the pinnipeds found along this coast that can swim at birth.
Enemies:	In addition to many adults being killed by fishermen, killer whales, and sharks, Arctic birds of prey such as eagles kill many pups.

STELLER'S SEA LION

This is the largest of the sea lions and is one of the most acrobatic of the pinnipeds, making its underwater ballet an unforgettable sight. When swimming its small ears are laid flat on the sides of its head.

Length:	Males reach over 3m (10'); females are about 2/3 of the size of males
Weight:	Males may reach over 900 kg. (one ton)
Color:	Yellow-brown; mature males grow a mane of long, coarse hair.
Range:	Widely distributed along Canadian, U.S., and Russian coasts.
Population:	World population numbers about 300,000
Food:	Squid, lamprey, and only small amounts of fish
Pups:	Born early in July, remain on land but by August they are swimming. They nurse for up to 2 years; occasionally a pup may be seen nursing along with his one- and two-year old siblings!
Enemies:	They are still being slaughtered by man to provide pet food.

CALIFORNIA SEA LION

This is the best known of pinnipeds for it is featured in circus acts, zoos and aquariums because it is easily trained and has an excellent memory. Because it is very responsive to learned behavior it is also used for U.S. military purposes.

Length:	Males often exceed 2m (7 ft.), while females are slightly smaller.
Weight:	Males weigh over 270 kg. (600 lbs.), females are smaller. Males have a prominent bony crest on their heads and they have shorter, more rounded snouts than fur seals.
Color:	Deep brown.
ID:	They are easily distinguished from harbor seals by noticeable ear lobes.
Range:	They live within 15 km (9 miles) the California coast and Galapagos Islands. The Japanese subspecies is almost extinct as a result of continued depredation.
Population:	In the tens of thousands
Food:	Squid, octopus and some fish
Pups:	Nurse for about 3 months, remain on land until 2 weeks old when they take easily to the water.

INDEX

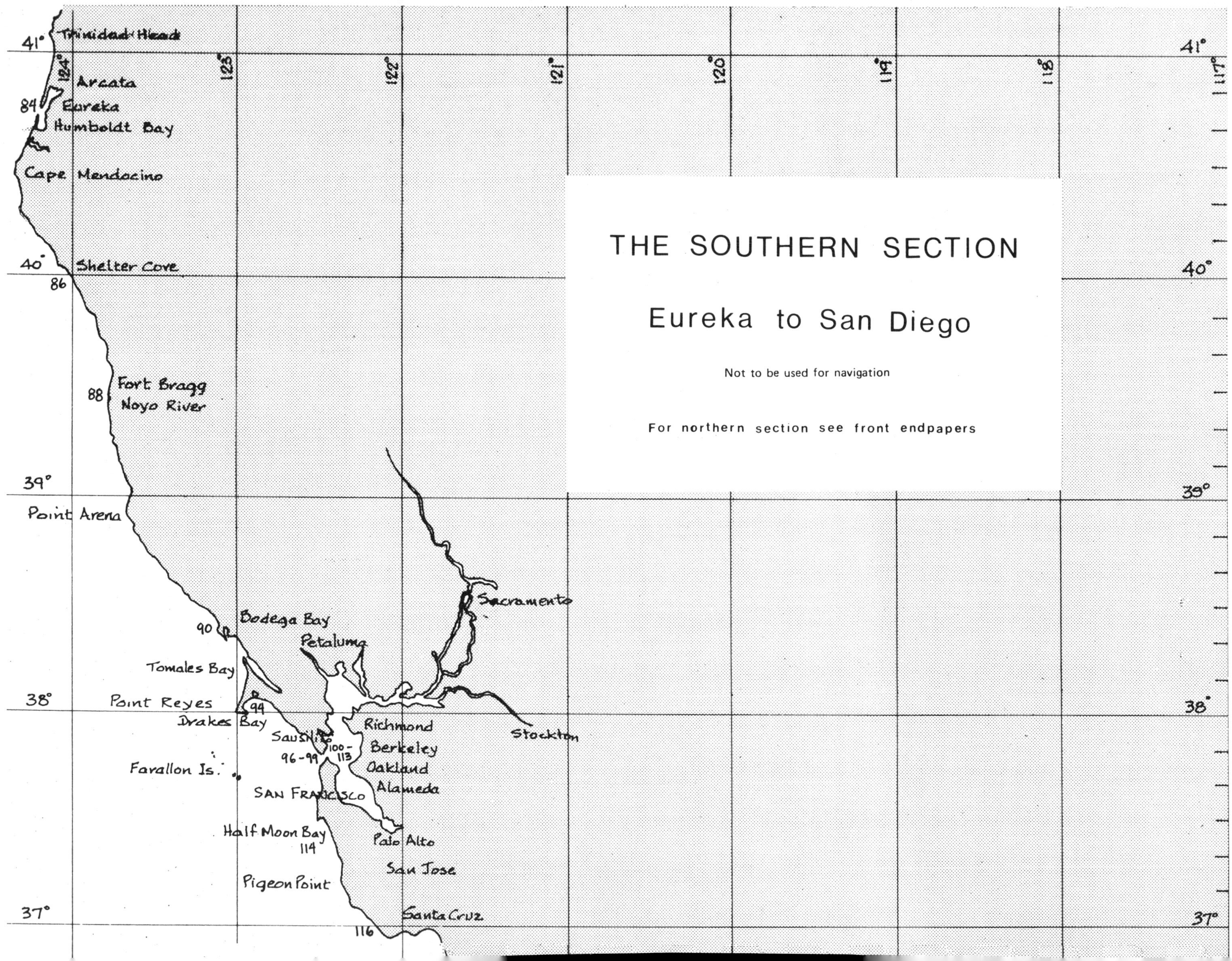
THE SOUTHERN SECTION
Eureka to San Diego
Not to be used for navigation
For northern section see front endpapers
41°
40°
39°
38°
37°
124°
123°
122°
121°
120°
119°
118°
117°
Trinidad Head
Arcata
84
Eureka
Humboldt Bay
Cape Mendocino
Shelter Cove
86
Fort Bragg
88
Noyo River
Point Arena
90
Bodega Bay
Petaluma
Sacramento
Tomales Bay
Point Reyes
94
Drakes Bay
Sausalito
96-99
100-113
Richmond
Berkeley
Oakland
Alameda
Stockton
Farallon Is.
SAN FRANCISCO
Half Moon Bay
114
Palo Alto
San Jose
Pigeon Point
Santa Cruz
116